ESO ASTROPHYSICS SYMPOSIA
European Southern Observatory

Series Editor: Bruno Leibundgut

L. Stanghellini J.R. Walsh N.G. Douglas (Eds.)

Planetary Nebulae Beyond the Milky Way

Proceedings
of the ESO Workshop held
at Garching, Germany, 19-21 May, 2004

Springer

Volume Editors

Letizia Stanghellini
NOAO
950 N Cherry Ave
Tucson AZ 85719 USA
E-mail: letizia@noao.edu

Nigel G. Douglas
Kapteyn Astronomical Institute
University of Groningen
P.O. Box 800
9700 AV Groningen
The Netherlands
E-mail: ndouglas@astro.rug.nl

Jeremy R. Walsh
European Southern Observatory
Karl-Schwarzschild-Str. 2
85748 Garching
Germany
E-mail: jwalsh@eso.org

Series Editor

Bruno Leibundgut
European Southern Observatory
Karl-Schwarzschild-Str. 2
85748 Garching
Germany

ISBN-10 3-540-31011-8 Springer Berlin Heidelberg New York
ISBN-13 978-3-540-31011-2 Springer Berlin Heidelberg New York

Library of Congress Control Number: 2005939176

Springer is a part of Springer Science+Business Media
springer.com

Printed in The Netherlands

Typesetting: by the authors and TechBooks using a Springer LaTeX macro package

Cover design: *design & production* GmbH, Heidelberg

Printed on acid-free paper SPIN: 11604792 55/TechBooks 5 4 3 2 1 0

Preface

In the last decade extra-galactic planetary nebulae (PNe) have gained increasing importance. Improved observational capabilities have allowed fainter and fainter PNe to be studied in galaxies well beyond the Milky Way. Planetary nebulae, characterized by their emission line spectrum, in particular the strong [O III] 5007Å line, can be detected to at least 30 Mpc. They are found in galaxies of all types and also between the galaxies in nearby galaxy clusters. They are valuable as probes, for providing both the velocity of their host stars and also the evolutionary status and relation to the stellar population from which they formed.

Whilst there had been a 1-day meeting devoted to extra-galactic PNe at the IAU General Assembly in the Hague in 1994, it was apparent to us that the topic had expanded to the extent that a full conference was well justified. The last day of the IAU Symposia on Planetary Nebulae had typically been allocated to extra-galactic PNe, but it was clear that it could not do full justice to the developments in the field. The resulting workshop was held at ESO Headquarters from 19 to 21 May 2004 and was devoted to exploring the potential for PN in the extra-galactic context.

The Scientific Organizing Committee (SOC) was formed to include scientists covering all the major areas of extra-galactic PN research, from stellar evolution to extra-galactic distance scale to dynamical studies. The SOC consisted of Magda Arnaboldi (Italy), Robin Ciardullo (USA), Nigel Douglas (The Netherlands), Ken Freeman (Australia), George Jacoby (USA), Roberto Mendez (USA), Richard Shaw (USA), Grażyna Stasińska (France) and was co-chaired by two of us (Stanghellini and Walsh).

About 65 scientists from around the world participated in the workshop which consisted of a mix of invited and contributed talks and a display of posters. The authors of poster papers presented their work during a short oral presentation. There were also two discussion sessions: one on "Observational Challenges," chaired by Quentin Parker, and another on "Future Challenges," chaired by Richard Shaw. Alan Moorwood gave the welcoming address and Holland Ford summarized the workshop with some witty collages of speakers slides and digital photographs taken at the conference. During the busy 3 days in Garching, we found that many scientific collaborations, and new friendships, were born, and old ones renewed.

During the preparations for the conference a German amateur astronomer, Gunther Cibis, asked us if we were going to tackle the nomenclature for extragalactic PNe. None of us had thought carefully about this and we were forced to admit that this was now becoming important with the burgeoning number of PNe being discovered in nearby galaxies. Much of one of the discussion sessions was devoted to this subject and Agnes Acker and Quentin Parker contributed a proceedings article. Whilst no clear consensus emerged, we hope that the subject will be considered at the next IAU meeting on PNe in Hawaii in 2006.

Even though one of the days of the conference was scheduled on a Bavarian holiday, the local organization went flawlessly, thanks primarily to Christina Stoffer. Britt Sjoeberg also helped out during the conference. Marina Rejkuba provided the mainstay of the Local Organizing Committee, Nausicaa Delmotte helped with the web pages, and Ed Janssen designed the conference poster. Pamela Bristow very ably prepared the proceedings for publication.

This meeting was the first full-scale conference devoted to the subject of planetary nebulae beyond the Milky Way. It is apparent from these proceedings that the field is maturing rapidly and is multi-facetted enough that, within a few years, a sequel to this meeting can be considered.

Tucson, Garching, Groningen
January 2005

Letizia Stanghellini
Jeremy Walsh
Nigel Douglas

Contents

Part IX Prospects

List of Participants

ARNABOLDI, Magda
INAF, Obs. of Turin
arnaboldi@to.astro.it

ARRIETA, Anabel
Univ. Iberoamericana,
Fisica y Matematicas, Mexico
anabel.arrieta@uia.mx

BARLOW, Michael
University College London
mjb@star.ucl.ac.uk

BUZZONI, Alberto
INAF - Oss. Astronomico di Bologna
buzzoni@bo.astro.it

CHITA, Sabina
MPI for Radioastronomy, Bonn
SCHITA@MPIFR-BONN.MPG.DE

CIARDULLO, Robin
The Pennsylvania State University
rbc@astro.psu.edu

CIONI, Maria-Rosa
ESO-Garching
mcioni@eso.org

CORRADI, Romano
Isaac Newton Group
S. Cruz de la Palma
rcorradi@ing.iac.es

DE LORENZI, Flavio
University of Basel
Astronomical Institute
lorenzi@astro.unibas.ch

DOUGLAS, Nigel
Kapteyn Astronomical Institute
Groningen
ndouglas@astro.rug.nl

DUFOUR, Reginald
Rice University, Physics & Astronomy
rjd@rice.edu

ERCOLANO, Barbara
University College London
be@star.ucl.ac.uk

EXTER, Katrina
Instituto de Astrofisica de Canarias
katrina@ll.iac.es

FELDMEIER, John
Case Western Reserve Univ.
johnf@bottom.astr.cwru.edu

FERGUSON, Annette
MPI for Astrophysics, Garching
ferguson@mpa-garching.mpg.de

FORD, Holland
Johns Hopkins University
ford@pha.jhu.edu

GARCIA-LARIO, Pedro
ESA/ISO Data Centre
Pedro.Garcia.Lario@esa.int

GERHARD, Ortwin
University of Basel
Astronomical Institute
gerhard@astro.unibas.ch

GIRARDI, Leo
Osservatorio Astronomico di Trieste
lgirardi@ts.astro.it

GROENEWEGEN, Martin
Instituut voor Sterrenkunde, Leuven
groen@ster.kuleuven.ac.be

JACOBY, George
WIYN Observatory, Tucson
jacoby@wiyn.org

KÄUFL, Hans Ulrich
ESO-Garching
hukaufl@eso.org

KNIAZEV, Alexei
MPI for Astronomy, Heidelberg
kniazev@mpia.de

KÖPPEN, Joachim
Observatoire de Strasbourg
koppen@astro.u-strasbg.fr

KRAUS, Michaela
Astronomical Institute
Utrecht University
M.Kraus@phys.uu.nl

KUNTSCHNER, Harald
ST-ECF, Garching
hkuntsch@eso.org

KWOK, Sun
Inst. of Astronomy & Astrophysics
Acad. Sinica, Taipei
kwok@asiaa.sinica.edu.tw

LEE, Henry
University of Minnesota
hlee@astro.umn.edu

LEISY, Pierre
Isaac Newton Group
S. Cruz de la Palma
pleisy@ing.iac.es

LIU, Xiao-wei
Peking University
Department of Astronomy
liuxw@vega.bac.pku.edu.cn

MACIEL, Walter
IAG/Univ. of São Paulo
maciel@astro.iag.usp.br

MAGRINI, Laura
Dept. of Astronomy, Florence
laura@arcetri.astro.it

MANCHADO, Arturo
Instituto de Astrofisica de Canarias
amt@iac.es

MATSUURA, Mikako
Department of Physics, UMIST
m.matsuura@umist.ac.uk

MENDEZ, Roberto
Institute for Astronomy
University of Hawaii
mendez@ifa.hawaii.edu

MERRETT, Helen
University of Nottingham
ppxhm@nottingham.ac.uk

NAPOLITANO, Nicola Rosario
Kapteyn Astronomical Institute
Groningen
nicola@astro.rug.nl

NICKELER, Dieter
Astronomical Institute
Utrecht University
D.H.Nickeler@phys.uu.nl

PARKER, Quentin
Macquarie Univ., Sydney/
Anglo-Australian Obs.
qap@ics.mq.edu.au

PATRIARCHI, Patrizio
Istituto di Radioastronomia - CNR
Florence
pat@arcetri.astro.it

PEÑA, Miriam
Instituto de Astronomia, UNAM
Mexico
miriam@astroscu.unam.mx

PENG, Eric
Rutgers University
ericpeng@physics.rutgers.edu

PERINOTTO, Mario
Dept. of Astronomy, Florence
mariop@arcetri.astro.it

REID, Warren
Macquarie University, Sydney
warren@ics.mq.edu.au

REJKUBA, Marina
ESO-Garching
mrejkuba@eso.org

RICHER, Michael
Instituto de Astronomia, UNAM
Mexico
richer@astrosen.unam.mx

ROMANOWSY, Aaron
University of Nottingham
aaron.romanowsky
@nottingham.ac.uk

ROTH, Martin
Astrophysikalisches Institut Potsdam
mmroth@aip.de

SAMBHUS, Niranjan
University of Basel
Astronomical Institute
sambhus@astro.unibas.ch

SCHOENBERNER, Detlef
Astrophysikalisches Institut Potsdam
deschoenberner@aip.de

SCHUBERTH, Ylva
Sternwarte der Universität Bonn
ylva@astro.uni-bonn.de

SHAW, Richard
NOAO, Tucson
shaw@noao.edu

STANGHELLINI, Letizia
NOAO, Tucson
letizia@noao.edu

STASINSKA, Grazyna
LUTH, Observatoire de Paris-Meudon
grazyna.stasinska@obspm.fr

TSAMIS, Yiannis
LUTH, Observatoire de Paris-Meudon
Yiannis.Tsamis@obspm.fr

VAN DE STEENE, Griet C.
Royal Observatory of Belgium
Brussels
gsteene@oma.be

VILLAVER, Eva
Space Telescope Science Institute
villaver@stsci.edu

WALSH, Jeremy
ST-ECF, Garching
jwalsh@eso.org

WANG, Wei
Peking University
Department of Astronomy
bwaw@vega.bac.pku.edu.cn

WILLSON, Lee Anne
Iowa State University
lwillson@iastate.edu

ZHANG, Yong
Peking University
Department of Astronomy
zhangy@bac.pku.edu.cn

ZIJLSTRA, Albert
Department of Physics, UMIST
a.zijlstra@umist.ac.uk

Part I

Historical Overview

Planetary Nebulae Beyond the Milky Way – Historical Overview

Michael J. Barlow

Department of Physics & Astronomy,
University College London,
Gower Street, London WC1E 6BT, U.K.

1 Introduction

Up to 10% of the total luminosity of a planetary nebula, $\sim$500 $L_{\odot}$, can be emitted in the dominant cooling line, [O III] λ5007. This, coupled with the narrowness of the line ($\sim$15-25 km s^{-1}), makes it extremely easy to detect PNe in external galaxies using a narrow-band filter tuned to the galaxy redshift. The availability of multiple independent distance indicators for our closest neighbouring galaxies, the Magellanic Clouds and M 31, means that the luminosities of the PNe in these galaxies (in particular the large numbers of PNe in the LMC) can be used to calibrate PN luminosity functions, which have been used over the past 15 years as probes of the Hubble constant. Given the fact that we still do not have accurate distances for most planetary nebulae in the Milky Way, this has been an astonishing development. Unlike H II regions, which cannot be used to probe elliptical galaxies, planetary nebulae can be used to probe the dynamics and metallicity of any type of galaxy. Today, via accurate radial velocity measurements, extragalactic PNe are being used as dynamical mass probes of galaxies, and even of the stellar mass content of the intracluster regions of galaxy clusters, with dedicated 'planetary nebula spectrographs' being built to further such studies. The angular resolution of the Hubble Space Telescope has turned out to be ideally suited to the study of PNe in the LMC and SMC, and the exploitation of the accurately known distances to these two galaxies has allowed reliable luminosities and masses to be derived for the central stars and nebulae, thereby putting PN research on a much more quantitative footing.

In this overview of past and recent studies of extragalactic planetary nebulae (PNe), I have divided the research into two main areas: (a) surveys for planetary nebulae in other galaxies, PN luminosity functions and intracluster PNe; (b) detailed studies of planetary nebulae in the Magellanic Clouds, particularly studies carried out with the HST.

2 Surveys for Extragalactic Planetary Nebulae

The low dust extinctions to the LMC and SMC mean that virtually all nebulae present can be detected by a survey down to a given intrinsic luminosity limit. The low reddening also makes them well suited to study in the ultraviolet. Sanduleak, MacConnell & Philip (1978: SMP) published a catalogue of all PNe that had been discovered in the Magellanic Clouds up till then. Of the 28 SMC PNe in their list, 18 had been discovered by Henize (1956), 7 were from Lindsay (1961) and 3 came from SMP's own objective prism survey. Of the 102 LMC PNe in their list, 59 were from Henize (1956), 21 came from Lindsay & Mullan (1963) and 22 came from their own objective prism survey.

Ford et al. (1973) detected 26 PNe in the dwarf elliptical companion galaxies to M 31: NGC 185, NGC 205 and NGC 221, via filter imaging in [O III] and Hα. Ford & Jenner (1975) followed up with deeper observations of NGC 221, detecting 21 PNe in its field (see Fig. 1). Three of the 14 PNe with observed radial velocities belonged to M 31, the rest to NGC 221 (they were distinguished by means of the 200 km s^{-1} difference between the two galaxies).

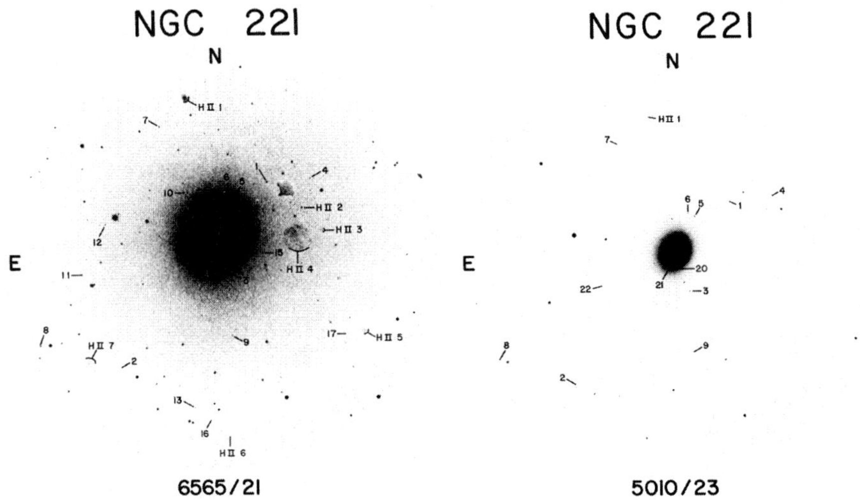

Fig. 1. Hα (left) and [O III] (right) images of NGC 221 (= M 32). The central wavelengths and widths (in Å) of the filters are indicated at the bottom of the figures. From Ford & Jenner (1975).

A deep narrow-band [O III] and Hα filter survey by Jacoby (1980) led to the detection of 53 faint new SMC and LMC PNe in selected fields and the derivation of a PN luminosity function (PNLF) that reached six magnitudes below the brightest. This work led the foundations for subsequent studies that exploited PNLF's for more distant galaxies. Ciardullo et al. (1989a) detected 429 PNe in the bulge of M 31, of which 104 were in the first 2.5 mags of its PNLF (Fig. 2). A steep turnover observed at the high luminosity end was interpreted as corresponding to a sharp cutoff in the upper mass limit of PN central stars, consistent with the expectation that no central star should exceed the Chandrasekhar mass limit and that they should evolve increasingly quickly as this mass is approached.

Ciardullo, Jacoby & Ford (1989b) detected 249 PNe in three early type galaxies in the Leo I cluster: NGC 3377 (E6), NGC 3379 (E0) and NGC 3384 (S0).

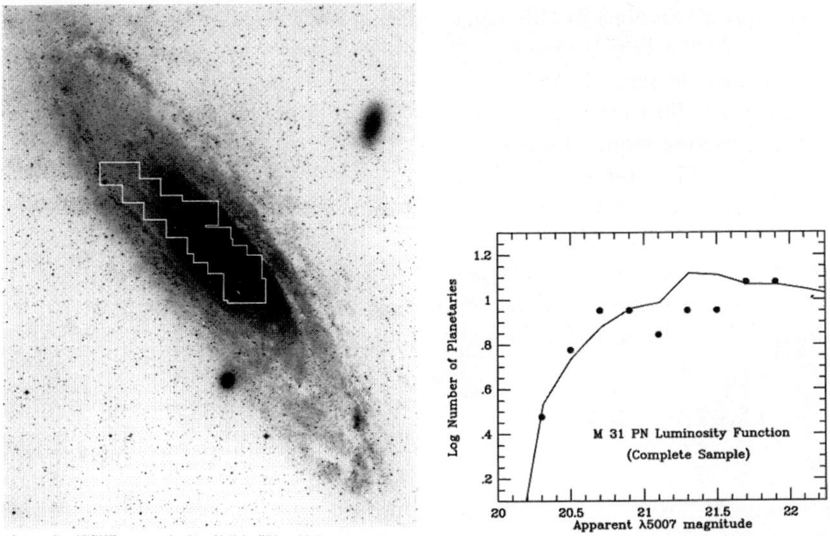

Fig. 2. Image of M 31 (left), showing the region surveyed for PNe using an [O III] 5007 Å filter. The plot on the right shows the resulting PN luminosity function for the 429 PNe discovered, from Ciardullo et al. (1989a)

All had similar PN [O III] luminosity functions, indistinguishable from those observed in the spiral galaxies M 31 and M 81. The PNLF invariance with galaxy type and metallicity demonstrated the PNLF to be an excellent standard candle. Model simulations indicated a central star mass distribution highly peaked at ~0.6 M⊙, similar to that of nearby white dwarfs.

PNe are now routinely used to study the internal dynamics and mass distributions of both early and late type galaxies, e.g. Arnaboldi et al. (1998) have measured the rotation curve and velocity dispersions along two axes of the early type galaxy NGC 1316 (Fornax A) from radial velocity measurements of 43 PNe (Fig. 4). The total galaxy mass inside 16 kpc radius was derived to be 2.9×10^{11} solar masses.

Méndez et al. (2001) detected 535 PNe in the elliptical galaxy NGC 4697 using filter images and slitless spectra. The former gave the [O III] PNLF (Fig. 3, right), which yielded a distance of 10.5 Mpc. The latter gave radial velocities for 531 of the PNe, allowing an accurate mass to luminosity ratio to be derived for the galaxy.

Turning to more local regions, Magrini et al. (2003) have surveyed a number of Local Group dwarf galaxies for PNe. They found that the number of PN per unit stellar luminosity appears to decline steeply for metallicities below 0.1 solar (Fig. 5).

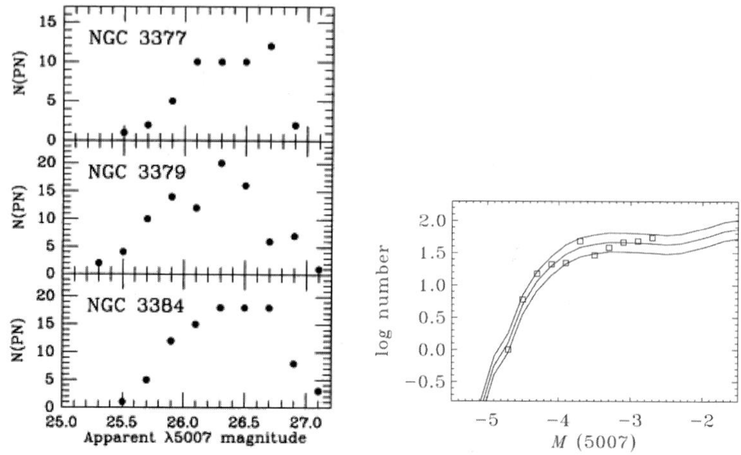

Fig. 3. [O III] PNLF's for three Leo I galaxies (left), derived by Ciardullo et al. (1989b). The plot on the right shows the [O III] PNLF derived by Méndez et al. (2001) for 535 PNe in the elliptical galaxy NGC 4697, which is located in the Virgo southern extension. The three lines show PNLF simulations from Méndez & Soffner (1997), for three different sample sizes: 2500, 3500, and 4900 PNe.

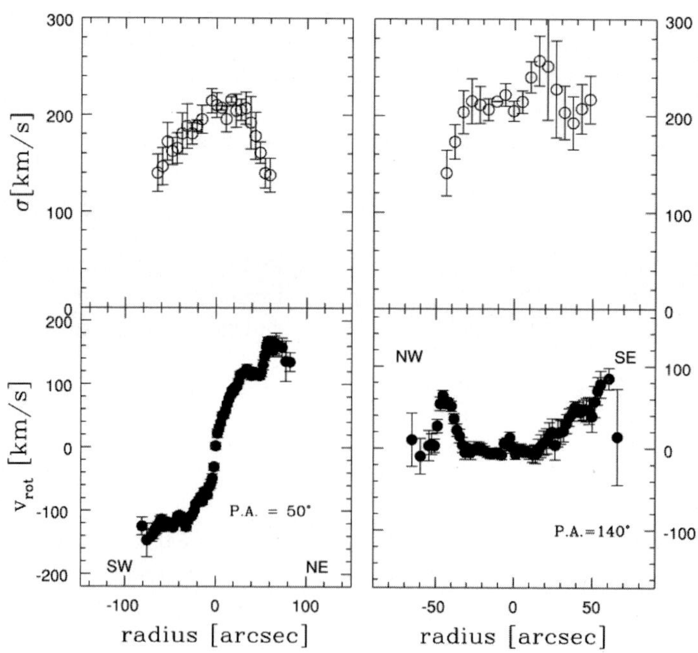

Fig. 4. Velocity dispersions (upper) and rotation curves (lower) for NGC 1316. From Arnaboldi et al. (1998).

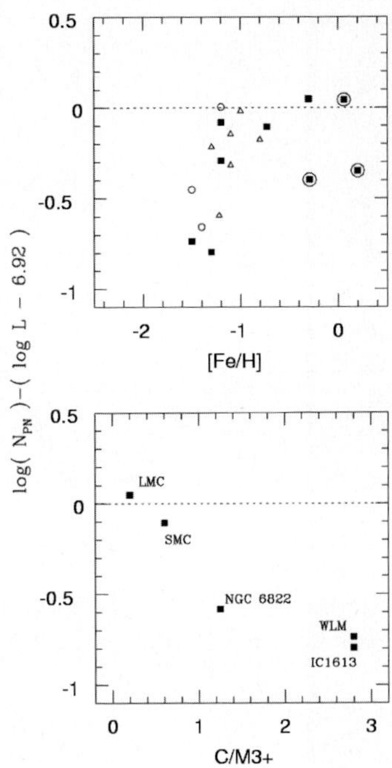

Fig. 5. Number of planetary nebulae per unit galactic luminosity as a function of galaxy metallicity (upper) and as a function of the ratio of carbon stars to O-rich late-type AGB stars (lower). From Magrini et al. (2003).

3 Planetary Nebulae in the Intracluster Regions of Galaxy Clusters

Arnaboldi et al. (1996) measured radial velocities for 19 PNe in the outer regions of the giant elliptical galaxy NGC 4406, in the southern Virgo extension region. Although this galaxy has a radial velocity of –227 km s^{-1}, three of the PNe had radial velocities close to 1400 km s^{-1}, the mean radial velocity of the Virgo cluster. It was concluded that they were intracluster PNe. Theuns & Warren (1997) discovered ten PN candidates in the Fornax cluster, in fields well away from any Fornax galaxy – consistent with tidal stripping of cluster galaxies. They estimated that intracluster stars could account for up to 40% of all the stars in the Fornax cluster.

Méndez et al. (1997) surveyed a 50 arcmin^2 area near the centre of the Virgo cluster, detecting 11 PN candidates. They estimated that a stellar population comprising 4×10^9 $M_\odot$ lay in their survey area and that such a population could account for up to 50% of the total stellar mass in the Virgo cluster. Feldmeier, Ciardullo & Jacoby (1998) searched for intracluster PNe in several Virgo Cluster fields. Comparison with the normalised M 87 PNLF showed an excess at the bright end, attributed by them to intracluster PNe (see Fig. 6).

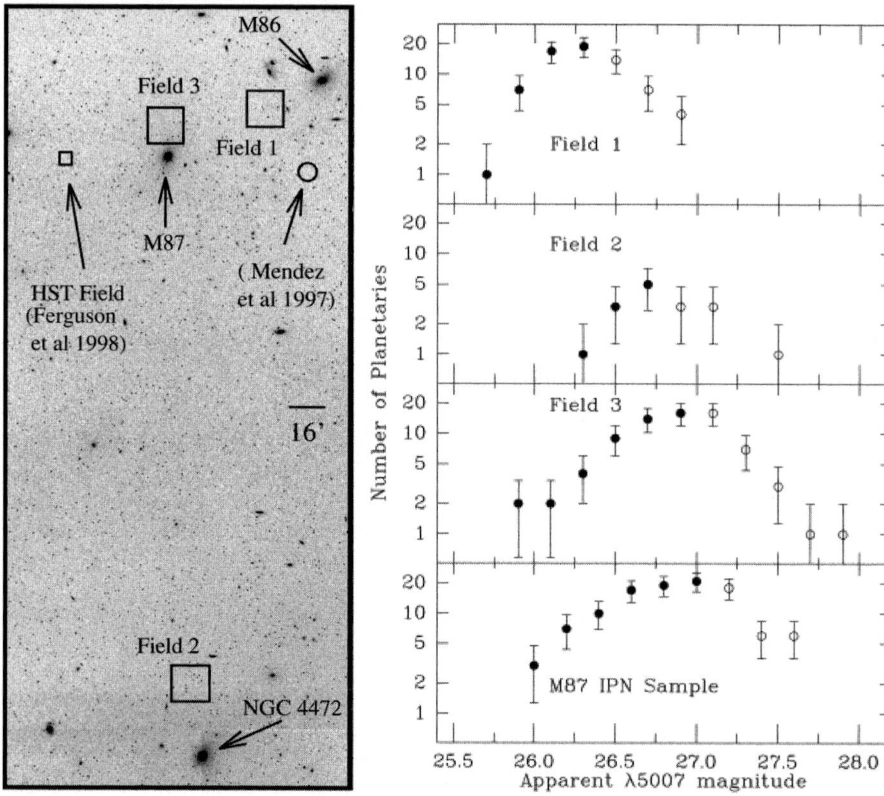

Fig. 6. Left: Virgo cluster fields searched for PNe by Feldmeier et al. (1998). Right: the resulting PN luminosity functions.

4 Observations of Magellanic Cloud Planetary Nebulae

Magellanic Cloud PNe are bright enough that Dopita et al. (1988) were able to obtain echelle spectra with 11.5 km s $^{-1}$ resolution of 94 LMC PNe in the [O III] 5007 Å line, using 1.0 m and 2.3 m telescopes. The nebulae were spatially unresolved from the ground but those with asymmetric velocity profiles (Fig. 7) indicate non-spherically symmetric nebulae. The radial velocities were used by Meatheringham et al. (1988) to conduct a dynamical study of the PN population in the LMC.

Monk et al. (1988) analysed 3.9-m AAT spectra of 21 SMC and 50 LMC PNe. They found that the PN and H II region abundances of oxygen and neon agreed within the same galaxy but that nitrogen in the planetary nebulae was enhanced by 0.9 dex in both the SMC and LMC. This was interpreted by them as due to CN cycle processing of almost all of the original carbon into nitrogen at the time of the first dredge-up. This process appears to have operated much

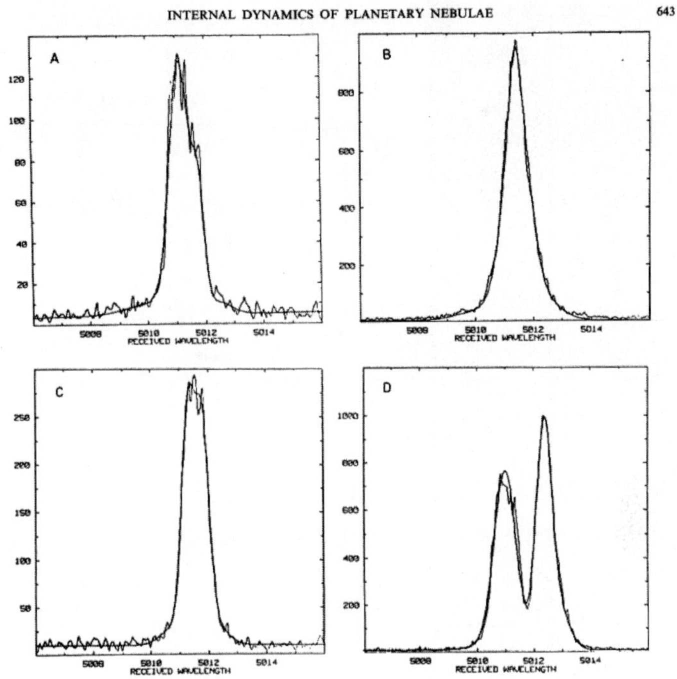

Fig. 7. High spectra resolution profiles of the [O III] 5007 Å line for four LMC PNe. Asymmetric or double-peaked profiles indicate non-spherically symmetric nebulae. From Dopita et al. (1988).

more efficiently in the metal-poor Clouds than in the Galaxy, in agreement with theoretical expectations (e.g. Becker & Iben 1980).

Meatheringham & Dopita (1991) obtained optical spectra of 73 LMC and SMC PNe and Dopita & Meatheringham (1991) applied a grid of photoionization models to these observations to derive nebular abundances and central star effective temperatures, luminosities and masses (Fig. 8).

4.1 The HST Era

The first spatially resolved images of Magellanic Cloud PNe were published by Blades et al. (1992), who presented pre-COSTAR Faint Object Camera [O III] 5007 Å images of two SMC PNe and one LMC PN, with 50 milli-arcsec resolution (the FOC was the only instrument that reached the HST's optical diffraction-limited resolution). One LMC nebula N 66 (=WS 35), was found by them to have a particularly unusual morphology, showing multiple lobes and a bright central star.

In a series of papers, Dopita, Vassiliadis and collaborators have published WFPC and WFPC-2 [O III] images of Magellanic Cloud PNe obtained over several HST cycles, mostly pre-COSTAR, while Stanghellini et al. (1999) have

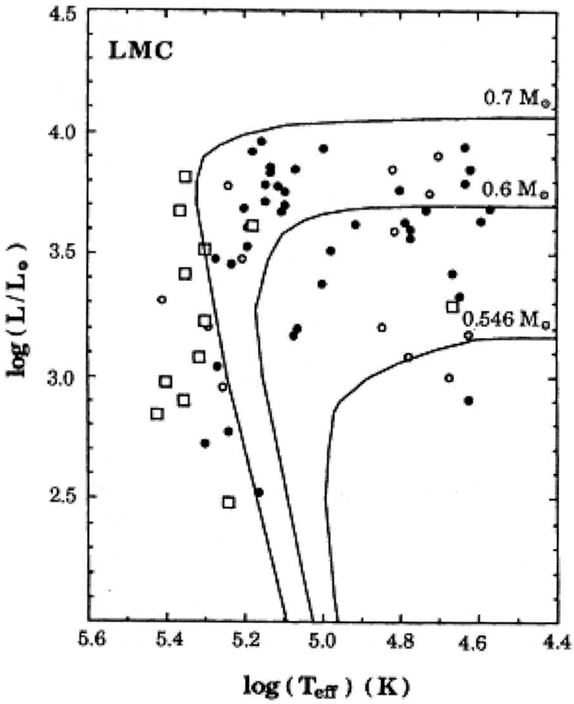

Fig. 8. Hertzsprung-Russell diagram for LMC PN central stars, together with H-burning evolutionary tracks for three different core masses. From Dopita & Meatheringham (1991).

published pre-COSTAR (and some post-COSTAR) HST FOC images of 27 Magellanic Cloud PNe taken in the [O III] line, some examples of which are shown in Fig. 9.

Vassiliadis et al. (1996) presented HST FOS UV spectra for a number of Magellanic Cloud PNe, while Dopita et al. (1997) have fitted photoionization models to the combined UV+optical spectra of 10 of these PNe in the LMC. Central star effective temperatures and luminosities were derived, with a star being classed as an H-burner or a He-burner according to which evolutionary track gave an age in agreement with the nebular expansion age (given by the measured nebular radius and expansion velocity). Fig. 10 shows one of their spectral fits, as well as the HR diagram for the objects assigned as He-burners.

The addition of STIS to its instrument suite powerfully increased the HST's capabilities for studies of Magellanic Cloud PNe, with STIS's ability to obtain slitless spectral imaging in multiple emission lines making a particularly strong impact. Shaw et al. (2001) have presented STIS images for 29 LMC PNe, obtained as part of an HST Snapshot survey. The overall sample of 60 LMC PNe which had HST images showed a much higher incidence of non-symmetric neb-

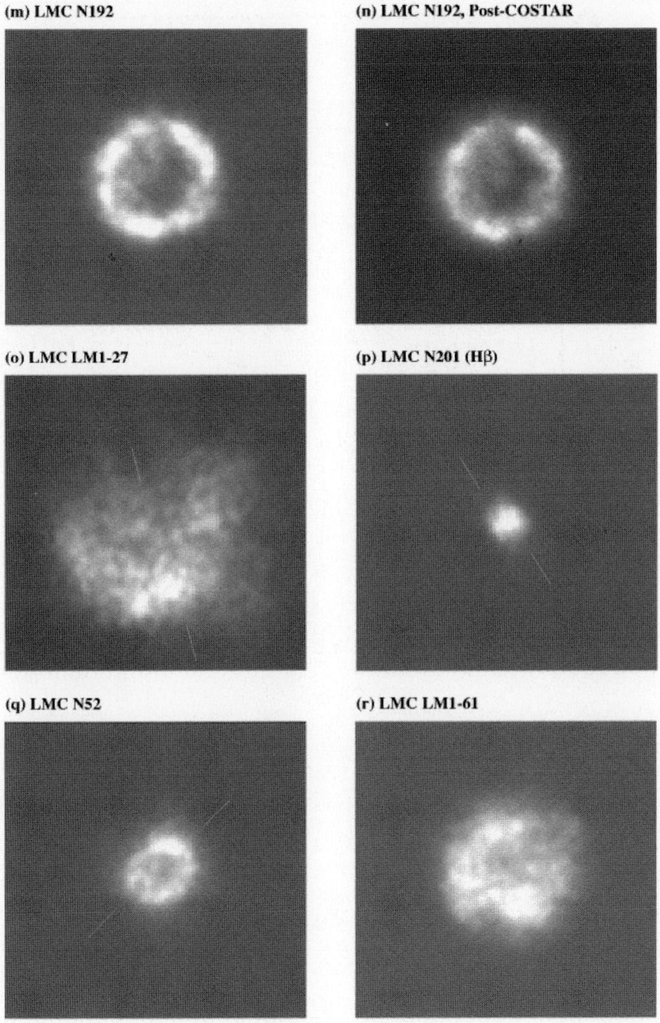

Fig. 9. HST Faint Object Camera images of five LMC PNe taken in the [O III] 5007 Å line. Each image is 2.2 arcsec on a side. From Stanghellini et al. (1999).

ulae than do comparable Galactic PN samples. Stanghellini et al. (2002) have presented STIS slitless spectra of 29 more LMC PNe (some examples are shown in Fig. 11), while Stanghellini et al. (2003) presented STIS images and slitless spectra for 27 more SMC PNe.

Pẽna et al. (1994) discovered that the central star of LMC N66 was much brighter at IUE UV wavelengths in the 1990's than it was in the 1980's. Since then, it has been monitored closely. It reached maximum brightness in 1994 and by 2000 had faded to a fainter level than in 1983. Since 1994 its optical and UV

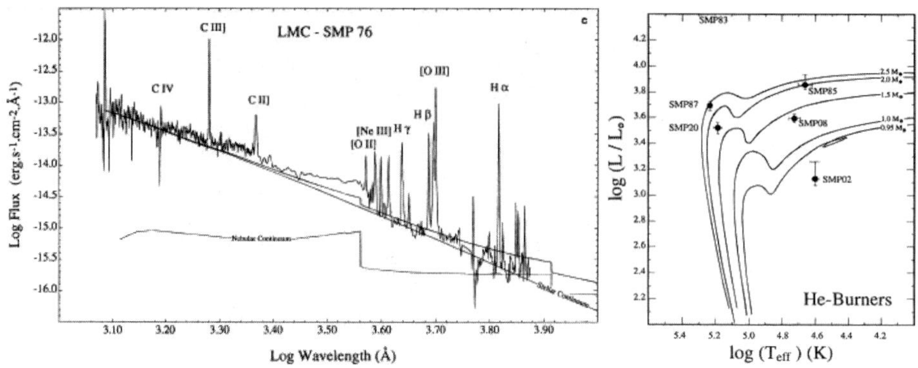

Fig. 10. Left: model stellar + nebular continuum fit to the UV+optical energy distribution of LMC SMP 76. Right: Hertzsprung-Russell diagram showing central star locations, together with He-burning stellar evolutionary tracks that are labelled by their initial main sequence masses. From Dopita et al. (1997).

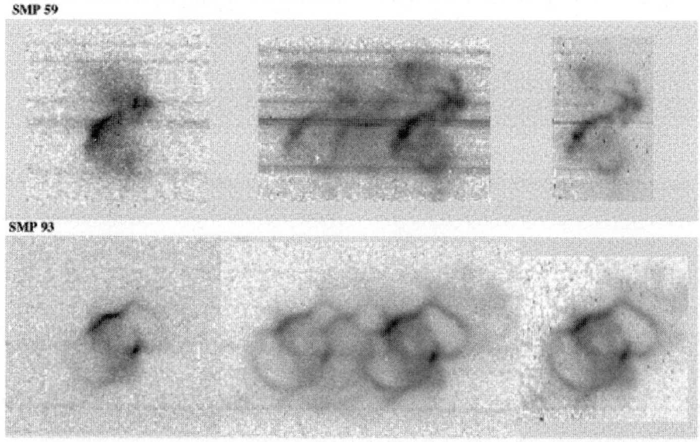

Fig. 11. HST STIS slitless spectra of the LMC PNe SMP 59 (LM 2-25) and SMP 93 (N 181). The images on the left are in the [O III] 5007 Å line, those in the middle show the Hα plus [N II] 6548, 6584 Å blend, and those on the right are deblended [N II] images. From Stanghellini et al. (2002).

spectrum has been that of a WN star, the only known case amongst PN nuclei. Fig. 12 shows an HST FOC [O III] image of LMC N66, from Pẽna et al. (2004), while Fig. 13 shows a sequence of IUE and HST UV spectra, from Hamann et al. (2003), which nicely illustrates the rise and fall of the central star's brightness. Hamann et al. suggested that rather than this being a final thermal pulse event, its nucleus may be a white dwarf accreting matter from a companion and could be the precursor to a Type Ia supernova.

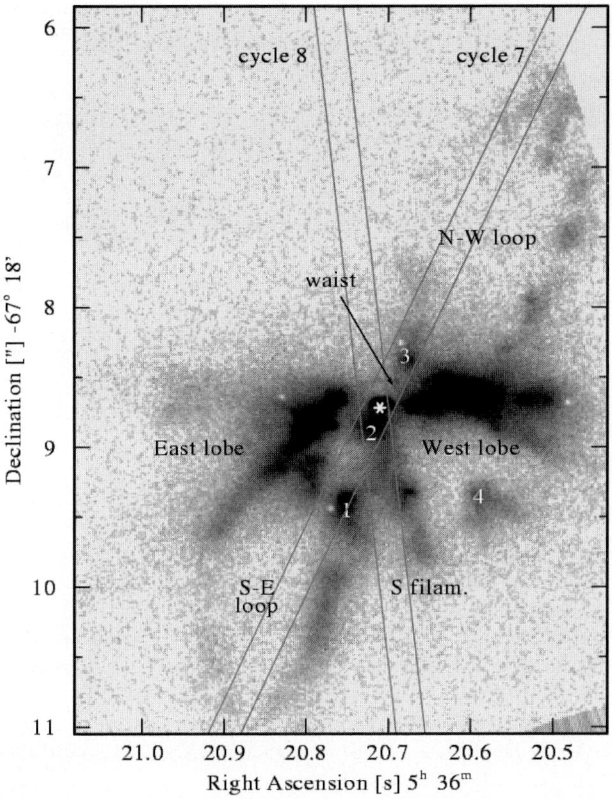

Fig. 12. HST Faint Object Camera (not WFPC) image of LMC N66 (=WS35) in the [O III] 5007 Å line, acquired in February 1994 when the central star was near peak brightness. From Pẽna et al. (2004).

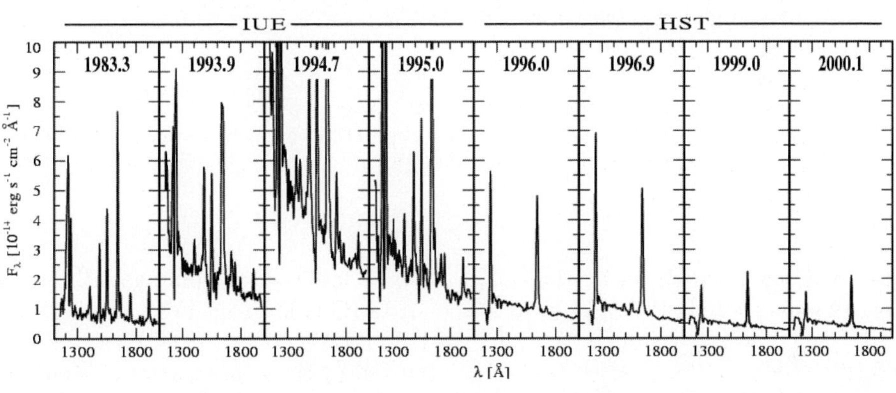

Fig. 13. A time sequence of 1200-2000 Å IUE and HST spectra of the central star of LMC N66, showing the transient brightening event that peaked in 1993-4. From Hamann et al. (2003).

References

1. M. Arnaboldi et al.: ApJ, **472**, 145 (1996)
2. M. Arnaboldi et al.: ApJ, **507**, 759 (1998)
3. S. A. Becker, I. Iben Jr.: ApJ, **237**, 129 (1980)
4. J. C. Blades et al.: ApJ, **398**, L41 (1992)
5. R. Ciardullo et al.: ApJ, **339**, 53 (1989a)
6. R. Ciardullo, G. H. Jacoby, H. C. Ford: ApJ, **344**, 715 (1989b)
7. M. A. Dopita et al.: ApJ, **327**, 639 (1988)
8. M. A. Dopita, S. J. Meatheringham: ApJ, **377**, 480 (1991)
9. M. A. Dopita et al.: ApJ, **474**, 188 (1997)
10. J. J. Feldmeier, R. Ciardullo, G. H. Jacoby: ApJ, **503**, 109 (1998)
11. H. C. Ford, D. C. Jenner, H. W. Epps: ApJ, **183**, L73 (1973)
12. H. C. Ford, D. C. Jenner: ApJ, **202**, 365 (1975)
13. W. -R. Hamann et al.: A&A, **409**, 969 (2003)
14. K. G. Henize: ApJS, **2**, 315 (1956)
15. G. H. Jacoby: ApJS, **42**, 1 (1980)
16. E. M. Lindsay: AJ, **66**, 169 (1961)
17. E. M. Lindsay, D. J. Mullan: Irish AJ, **6**, 51 (1963)
18. L. Magrini et al.: A&A, **407**, 51 (2003)
19. S. J. Meatheringham et al.: ApJ, **327**, 651 (1988)
20. S. J. Meatheringham, M. A. Dopita: ApJS, **75**, 407 (1991)
21. R. H. Méndez et al.: ApJ, **491**, L23 (1997)
22. R. H. Méndez, T. Soffner: A&A, **321**, 898 (1997)
23. R. H. Méndez et al.: ApJ, **563**, 135 (2001)
24. D. J. Monk, M. J. Barlow, R. E. S. Clegg, MNRAS, **234**, 583 (1988)
25. M. Pẽna et al.: ApJ, **428**, L9 (1994)
26. M. Pẽna et al.: A&A, **419**, 583 (2004)
27. N. Sanduleak, D. J. MacConnell, A. G. D. Philip: PASP. **90**, 621 (1978)
28. R. A. Shaw et al.: ApJ, **548**, 727 (2001)
29. L. Stanghellini et al.: ApJ, **510**, 687 (1999)
30. L. Stanghellini et al.: ApJ, **575**, 178 (2002)
31. L. Stanghellini et al.: ApJ, **596**, 997 (2003)
32. T. Theuns, S. J. Warren: MNRAS, **284**, L11 (1997)
33. E. Vassiliadis et al.: ApJS, **105**, 375 (1996)

Part II

Planetary Nebula Surveys

Surveys for Planetary Nebulae in the Magellanic Clouds

George H. Jacoby

WIYN Observatory
950 N. Cherry Avenue
Tucson, Arizona 85719

Abstract. As the nearest major galaxies to the Milky Way, the Magellanic Clouds are a highly productive focal point to study extragalactic PN. These PN are bright and spatially resolved with modern telescopes, providing a single environment where individual PN can be studied in the contexts of both stellar and galaxy evolution. Also, it is possible only in the Clouds to directly observe a large number of central stars at a known distance.

Surveys are approaching completeness: ∼65% in the smaller (∼ 20 sq. deg.) SMC and ∼35% in the (∼ 130 sq. deg.) LMC. The "bottom" of the PN luminosity function can now be reached with mosaic CCD imagers, allowing complete PN counts in major galaxies for the first time. Surveys, though, may always be incomplete due to a variety of factors.

The SMC is host to one of the most intriguing new results about PN – that there are gaps (or deficits) in the luminosity function. A gap may indicate that the progenitor stars are drawn from a few discrete stellar populations, or that there are two or more routes to producing PN (e.g., high mass single stars and low mass interacting binary stars). The Clouds are a unique location to investigate the origin of the deficit.

1 Motivation for Surveying the Clouds

The Magellanic Clouds are the closest major galaxies to the Milky Way. Although there are dwarf galaxies that are nearer, they are roughly 100 times lower luminosity and so, their planetary nebula (PN) populations are extremely sparse or nonexistent. Consequently, the study of extragalactic PN began in the Clouds with the pioneering surveys by Henize (1956), Lindsay (1961), Henize & Westerlund (1963), and Westerlund & Smith (1964). As a historical note, Baade and Swope (1955) identified 5 PN in M31 at about the same time.

The appeal of the Cloud PN is driven by the unique astrophysical advantages that they offer – each Cloud provides a large collection of PN at a common distance that is reasonably well-determined and relatively small. Webster (1976) was among the first to capitalize on the merits of the Cloud PN. The only comparable sample is the Milky Way's galactic center, and while they are 7 times closer, those PN suffer from severe interstellar extinction. Thus, the Cloud PN excel in the following areas:

- Proximity for brightness and spatial resolution
- Good sample size for statistical studies

- Low extinction/reddening for broad spectral coverage
- Low and intermediate metallicity samples
- Access to central stars at a known distance

Surveys provide a "kickstart" for the astrophysics. From the survey data alone, we can measure emission-line luminosities very accurately, usually in the [O III] λ5007 line, as well as the statistics of the PN counts. The latter provide a direct measure of the stellar death rates, and consequently, the rate of return of processed material back into the ISM.

With HST imaging (e.g., Stanghellini et al 2003) and UV spectroscopy, combined with ground-based spectroscopy (e.g., Meatheringham & Dopita 1991), it is straight-forward to measure an unequalled number of parameters directly, including those for the central stars, allowing one to build extremely accurate nebular photoionization models. These, in turn, allow the most accurate estimates for chemical composition diversity of the PN (Dopita & Meatheringham 1991), their central star ages and masses, and the nebular sizes and ages. By comparing the age of the nebula to that of the central star, we can test models of the late stages of stellar evolution in ways that are not possible anywhere else.

The large number ($\sim$ 10) of Cloud PN papers at this conference demonstrates their continuing attraction. Here, we discuss the current state-of-the-art in the surveys needed to find those PN.

2 Survey Techniques

In this paper, the term "survey" refers to the process of identifying previously unknown PN. There are two principal techniques that are used extensively – direct imaging through filters that isolate or reject specific emission lines (e.g., on-band/off-band [O III] imaging) and wide-field slitless spectroscopy (e.g., objective prism imaging). One can imagine applying other methods that were used in the Galaxy, for example, IRAS-like color surveys (van de Steene & Pottasch 1993).

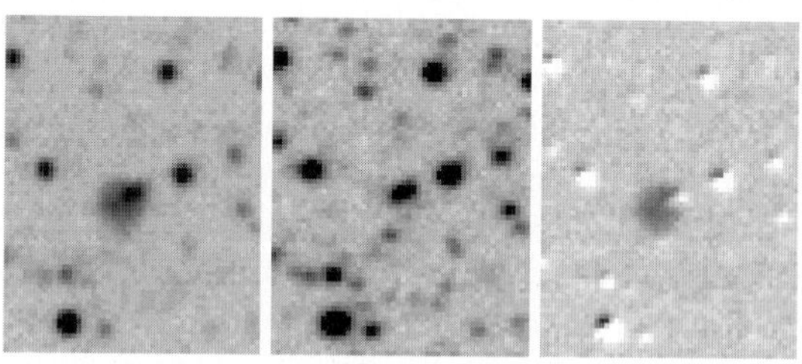

Fig. 1. An SMC PN identified by Jacoby & De Marco with the CTIO 4-m showing the on-band (left) and off-band (center) images. The right panel shows the "difference" image (with aliasing from imperfect subtractions). This candidate is not yet confirmed, but has coordinates α_{2000} = 00 49 53.8, δ_{2000} = −71 47 33 and is ~4 arcsec in diameter.

The on-band/off-band method has been extremely effective for finding PN in larger, more distant, galaxies where suppressing the background galaxy light is essential. With narrow-band filters centered on the doppler-shifted emission line of interest (usually λ5007 because it tends to be the brightest line), PN are identified from their presence in the on-band frame and their relative faintness in the off-band frame (Figure 1), either by visually scanning the difference image (subtraction of the off-band from the on-band), or by automated photometry, looking for large discrepancies between the objects on the 2 images. For the Cloud PN, automated methods are complicated by the likelihood that faint PN can have diameters of several arcsec or be completely stellar, or they may fall in the presence of clumpy diffuse HII emission. To date, automated methods (Feldmeier et al 2003) have not been used in the Clouds.

The use of an objective prism on a wide-field telescope with a medium band-pass filter offers important advantages. These were exploited famously by Sanduleak, MacConnell, & Philip (1978, SMP). The wide-field of the CTIO Curtis Schmidt greatly improved the survey efficiency, providing nearly complete coverage of each Cloud in one exposure. More importantly, a survey at either [O III] λ5007 or at Hα includes emission from nearby lines (either [O III] λ4959 or [N II] $\lambda\lambda$6548, 6583) and continuum to corroborate the classification of an object.

An exciting new variant on the latter method is the planetary nebula spectrograph, or PNS, as pioneered by Douglas et al (2002). It has demonstrated that PN can be found by the thousands when a special purpose instrument is designed and applied to a rich galaxy (Merrett et al 2003). This camera also yields kinematics for each PN, allowing a measurement of the host galaxy's dark matter content directly from the survey material (Romanowsky et al 2003, Romanowsky, these proceedings).

A survey technique that has been used extensively in the Milky Way, and potentially could be used in the Clouds, is high spatial resolution imaging. That is, one visually inspects deep images for objects that have the morphological appearance of PN. Abell (1966) found 85 old galactic PN in this way. One could imagine a small space-based imager doing this sort of survey, as is being done in the Galaxy by Parker et al (2003) from the ground.

While a large CCD mosaic is generally the preferred survey instrument, modern photographic material can be competitive when field of view is the primary metric. As reported at this conference, Reid & Parker (these proceedings) are surveying the LMC by stacking digitized Hα photographic images. They report finding $\sim$ 1000 candidate PN, but spectroscopic verification is not yet complete. As the next generation of CCD mosaic cameras come on-line (e.g. the 26x26 cm Omegacam on the VST, Kuiken et al 2004), the photographic advantage is marginalized, and one can imagine a VST survey of the Clouds in the future.

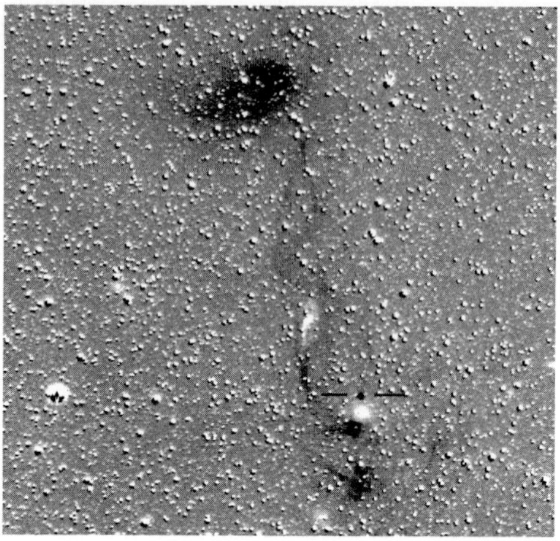

Fig. 2. An unusual example of a background galaxy cluster, probably suffering interactions, and whose redshift pushes an emission-line (probably Hβ λ4861) into the CTIO 4-m [O III] bandpass. The image represents 3.6 arcmin on the sky

3 Challenges to Surveys in the Clouds

Despite the ease with which PN can be found in the Clouds, finding all of them is a daunting challenge, especially the faintest. Here are the main stumbling blocks to a complete and reliable survey of PN in the Clouds.

3.1 Choice of Emission-Lines

For the SMC, the [O III] λ5007 line may be sub-optimal because the oxygen abundance is so low. At metallicities approaching 0.1 Solar, the λ5007 line is, on average, only slightly stronger than Hα. Other considerations for choosing an emission-line are (a) the sky is darker near λ5007, (b) confusion from HII regions and the brightness in diffuse emission is more serious in Hα, and (c) atmospheric seeing is better in the red. Ideally, one should perform a survey in several emission lines, but telescope time becomes a limiting factor.

3.2 Angular Extent

The proximity of the Clouds to the Milky Way (50 kpc to the LMC; 62 kpc to the SMC) is not all good news – proximity brings the burden of large angular extent on the sky. If one includes the Magellanic Stream, parts of these galaxies extend more than 90 deg across the sky (Putman et al 2003). The stream, though, contains very few stars and is not likely to be an important source of PN.

SMP recognized that PN in the LMC covered a vast area on the sky, at least 11 deg in diameter. Recent RGB and AGB star counts (van der Marel 2001) show that the stellar component of the LMC extends at least that far, covering ∼130 sq. deg. on the sky. To survey that field with the largest cameras available today would require ∼200 pointings. The SMC is not nearly as extended, covering only 20 sq. deg., but it still represents an observational and data handling challenge.

Neither Cloud has been 100% surveyed. Because the great majority of PN are close to the centrally luminous galaxy centers, though, roughly half of the PN in each galaxy are known. We estimate this fraction more precisely in Section 4.

3.3 Contamination

HII regions, supernova remnants, novae, and background galaxies are also strong line emitters. Thus, all PN candidates should be screened spectroscopically before publication. Several candidates in my first Cloud catalog (Jacoby 1980), for example, proved to be invalid after spectroscopy (Boroson & Liebert 1989).

Compact HII regions have always been a problem and some can mimic the spectra of a PN. This can be a serious problem in distant spiral galaxies and steps must be taken to minimize the impact of HII contamination (Feldmeier, Ciardullo, & Jacoby 1997) such as obtaining Hα images and avoiding spiral arms. The problem is much less severe in the Magellanic Clouds - the proximity allows the surveyer to spot HII regions from the strong continuum emitted by the O and B stars that provide the ionizing flux.

Novae were an unexpected contaminant. Novae are well-known to be strong Hα sources, but they also are weak [O III] emitters at certain phases. Two novae were recovered in the SMC and one in the LMC as potential PN candidates. Checking coordinates via SIMBAD revealed their true nature in these cases.

Background galaxies enter into the candidate list in several ways. The common scenario for more distant galaxies is that a Lyα emitter at redshift 3.1 falls into the [O III] filter bandpass. These galaxies are highly clustered according to their large-scale structure in the early universe, and so, there is no reliable estimate for the probable detection rate in a specific piece of sky. With a large uncertainty, rates are $\sim$ 70 per square degree for m(5007) $\sim$ 25.5-27.0. These magnitudes, where Lyα galaxies begin to emerge, is nearly 10 mag below the brightest PN in the Clouds and should never be a contaminant there.

One can also detect the nearer [O II] emitters entering the filter bandpass from a redshift of 0.34, but these are relatively rare because the volume of space at low redshift is so small. Similarly, Hβ emitters at a redshift of 0.03 can be detected on rare occasions such as the bizarre situation shown in Figure 2. This emission-line complex behind the SMC is interesting in its own right. Components of the complex could easily have been mistaken for 3-4 PN candidates, especially if only the brighter patches were imaged. The excellent depth of this CTIO 4-m image shows strands of connecting material in what may be an interacting group of galaxies. Coincidentally, there is one bona fide PN among the emission (MA-1682), as shown straddled by the horizontal lines.

Variable stars also can enter as a contaminant when the on-band/off-band method is used. A star may be in a dim cycle when the off-band image is taken.

3.4 Validation

Once a PN candidate has been identified, it must be confirmed to have the spectral signature of a true PN. Even if that test is passed, the object is not guaranteed to be a PN. This caveat is true even in our own galaxy where some objects have PN morphology and PN spectral properties but turn out to be something else (e.g., supernova or nova remnant, symbiotic star, HII region).

Nevertheless, without spectral evidence, an object should not be classified as a PN. For the more distant galaxies, this requirement is not practical because those objects are 5,000 times fainter; the reader is cautioned to consider a candidate as simply that - a candidate.

Even for nearby galaxies like the Clouds, spectroscopy is a challenge. The important diagnostic lines cannot be observed in very faint PN. Also, the Clouds present their own special conditions. Exacerbating their faintness, Cloud PN are generally spatially extended and only a fraction of their total flux makes it through the slit of most spectrographs. In addition, improper "sky" subtraction due to the ever-present diffuse emission and crowding by stars (Figure 3) complicates spectroscopy (but see Roth et al 2005).

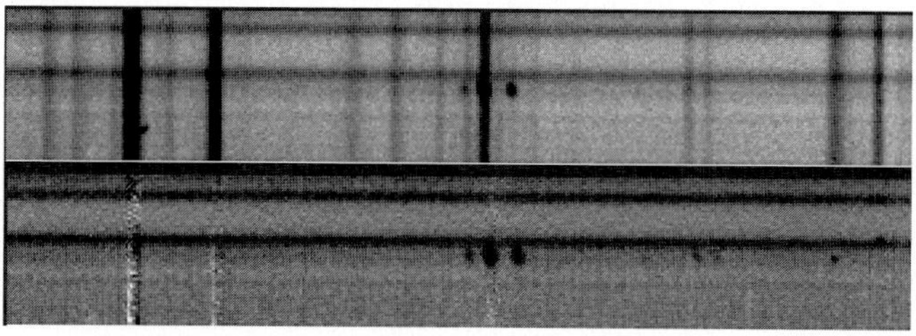

Fig. 3. The Hα region of SMC PN JD-17 from Jacoby & De Marco (2002). The upper panel shows the original spectrum and the lower panel shows the result after "sky subtraction". Emission from diffuse HII and night sky is reasonably well removed, but one of the many stars that fell along the slit did not subtract well. Subsequent emission-line analysis is partially contaminated by the stellar spectrum.

3.5 Nomenclature

PN in the Clouds have been studied by so many different individuals while the field was maturing that numerous naming conventions exist. The current state of Magellanic Cloud PN names is chaotic, with objects appearing in multiple catalogs under different names and at different coordinates. See Parker & Acker (2005) for suggestions to bring order to extragalactic PN cataloging schemes.

4 Estimated Number of PN in the Clouds

To estimate the total number of PN in each Cloud, I start with a summary of "modern" surveys in Tables 1 and 2.

Table 1. SMC Surveys

Survey	Number Found	Number New	Depth mags
SMP 1978	28	3	3
Jacoby 1980	27	19	5
Sanduleak & Pesch 1981	6	6	3
Morgan & Good 1985	13	10	3
Meyssonnier & Azzopardi 1993	62	18	4
Morgan 1995	62	9	3
Murphy & Bessel 2000	131	108	?
Jacoby & De Marco 2002	59	25	6
Galle, Winkler, & Smith[a]	69	13	4?
Jacoby & De Marco[a]	15	4	7

[a] Not yet published

Table 2. LMC Surveys

Survey	Number Found	Number New	Depth mags
SMP 1978	28	3	3
Jacoby 1980	27	19	5
Sanduleak 1984	25	13	3
Morgan & Good 1992	98	86	5
Morgan 1994	265	54	5
Leisy, Francois, & Fouqué[a]	10	4	9
Jacoby & De Marco[a]	15	10	7
Reid & Parker[a]	~ 1000	136	7?

[a] Not yet published

4.1 Survey Depth

Estimates for survey depth in the tables (usually in [O III]) are not very accurate; rarely is the topic discussed quantitatively in the Cloud PN literature. The second survey by Jacoby & De Marco (hereafter, JD2) used the CTIO 4-m, but is not published. It extends about 1 mag deeper than their 2002 survey (hereafter JD1) using the ESO 2.2-m. Consequently, we can see how many PN are found by extending a survey 1 or more mags deeper.

JD2 covered 6 SMC fields of about 0.36 square degrees each – 3 that were included in JD1 and 3 new fields a bit further out in the galaxy (Figure 4). The JD1 fields had 10 PN previously known; the deeper re-survey found 2 more PN, both in a single field having only 2 previously known PN (1 from SMP, 1 from JD1). Of the 3 fields not in JD1, 4 PN were previously known from Mayssonier & Azzopardi (1993), and 2 additional candidates were found in 2 different fields. Thus, by going 1 mag deeper, the yield increased by 20%; by going 3 mag deeper, the yield doubled. These numbers have statistically low weight.

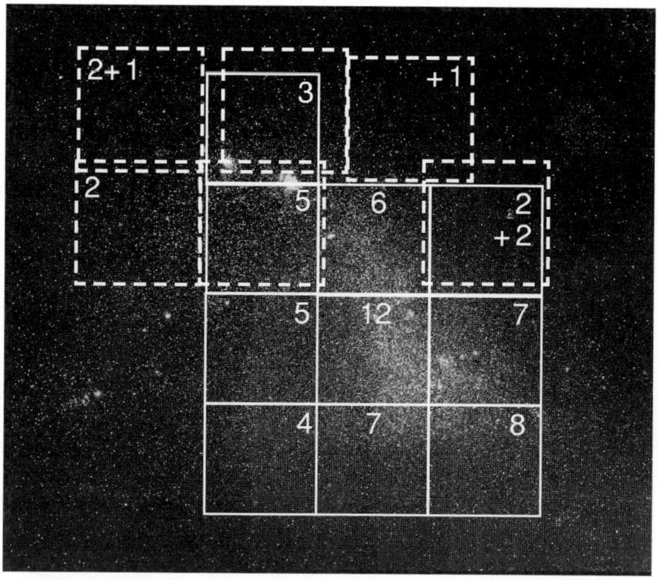

Fig. 4. The SMC fields of the first Jacoby & De Marco (2002) survey are shown as solid boxes; those from the second survey are shown with dashed lines. The numbers in the boxes indicate the number of PN in each field; those with a plus sign ("+") in the second survey fields indicate the number of candidate PN (but unconfirmed).

The single JD2 LMC field is at the eastern end of the bar where 5 PN were previously known (4 from SMP, 1 from Morgan 1994). JD2 identified 10 new candidates. By extending 2 mags, the number of PN was tripled. However, the statistics are still weak and conflict with the SMC numbers. A deep CCD survey of the LMC is an obvious project that would greatly improve the statistics.

4.2 Spatial Completeness

The SMC has been observed to faint magnitudes in JD1, JD2, and Jacoby (1980) to cover ∼ 65% of its surface. The remaining 35% is not likely to yield a large number of PN, given the fall-off seen in Figure 4 from the galaxy center.

The LMC has not been surveyed very deeply over much of the huge angular area required. Nevertheless, roughly 40% has been surveyed at depths of 4-6 mags down the PNLF, and nearly all of the LMC has been surveyed to ∼3 mags down by SMP and Sanduleak (1984).

4.3 Using the PN Luminosity Function

The PN luminosity function (PNLF) provides a guide for extrapolating shallow surveys to deeper limits. We begin with the SMC PNLF because the LMC has not been surveyed as widely or as deeply as the SMC even though the LMC has far more PN. The JD1 and JD2 surveys show that increasing the depth doesn't add a large number of PN, nor does extending the survey beyond the main body of the galaxy. Thus, the PN luminosity function in the SMC (Figure 5) is drawn from a reasonably complete and homogeneous sample and represents the state-of-the-art in terms of depth.

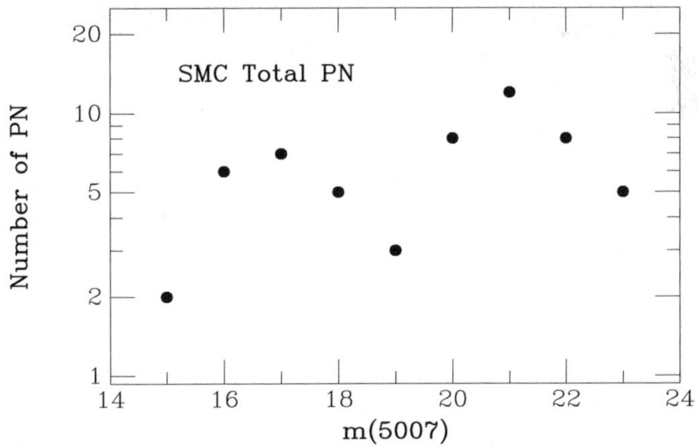

Fig. 5. The SMC PNLF from Jacoby & De Marco (2002). Note the pronounced deficit at m_{5007} = 19, and that the function drops beyond m_{5007} = 21. Both features are statistically significant with survey completeness to $m_{5007} \sim 22$.

Figure 5 suggests that completeness within the central survey regions are deep enough that the "bottom" of the PNLF may have been reached, where fainter PN can no longer be recognized. Jacoby (1980) defined completeness as 8 mag below the brightest, based on Galactic samples. While this definition should be revisited, it is adequate for now. Given the limited spatial survey, a total PN count requires a modest, but uncertain, correction for spatial incompleteness. On the other hand, the total number within the main body of the SMC has essentially been measured directly. From JD1 and JD2, this number is 67 PN. In all, there are 84 PN known in the SMC; thus we only need to correct the outer regions (17 PN) for depth. The outer surveys extend about 4 mags down the PNLF, representing 35.7% of the total. Thus, there are about 84 + 17/0.357, or ∼132, PN in the SMC.

The LMC is a greater challenge because of the diversity of survey material. There are ∼ 350 known PN in the LMC. If we also use the 4 mag survey limit as above, then the depth correction leads to ∼980 total PN. This is consistent with the Reid & Parker report of ∼ 1000, as well as the early statistics in the JD2 field.

5 The PNLF Deficit

Early studies of the PNLF (Jacoby 1980, Jacoby 1989) predicted that the number per magnitude should monotonically increase. In hindsight, those analyses were naive and obviously were wrong. Consequently, predictions of total numbers of PN based on extrapolating those PNLFs are inaccurate.

The implications of a PNLF deficit at ∼ 4 mag down the PNLF (Jacoby & De Marco 2002) may have profound impact on the entire PN field. This is very nearly the luminosity where PN derived from an old stellar population with low mass PN central stars would attain their limiting brightness (Marigo et al 2004). Figure 26 of Marigo et al (2004) is especially illuminating, as it shows the PNLF expected from a dual age population. That PNLF is effectively a composite of 2 PNLFs and looks remarkably similar to the SMC PNLF.

The Marigo model, though, fails dramatically in old elliptical galaxies. It predicts a peak PN luminosity about 4-5 mag fainter than is observed (e.g., M87 in Virgo). Thus, we have a puzzle - how can an old progenitor population produce a large PN sample with luminosities representative of 1 Gyr stars?

Perhaps the most promising avenue is a pathway through merged binary stars. A similar argument has been made to explain blue stragglers. For now, this is speculation – it is a challenge to get enough binary stars to merge in order to explain the large number of bright PN in elliptical galaxies. Nevertheless, it may be that most PN, if not all, form from binary stars, either as mergers or mass transfer recipients. Recent kinematic evidence suggests that a majority of galactic central stars are, in fact, binaries (De Marco et al 2004).

6 Astrophysical Questions for PN Enthusiasts

Almost any problem in the field of planetary nebulae can be studied better in the Clouds than anywhere else because these objects are at a known distance, many are very bright, they suffer little interstellar reddening, and most are spatially extended with space-based imagers. Thus, we can derive PN properties with great accuracy: e.g., absolute fluxes, physical sizes, and evolutionary ages.

The Clouds offer unique opportunities for studying the physics of PN, their role in galaxy evolution, and the clues they provide into the mysteries of stellar evolution. Some questions where Cloud PN have a special role are:

- How many PN are in the SMC and LMC? How do these counts compare to galaxy evolution models? Only in the Clouds can a complete count be made.
- What fraction of PN have binary central stars? Cloud PN are close enough together on the sky that multi-object spectrographs can be used to efficiently monitor central star kinematics. In the Galaxy, these must be observed one at a time; in more distant galaxies, the central stars cannot be seen.
- Can we map the dark matter in the Clouds? Velocities are easy to measure to 1 km/s. In the LMC, we have a sample of $\sim$ 1000 PN. A comparison between PN and stellar kinematics for different age stars (e.g., A-G stars) will yield clues to the statistically dominant progenitors of the PN. The result may implicate binaries as the source of many PN if their kinematics are consistent with an old population but they have high mass central stars. Only in the Clouds can the stellar and PN kinematics be measured.
- What is the distribution of central star masses? Does this depend on metallicity (comparing the SMC and LMC)? The Clouds provide the only large sample of observable central stars at a known distance.
- Can we be more quantitative about the correlations between morphology and PN properties (e.g., central star mass, age)? Only in the Clouds can morphology correlations be made.

Several projects emerge as obvious items on the PN "to do" list. There is still plenty of work remaining in the Clouds and much of it begins with surveys!

1. Complete a deep CCD survey of the LMC; finish the survey in the SMC.
2. Measure velocities for all the PN; this is a by-product of the follow-up spectroscopy needed to confirm the survey candidates.
3. Obtain high resolution images of as many PN as possible to measure their morphologies, radii, and central star luminosities. This should be possible with ground-based telescopes having AO systems.
4. Further develop models for the PNLF, given multiple age populations and the possibility of binary star interactions to enhance the bright end.

Acknowledgements

Orsola De Marco supported the CTIO 4-m survey observations and provided valuable ideas about PN formation from binary stars.

References

1. G. O. Abell: ApJ **144** 259 (1966)
2. W. Baade, H. Swope: A.J. **60**, 149 (1955)
3. T. A. Boroson & J. Liebert: ApJ **339** 844 (1989)
4. O. De Marco, H. E. Bond, D. Harmer, & A. J. Fleming: ApJL **602** 93 (2004)
5. M. A. Dopita & S. J. Meatheringham: ApJ **467** 115 (1991)
6. N. G. Douglas et al: PASP **114**, 1234 (2002)
7. J. J. Feldmeier: 'Intracluster Planetary Nebulae as Probes of Intracluster Starlight'. In: *Planetary Nebulae beyond the Milky Way at Garching, Germany, May 19–21, 2004*, ed. by J.R. Walsh, L. Stanghellini & N. Douglas (Springer, Heidelberg 2005), in press
8. J. J. Feldmeier, R. Ciardullo, & G. H. Jacoby: ApJ **479** 231 (1997)
9. J. J. Feldmeier, R. Ciardullo, G. H. Jacoby, & P. R. Durrell: ApJS **145** 65 (2003)
10. K. G. Henize: ApJS **2**, 315 (1956)
11. K. G. Henize & B. E. Westerlund: ApJ **137** 747 (1963)
12. G. H. Jacoby: ApJS **42** 1 (1980)
13. G. H. Jacoby: ApJ **339** 39 (1989)
14. G. H. Jacoby & O. De Marco: AJ **123** 269 (2002)
15. K. Kuiken et al: 'OmegaCAM: Wide-field imaging with fine spatial resolution'. In: *SPIE Conference Astronomical Telescopes and Instrumentation at Glasgow, Scotland, June 21-25, 2004*
16. E. M. Lindsay: AJ **66** 169 (1961)
17. P. Marigo et al: A&A **423** 995 (2004)
18. S. J. Meatheringham & M. A. Dopita: ApJS **75** 407 (1991)
19. H. R. Merrett et al: MNRAS **346**, 62 (2003)
20. N. Meyssonnier & M. Azzopardi: A&AS **102** 451 (1993)
21. D. H. Morgan: A&AS **103** 235 (1994)
22. D. H. Morgan: A&AS **112** 445 (1995)
23. D. H. Morgan & R. A. Good: MNRAS **213** 491 (1985)
24. M. T. Murphy & M. S. Bessell: MNRAS **311** 741 (2000)
25. Q. A. Parker et al: 'A Rich New Vein of Planetary Nebulae From the AAO/UKST Hα Survey'. In: *IAU Symposium No. 209, Planetary Nebulae: Their Evolution and Role in the Universe* ed. S. Kwok, M. Dopita, & P. Sutherland (ASP, San Francisco 2003) pp. 25–32
26. Q. A. Parker & A. Acker: 'Extragalactic PN Nomenclature'. In: *Planetary Nebulae beyond the Milky Way at Garching, Germany, May 19–21, 2004*, ed. by J.R. Walsh, L. Stanghellini & N. Douglas (Springer, Heidelberg 2005), in press
27. M. E. Putman et al: ApJ **597** 948 (2003)
28. A. I. Romanowsky et al: Sci **301**, 1696 (2003)
29. M. M. Roth, D. Schoenberner, T. Becker, & M. Steffen: 'Crowded-field 3D Spectrophotometry of PNe in Local Group Galaxies'. In: *Planetary Nebulae beyond the Milky Way at Garching, Germany, May 19–21, 2004*, ed. by J.R. Walsh, L. Stanghellini & N. Douglas (Springer, Heidelberg 2005), in press
30. N. Sanduleak: in *IAU Symp. 108: Structure and Evolution of the Magellanic Clouds, Tubingen, Germany, Sept 5-8, 1983*, ed. by S. van den Bergh (Reidel, Dordrecht 1984), p. 231 (1984)
31. N. Sanduleak, D. J. MacConnell, & A. G. D. Philip: PASP **90**, 621 (1978)
32. N. Sanduleak & P. Pesch: PASP **93** 431 (1981)

33. L. Stanghellini, R. A. Shaw, B. Balick, M. Mutchler, J. C. Blades, E. Villaver: ApJ **596** 997 (2003)
34. B. L. Webster: MNRAS **174** 513 (1976)
35. B. E. Westerlund & L. F. Smith: MNRAS **127** 449 (1964)
36. G. C. M. van de Steene & S. R. Pottasch: A&A **274** 895 (1993)
37. R. P. van der Marel: AJ **122** 1827 (2001)

H-alpha Stacked Images Reveal Large Numbers of PNe in the LMC

Warrenk Reid[1] and Quentin Parker[1,2]

[1] Macquarie University, Sydney, Australia
[2] Anglo-Australian Observatory, Sydney, Australia

Abstract. Our new, deep, high resolution Hα and matching R-band UKST multi-exposure stack of the central 25 sq. degrees of the LMC promises to provide an unprecedented homogeneous sample of >1,000 new PNe. Our preliminary 2dF spectroscopy on the AAT has vindicated our selection process and confirmed 136 new PNe and 57 emission-line stars out of a sample of 263 candidate sources within an initial 2.5 sq. deg. area. To date approximately one third of the entire LMC has been scanned for candidates (~7.5 sq. deg.). More than 750 new emission sources have been catalogued so far along with independent re-identification of all known and possible PNe found from other surveys.

Once our image analysis is complete, we plan comprehensive spectroscopic follow-up of the whole sample, not only to confirm our PN candidates but also to derive nebula temperatures and densities which, with the aid of photoionization modeling, will yield stellar parameters which are vital for constructing H-R diagrams for these objects. A prime objective of the survey is to produce a Luminosity Function which will be the most accurate and comprehensive ever derived in terms of numbers, magnitude range and evolutionary state; offering significant new insights into the LMC's evolutionary history. The observation and measurement of our newly discovered AGB halos around 60% of these PN will also assist in determining the initial-to-final mass ratios for this phase of stellar evolution.

1 Background

Since the Hα survey of the Magellanic Clouds by Henize (1956), subsequent Hα surveys have gone progressively deeper by increasing resolution and improving observational configurations. Although this resulted in improved sensitivity and increased the number of emission-line detections, observers continued to use single objective-prism plates and separated emission-line stars from nebulae according to whether a continuum could be seen adjacent to the Hα line.

The dominant method for the discovery of LMC PNe has been the identification of the [*OIII*] 5007, 4959 lines on objective-prism plates. The high number of candidates rejected through follow-up spectroscopy (Morgan 1995) shows that secondary plate images may play an important role in eliminating spurious identifications prior to spectroscopy. The need for a thorough, deep, Hα survey of the LMC to detect PNe became clear. As part of the AAO/UKST Hα survey of the South galactic plane (Parker & Phillips 2003), an equivalent mini-survey of the entire LMC and surrounding regions was also undertaken. This included an additional 12x 2-hour exposures and 6x 15-minute exposures on the central LMC field. The intention was to form a multi-exposure stack of these exposures to gain significant additional depth in order to search for faint emission-line sources.

1.1 The H-alpha Filter

The filter is effectively the world's largest monolithic interference filter to be used in astronomy. The central wavelength of 6590Å and bandpass 70Å work effectively in the UKST's fast f/2.48 converging beam (the 305mm diameter filter gives a 5.5deg. field with a peak transmission of 85%; Parker & Bland-Hawthorn, 1998).

1.2 The Imaging Detector: Kodak, Tech-Pan Emulsion

Generally speaking, in good seeing conditions, Tech-Pan UKST R- band exposures go about 1 magnitude deeper than the equivalent standard IIIaF R band images with better imaging, improved resolution and lower noise characteristics. Both the Hα and contemporaneous red exposures were made on Tech-Pan emulsion. The use of the same emulsion for both Hα and Short Red (SR) exposures ensures an excellent correspondence of their image point spread functions when film pairs are taken under the same observing conditions (Parker and Phillipps 2001). The Schmidt Tech-Pan films (e.g. Parker & Malin 1999) are ideal tools to find large numbers of PNe candidates within the LMC due to their inherent high resolution.

1.3 Data Reduction

The SuperCOSMOS machine in Edinburgh scanned and co-added the individual Hα/SR exposures on a pixel grid creating 0.67 arcsec pixels after 15-bit digitized plate scanning. The exposures reach depths of 21.5 for broad-band Red and $\sim$22 (4.5×10^{-15} ergs cm^2 s^{-1} Å^{-1}) for the Hα equivalent magnitude. This is much deeper than previous LMC Hα surveys such as Lindsay & Mullan (1963) R=$\sim$15 and Bohannan & Epps (1974) R=$\sim$14.5. Other more recent surveys have touched on the Magellanic Clouds in the process of surveying both hemispheres. The most contemporary of these is Southern H-alpha survey (SHASSA) undertaken at CTIO (Gaustad et.al., 2001). This is a full hemisphere survey however at 48 arcsec resolution, it doesn't offer the level of spatial resolution provided by the UKST H-alpha survey. The UM/CTIO Magellanic Cloud Emission-line Survey conducted by Chris Smith and the MCELS team (1998) uses a CCD detector on a Schmidt telescope covering the central 8 sq. deg. of the LMC. The instrument provides $2.035'$ pixel^{-1}, giving $\sim3''$ resolution with a field of view of 1.1 sq. deg.. Only the UKST survey can therefore detect and identify large numbers of faint point source emitters. Bland-Hawthorn, Shopbell and Malin (1993) have shown that digital stacking of UKST plates and films can achieve canonical poissonian depth gains. We estimate our stacked images reach $\sim$1.35 magnitudes deeper for Hα and 1 magnitudes fainter in Red than an individual exposure.

Another significant advantage of the stacking process is that small emulsion imperfections and adhering dust particles, scratches etc which can be problematic and lead to spurious detections are naturally eliminated as part of the combination process. Likewise the influence of variable stars is considerably alleviated by the time averaging process by stacking multiple exposures taken over a three year period.

2 Detection Technique

Candidate emission sources are found using an adaption of a technique whereby fits images from the deep SR stack are first coloured red and overlaid upon the matching stacked Hα images coloured blue. The coloured images are then merged and matched for point spread function (PSF) which is achieved both visually and through numerical intensity values, constantly displayed in the RGB program of KARMA. KARMA is a general-purpose image/movie display tool which can load multiple datasets, display multiple windows, overlay contours, annotations, show multiple overlaid profiles, etc. (Gooch, 1996). All continuum sources such as LMC Stars and background galaxies become a uniform colour but emission nebulosities and candidate compact emission sources develop a blue hue, making discovery and identification straightforward. Careful selection of software parameters allows the intensity of the matching Hα and SR images to be perfectly balanced allowing only peculiarities of one or other band-pass to be observed and measured. **All** previously known LMC PNe have been re-identified either by their extended halos surrounding the central PN or as areas of compact dense emission. In each case the Hα emission is seen as a bright blue aura.

Our preliminary application of this technique, applied to several 1-square degree sub-regions of the main LMC field, has produced extremely encouraging results. Areas were chosen which contained known emission sources and PNe, including some of the previous faintest detected sources such as PNeJ07 at m_B21.7, which was easily identified. Scrutiny of the combined, KARMA processed, 1-degree Hα/SR images revealed dozens of new PNe candidates in each sub-field. Several point sources exhibit very centrally concentrated emission, while others display rings and faint outer bubbles, halos or extensions. The enhanced angular widths of many of the emission sources revealed by our technique (up to 4 arcseconds in radius about the central star) strongly favour PNe in these cases.

3 Identification of New Candidate LMC Emission Sources

Only spectroscopic data will unambiguously determine the nature of the new candidate sources whilst their magnitude ($15 \leq$ R ≤ 21) and number density (100 sq/deg) make the 2dF spectrograph on the AAT the obvious choice for effective follow-up. All previously known PNe in the preliminary studied regions were successfully re-identified whilst the characteristics of the newly identified sources continue the trend to fainter magnitudes. We include every strong source that exhibits a halo whose diameter is >20% the diameter of the central source, regardless of the fact that many of them will be emission-line stars. The fact that Galactic Bulge Symbiotic stars and compact PNe look the same means we will not discard any star that appears to exhibit Hα emission. Therefore, we can be confident that we are extracting every unidentified PNe to the limit of the stacked data. Figure 1 shows the results of an initial AAT 2dF service time run where we were able to place almost 2/3 of the sources in a $1.25^{\circ 2}$ area on optical fibres for spectral confirmation. It can be seen that our data extends the LMC PNe range both in terms of number of detections and magnitude.

The apparent visual density of the Hα emission has proven to be a fair indication of the source type. All known PN are strong radiative emitters with thick emission halos that are more than double the visual diameter of the central source. By comparison with HST imaging of LMC PNe, we find film saturation increasing the diameter of the central PN source including shell structure by $\sim$4 times for an average magnitude 16 central PN at logarithmic intensity levels. This figure decreases to $\sim$2 times for the widest, large structure PNe such as SMP 93. Faint, wide- scale structures such as the AGB halos surrounding PNe will not suffer from the same PSF characteristics so we can be more confident about the large diameters we are observing.

As candidates are discovered, we record position, check data bases including (but not restricted to) SIMBAD for previous detections, allot an ID number and give each a probability rating. A note is made relating to any peculiarities associated with the source such as nearby bright or overlapping stars, bi-polar emission, shape, intensity, optical diameter, density, and proximity of the emission to any nearby *HII* regions. Magnitudes are found with the STARLINK

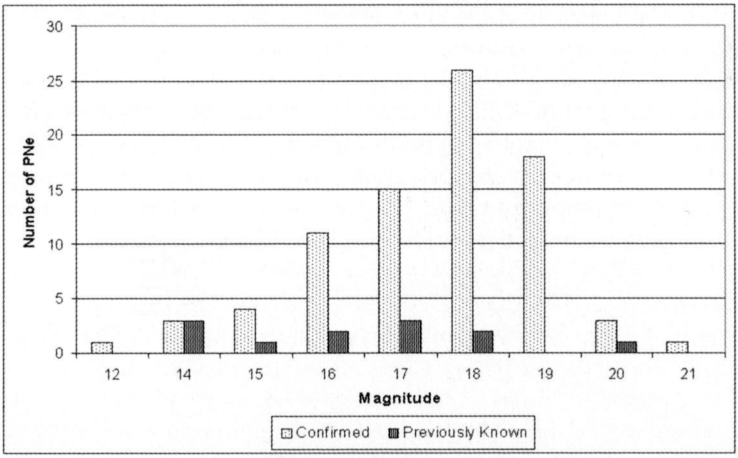

Fig. 1. Results of spectroscopic confirmation in a 1.25 sq. deg. area: number of PNe vs. magnitude.

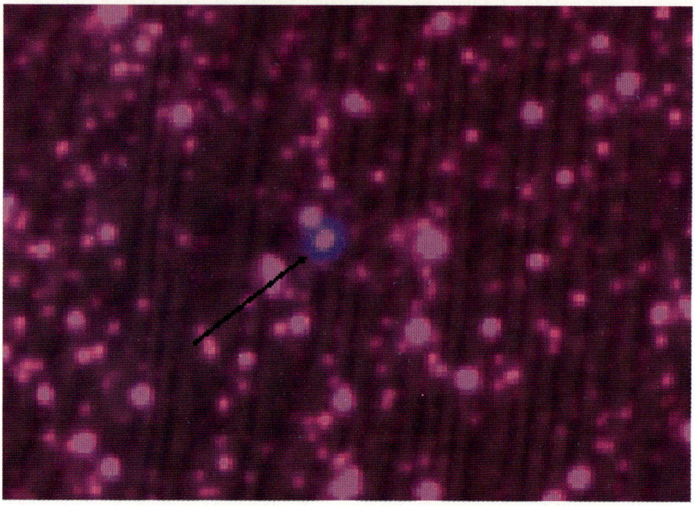

Fig. 2. Combined colour image of a newly discovered PN (shown arrowed).

PHOTOM package which has been calibrated to several catalogues and the SuperCOSMOS on-line UKST R-band catalogue.

4 Future Plans

With such a large number of PNe at a similar 50Kpc distance, the primary objective will be the creation of a comprehensive luminosity function. This will be achieved for the Hα line to allow us to better understand the evolutionary types in the LMC. It will also provide a probe into the structure and origin of the LMC as well as providing a distance probe when applied to other Hubble-type galaxies. Spectroscopic analysis will add further to our knowledge of AGB evolution as well as chemical enrichment in the LMC, especially when combined with morphology and evolutionary types. HR Diagrams are useful tools for examining the evolution of stars and understanding the different phases that evolve from differing stellar and nebula parameters. Extended late AGB phase Halos, which we have discovered for the first time in LMC PNe, will be examined, particularly their diameters and shapes with relation to the central PNe. The initial-to-final mass relation is another important area of stellar physics that can now be re-examined from measurements of halo parameters; since all of our sources are at a similar distance. Our data-base will provide a comprehensive catalogue which will be published and accessible on-line.

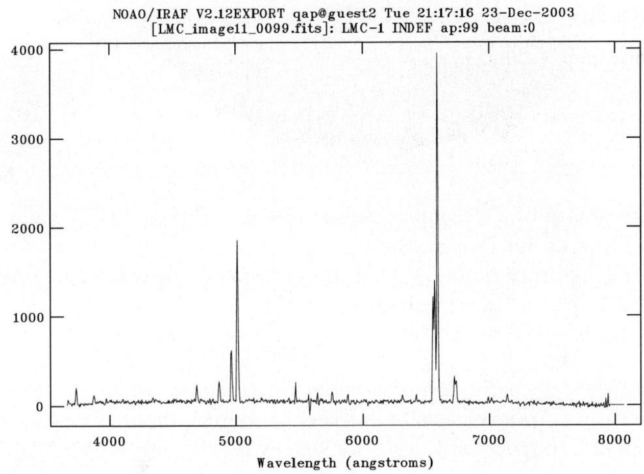

Fig. 3. Spectroscopically confirmed PN.

5 Conclusion

So far 6 out of 16 image cells have been analyzed. The detection rate increases 3-fold on the main bar. Approximately half the 263 candidates followed-up on 2dF have been confirmed as PN, 57 are emission-line stars, 33 require longer exposure times, while 33 are non-PNe detections which include late-type stars. At these rates, we expect to confirm more than 1,000 new PNe in December when we return to 2dF on the AAT.

References

1. Bland-Hawthorn J., Shopbell P.L. & Malin D.F. 1993, AJ., 106, 2154B
2. Bohannan, B.E. & Epps, H.W. 1974, A&AS, 18, 47
3. Gaustad, J.E., McCullough, P.R., Rosing, W. & Van Buren, D., 2001, PASP, 113, 1326
4. Gooch, R., 1996. Proc. ADASS. Ed. G. Jacoby & J. Barnes, 80 Henize, K.G. 1956, ApJS 42, 1
5. Henize, K.G. 1956, ApJS 42, 1
6. Morgan, D.H., 1995, A&AS, 112, 445
7. Lindsay, E.M. and Mullan D.J., 1963, Irish A.J., 6, 51
8. Parker, Q.A., Bland-Hawthorn, J., 1998 PASA, 15, 33p
9. Parker, Q.A. & Malin D., 1999, PASA 16, 288P
10. Parker, Q.A. & Phillipps, S. 2001, in ASP Conference Series No. 232. Eds. R. Clowes, A. Adamson & G. Bromage, 38P
11. Parker, Q.A. & Phillipps, S., 2003, in ASP Conference Series No. 289. Eds. S. Ikeuchi, J. Hearnshaw, T. Hanawa, 165P
12. Smith, C. & MCELS Team, 1998, Pub. Astron. Soc. Aust., 15, 163

The Census of Planetary Nebulae in the Local Group

Romano L.M. Corradi[1] and Laura Magrini[2]

[1] Isaac Newton Group of Telescopes, Apartado de Correos 321, 38700 Santa Cruz de La Palma, Spain
[2] Dipartimento di Astronomia e Scienza dello Spazio, Universitá di Firenze, L.go E. Fermi 2, 50125 Firenze, Italy

Abstract. A review of the imaging searches for planetary nebulae (PNe) in the Local Group (LG), with an updated list of references, is presented. Emphasis is put on the results of the Local Group Census, a recent survey carried out at the 2.5m INT telescope at La Palma.

PNe have been searched and found in all LG galaxies with $\log(L_V/L_\odot) > 6.7$; very few PNe are expected to be found in less luminous galaxies. The PN population size, when corrected for the different completeness limits of the various surveys, scales pretty well with the V luminosity of a galaxy, but it is suggested that it is somewhat sensitive to the relative amount of star formation in the last 2 to 8 Gyr. Some possible evidence of the effect of star formation in the last few Gyr on the absolute magnitude of the bright cutoff of the PN luminosity function is also presented.

1 An Overview of the Local Group

Table 1 presents a list of the LG galaxies included in the book of van den Berg ([34], hereafter vdB00). Their morphological type, absolute magnitude, and distance are indicated.

All galaxies within 1.3 Mpc from the Milky Way can be considered to belong to the LG. In particular, we can identify two main subgroups and a number of "free-floating" objects. The Milky Way subgroup includes our own galaxy and a number of irregular and spheroidal companions, the most massive ones being the Magellanic Clouds. The M31 subgroup includes all systems bound to the Andromeda galaxy, ranging from its relatively bright companions M32, NGC 205, NGC 147 and NGC 185 (the latter two located close to the centre of mass of the LG), and possibly the starburst irregular IC 10, to a number of much fainter dwarf galaxies. The third main spiral galaxy of the LG, M33, is instead quite isolated in spite of being seen projected in the sky not far from M31. In addition, a number of isolated, but quite interesting objects, are also found in the LG, like for example the irregular galaxies NGC 6822, IC 1613, WLM and Leo A. It is likely that the number of these isolated galaxies is larger than presently known, as searches for LG galaxies have been mostly concentrated around the Milky Way and in the direction of M31/M33.

In the outskirts of the LG, a few more dwarf galaxies are found (also listed in Table 1), among which NGC 3109, Antlia, Sextans A and B might form a physical association at about 1.3 Mpc from us. Beyond it, the closest systems of galaxies outside the LG are the IC 342/Maffei, South Polar (Sculptor), and M81 groups.

2 Imaging Searches for PNe in the LG

The chase for extragalactic PNe started with the discovery by W. Baade [2] of five PNe in M31. In the 60's, a hundred of candidate PNe were discovered in the Magellanic Clouds (see e.g. [35]), and one decade later the search was extended by H. Ford and collaborators to M31 and its companions [11] [12] [13] [14], and to other LG galaxies [17].

The number of candidate PNe in M31 increased to 429 in 1989 with the observations of [4], while we had to wait until 2000 to find a significant number (134) of these objects in M33 [22]. In more recent years, [19] made a very deep survey of the SMC, which is discussed in the review of G. Jacoby, these proceedings, while the advent of a dedicated spectrograph like the Planetary Nebula Spectrograph allowed to discover 2766 PNe in M31 [29], making the Andromeda galaxy the galactic system with the largest number of (candidate) PNe presently known.

2.1 The Local Group Census

A significant increase of the number of candidate PNe in several galaxies of the LG was recently achieved by the so-called "Local Group Census" (LGC), a narrowband imaging survey of the LG galaxies with Dec$\geq -30°$. Observations were carried out using the Wide Field Camera (WFC) of the 2.5m INT telescope of the Isaac Newton Group of Telescopes (ING) at La Palma. The WFC is equipped with a mosaic of four 2k$\times$4k EEV CCDs covering a field of view of $34' \times 34'$. The collaboration includes astronomers from various institutions in the UK, Spain, the Netherlands, as well as ESO in the person of the host of this conference, Jeremy Walsh, and it is on behalf of all them that we summarize here some details and results of the LGC.

The observations of the LGC have almost been completed during some 45 clear nights of observations in different runs from 2001 to 2003. All data are available to the astronomical community. Most galaxies were observed in four narrow filters which select the nebular emission lines [O III]500.7 nm, Hα+[N II], [S II]671.7,.673.1 nm, and He II468.6 nm. Additional images were taken in the broadband Sloan g', r', i' and the *Strömgren Y* filters, this latter one particularly suited to be used as a 'continuum' filter for [O III]. Exposure times were 1 hour for the narrowband filters, 30 min for *Strömgren Y*, and 20 min for the Sloan filters. This allowed us to reach, during photometric dark nights with good seeing conditions, a completeness limit corresponding to an equivalent V magnitude around 24.5 mag (using the standard definition of $m_{5007} = -2.5 \log F_{5007} - 13.74$ [18]) for unresolved emission–line objects in relatively crowded regions of galaxies.

Table 1. The galaxies in the Local Group and at its outer fringes

Name	Type	M_V	Dist [kpc]	Number of PNe	Reference
Local Group					
M31	Sb	−21.2	760	2766	[29]
Milky Way	Sbc	−20.9		∼2400	[1],[30]
M33	Sc	−18.9	795	152	[6]
LMC	Ir	−18.5	50	277	[20]
SMC	Ir	−17.1	59	105	[20]
M32	E2	−16.5	760	30	[4]
NGC 205	Sph	−16.4	760	35	[7]
IC 10	Ir	−16.3	660	16	[24]
NGC 6822	dIr	−16.0	500	17	[21]
NGC 185	Sph	−15.6	660	5	[7]
IC 1613	dIr	−15.3	725	3	[25]
NGC 147	Sph	−15.1	660	9	[7]
WLM	dIr	−14.4	925	1	[25]
Sagittarius	dSph	−13.8	24	3	[37]
Fornax	dSph	−13.1	138	1	[8]
Pegasus	dIr	−12.3	760	1	[17]
Leo I	dSph	−11.9	250		
And I	dSph	−11.8	810		
And II	dSph	−11.8	700		
Leo A	dIr	−11.5	690	1	[24]
DD 210	dIr	−11.3	1025		
Sag DIGD	dIr	−10.7	1300		
Pegasus II	dSph	−10.6	830		
Pisces	dIr	−10.4	810		
And V	dSph	−10.2	810		
And III	dSph	−10.2	760		
Leo II	dSph	−10.1	210		
Phoenix	dSph	−9.8	395		
Sculptor	dSph	−9.8	87		
Cassiopeia	dSph	−9.5	690		
Tucana	dSph	−9.6	870		
Sextans	dSph	−9.5	86		
Carina	dSph	−9.4	100		
Draco	dSph	−8.6	79		
Ursa Minor	dSph	−8.5	63		
LG outskirts					
GR 8	dSph	−11.6	2200	0	[25]
Antlia	dSph	−15.8	1330		
NGC 3109	dIr	−15.8	1330	18	[31]
Sextans B	dIr	−14.3	1600	5	[23]
Sextans A	dIr	−14.2	1320	1	[24]
EGB0427+63	dIr	−10.9	2200		

The galaxies observed by the LGC are (in brackets are the number of WFC fields when larger than one): Draco (4), DD210, EGB 0427+63, GR 8, IC 10, IC 1613, Leo A, Leo II, Leo I, M 33 (7), NGC 147, NGC 185, NGC 205 (2), NGC 6822 (4), Pegasus, Sextans A, Sextans B, Ursa Minor (4), and WLM. This list includes most galaxies in the LG in the Northern hemisphere, the most notable exception being M 31, for which observations aimed at detecting PNe were already being obtained by other ING programmes.

More details about the survey and a complete list of the data obtained can be found at

http://www.ing.iac.es/~rcorradi/LGC/

To date, data from the LGC have been analysed for the dwarf irregular galaxies Sextans A and B, IC 10, WLM, IC 1613, NGC 6822, Leo A, GR8 (see Table 1 for the list of references), as well as for the spheroidal companions to M31 – NGC 147, NGC 185, and NGC 205, significantly improving the number of candidate PNe known in these galaxies. Most of the detected PNe are found inside the $m_{pg} = 26.5$ mag arcsec^{-2} isophotal diameter of the galaxies, the most notable exceptions being several PNe in IC 10 and NGC 6822 which are found at large galactocentric distances. It should be noted that both IC 10 and NGC 6822 have the peculiarity, compared to other LG galaxies, that they are surrounded by huge neutral hydrogen envelopes [16] [9]. As PNe are excellent tracers of stellar populations in large volumes with low density of stars, whose integrated stellar light is hardly detectable, these 'isolated' PNe might reveal the presence of conspicuous stellar populations at galactocentric distances much larger than considered so far, possibly related to the large haloes of neutral gas. Spectroscopic confirmation of their nature as PNe and their association (via radial velocity studies) with the galaxies is needed to strengthen this possibility.

It should be mentioned here that another narrow- and broad-band survey of the LG, carried on by P. Massey and collaborators at the KPNO and CTIO 4-m telescopes, exists. As for the LGC, data are available to the whole astronomical community. Further information can be obtained at

http://www.lowell.edu/users/massey/lgsurvey.html

3 Discussion

3.1 The PN Population of the LG

Columns 5 and 6 of Table 1 list the number of candidate PNe known in each galaxy of the LG at the time this paper was written, and the references to the discovery paper. For all galaxies but the Milky Way, only the reference to the most recent and extensive work is given. The reader is referred to these articles for information about previous work. Some statistical properties of the PNe sample of the LG are presented below.

3.2 PN Population Size and Luminosity of LG Galaxies

Stellar population synthesis theories [32] predict the number of PNe in a single stellar population (a chemically homogeneous system of coeval stars) to be

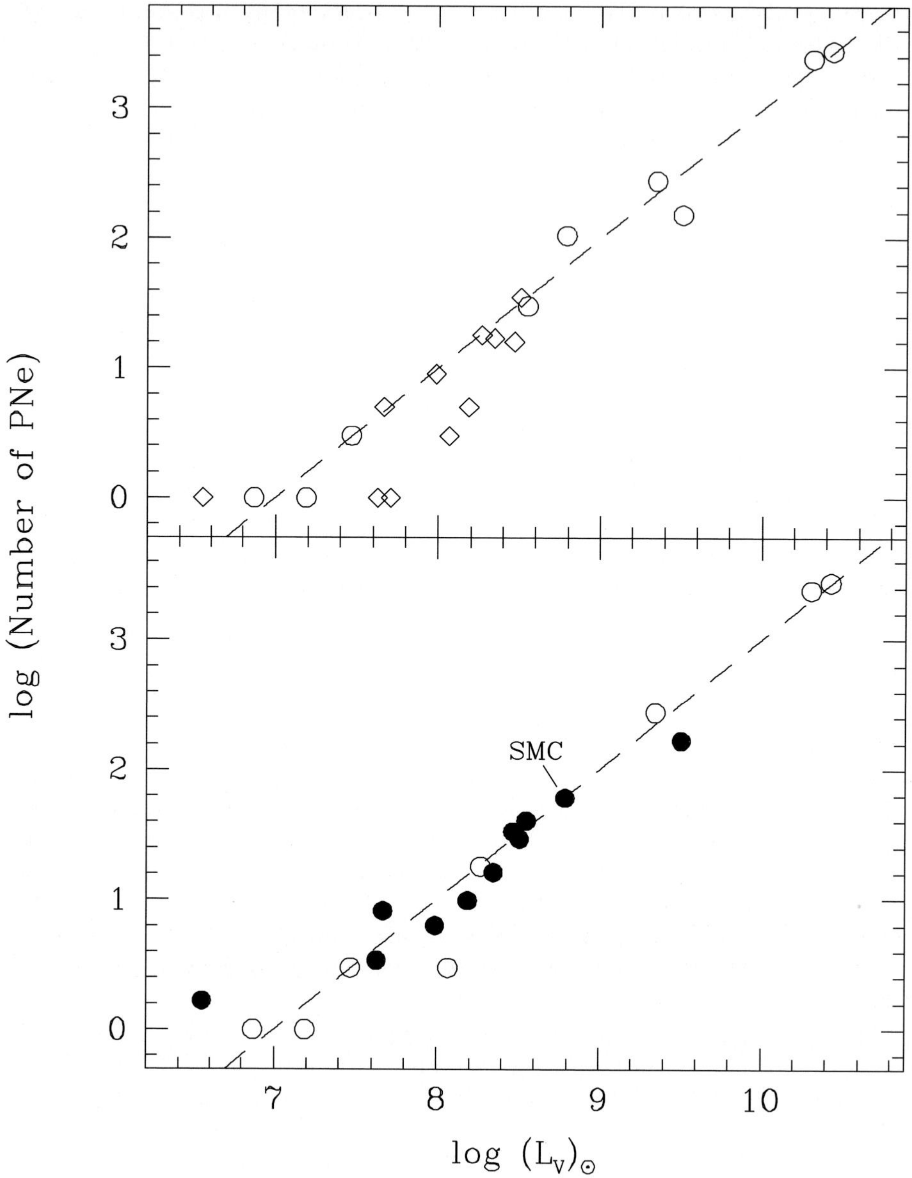

Fig. 1. Number of PNe as a function of the absolute V luminosity for LG galaxies. Top: observed number of PNe (from Table 1). Diamonds are data from the LGC, open circles are from other sources. Bottom: PN population size extrapolated to 4 mag below the bright end of the PNLF (filled circles). Open circles in this plot are points to which such a completeness correction cannot be applied. The dashed line is the scaling relation normalized using the observed population of the SMC, where the search for PNe is the most complete one [20].

directly proportional to the bolometric luminosity of the population, the proportionality coefficient being weakly dependent on age, metallicity and form of the initial mass function. For more complex systems like real galaxies, the proportionality coefficient, which is called the PN Specific Luminosity Rate (No. of PNe $L_{\odot\, bol}^{-1}$), is predicted to have a more significant dependence on the galaxian type [3]. Even so, a roughly linear correlation might be expected between the total number of PNe in a galaxy and its total luminosity. This was indeed shown by [24] for the LG galaxies, and an updated version of this relation is presented in Fig. 1. The V luminosity is used instead of the bolometric one, as the bolometric correction does not vary dramatically with the galaxian type [3]. In the top graph, the number of candidate PNe discovered by the various LG surveys is plotted. While a good correlation appears, it is definitely improved in the lower plot, where the number of PNe is extrapolated to the same completeness limit of 4 mag below the bright end of the PN luminosity function (PNLF) using its usual analytical representation [4], and for all cases in which the information of the completeness limit of the observations is available (filled circles).

Table 1 shows that PNe were searched and found in all LG galaxies with $M_V < -11.9$ ($\log L_V/L_\odot > 6.7$); considering the relation in Fig. 1, it is clear that only few PNe will be discovered in the relatively large number of lower luminosity galaxies of the LG.

3.3 Age Effects

In spite of the good correlation in Fig. 1, some age effects on the PN population size of LG galaxies are revealed if the star formation history at intermediate ages (2 to 8 Gyr) is considered. Fig. 2 shows the relation between the observed number of PNe and the total mass of LG galaxies. The solid line separates LG galaxies which had a stronger star formation in the last 2 to 8 Gyr (> 0.2 relative to the whole star formation history of the galaxy), from those where star formation was lower (data from [28]). Therefore the conclusion is that the PN population size of a galaxy is sensitive to the relative importance of star formation at these intermediate ages.

Another interesting suggestion about possible age effects on the [O III] luminosity of PNe comes from the comparison between NGC 147 and NGC 185 [7]. These two spheroidal galaxies are very similar, as far as morphological type, luminosity, and metallicity are concerned (vdB00). The main difference is their gas content and recent star formation history: while NGC 147 is gas free, and its star formation might have stopped some 2-3 Gyr ago, NGC 185 is gas rich, and active star formation in the last few Gyr is likely to have occurred in the inner region where all the five candidate PNe are located.

Fig. 3 shows the histogram of the absolute [O III] magnitudes of the candidate PNe in NGC 147 and NGC 185. In spite of the small number of objects involved, it seems that PNe in NGC 147 are systematically fainter than those of NGC 185, which populate the bright end of the "universal" PNLF. One possible explanation is that the bright PNe of NGC 185 are the product of stars formed in the last few Gyrs, belonging to a relatively young stellar population that is

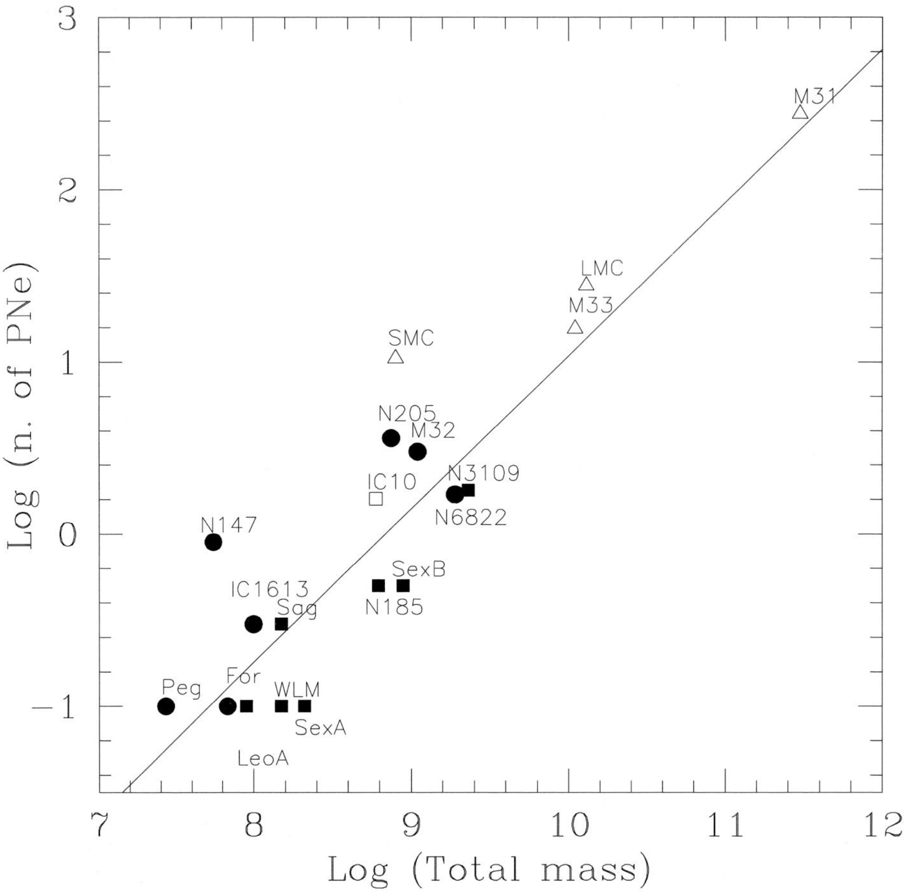

Fig. 2. The observed number of PNe vs. the total mass of LG galaxies. Filled circles are LG galaxies with relatively strong star formation between 2 and 8 Gyr ago (more than 20% of their whole star formation history), and filled squares those with lower star formation at these intermediate ages.

instead missing in NGC 147. This would support the results of [27] (see also [15]), who predicted the bright end of the PNLF to be populated by stars with an initial mass of 2 to 2.5 solar masses, whose lifetime is of 1-2 Gyr. This result, however, should be taken with extreme caution because of the small size of the PN samples of NGC 147 and NGC 185. A better knowledge of the recent star formation history of NGC 147 would also be needed.

3.4 Metallicity Effects

[24] suggested a possible fall in the observed number of PNe when [Fe/H]<<-1.0, which would indicate that below this metallicity the PN formation rate is quenched. This would be consistent with the fact that [Fe/H]=−1.0 corresponds

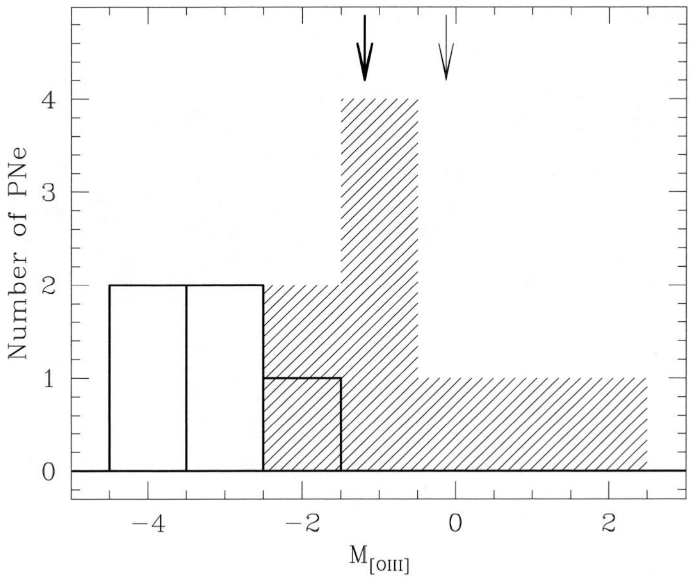

Fig. 3. Comparison of the absolute magnitude of the candidate PNe in NGC 147 (dashed histogram) and NGC 185 (thick solid line) [7]. The completeness limits of the LGC observations are indicated for each galaxy by the arrows in the upper part of the plot.

to the metallicity below which the AGB wind is expected to be driven by pulsations only, and no longer by radiation pressure on dust [36], which might in turn imply a reduced PN formation efficiency. More data are however needed to confirm this result, and other possible factors should also be considered (e.g. a reduced detectability of PNe via [O III] imaging at very low metallicities).

Fig. 4, an updated version of Fig. 4 in [24], shows the absolute [O III] magnitude of the brightest PN in the LG galaxies, before and after correction for the metallicity dependence on the bright cutoff of the PNLF using the prescription of [10]. After correcting for metallicity, points are more tightly grouped around the "universal" value of the PNLF cutoff. This seems to confirm that the metallicity correction by [10] is at least qualitatively correct, but it is difficult to disentangle this effect from the fact that the most metal poor galaxies are also the less massive ones, and owing to their small PN population sizes it is statistically more unlikely to observe PNe close to the PNLF cutoff.

4 Perspectives

The recent, new detections of PNe in many galaxies of the LG are clearly a starting point for future spectroscopical studies of individual objects, aimed at confirming their nature as PNe and, more important, at determining their

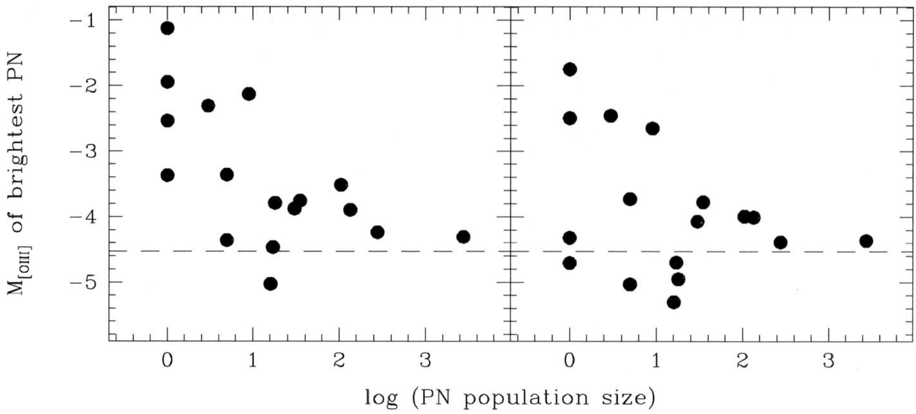

Fig. 4. The observed number of PNe vs. the absolute magnitude of the brightest PN for LG galaxies, before (left) and after (right) metallicity correction following [10]. The horizontal line shows the magnitude of the PNLF cutoff for solar metallicity [5].

physical and chemical properties and of their host galaxies. We have started such a programme at the observatories in Chile and La Palma (see e.g. [26] and [33]).

Concerning future imaging of the LG, Jacoby (see [20]) has opened an interesting route in trying to get to the faint end of the SMC PNLF. Similar works for other LG galaxies would result in accurate PNe counts which would allow a better understanding of the roles of metallicity and star formation history on the PN population sizes and their luminosity function.

References

1. Acker, A., Ochsenbein, F., Stenholm, B., Tylenda, R., Marcout, J., Schohn, C., 1992, 1996, Strasbourg–ESO catalogue of Galactic planetary nebulae, ESO
2. Baade, W., 1955, AJ, 60, 151
3. Buzzoni, A., Arnaboldi, M., this book
4. Ciardullo, R., Jacoby, G.H., Ford, H.C., Neill, J.D. 1989, ApJ, 339, 53
5. Ciardullo R., Feldmeier J.J., Jacoby G.H., Kuzio de Naray R., Laychak M., Durrell P.R., 2002, ApJ, 577, 31
6. Ciardullo R., et al. 2004, ApJ, in press
7. Corradi, R.L.M., Magrini, L., Greimel, R., Irwin, M., Leisy, P., Lennon, D.J., Mampaso, A., Perinotto, M., Pollacco, D.L., Walsh, J.R., Walton, N.A., Zijlstra, A.A., 2004, A&A, in press
8. Danziger, I.J., Webster, B.L., Dopita, M.A., Hawarden, T.G, 1978, ApJ, 220, 458
9. de Blok, W.J.G., Walter, F., 2000, ApJ 537, L95
10. Dopita, M. A., Jacoby, G. H., Vassiliadis, E. 1992, AJ, 389, 27
11. Ford, H.C., Jenner, D.C., Epps, H.W., 1973, AJ, 183, 73
12. Ford, H.C., Jenner, D.C., 1975, ApJ, 202, 365 (F75)
13. Ford, H.C., Jacoby, G.H., Jenner, D.C., 1977, ApJ, 213, 18

14. Ford, H.C., Jacoby, G.H. 1978, ApJS, 38, 351
15. Girardi, L., et al., this book
16. Huchtmeier, W.K., 1979 A&A, 75, 170
17. Jacoby, G.H., Lesser, M.P., 1981, AJ, 86, 185
18. Jacoby, G.H., 1989, ApJ, 339, 39
19. Jacoby, G.H., De Marco, O., 2002, AJ, 123, 269
20. Jacoby, G.H., this book
21. Leisy, P., Corradi, R.L.M., Magrini, L., Greimel, R., Mampaso, A., Dennefeld, M., 2004, A&A, submitted
22. Magrini L., Corradi, R.L.M., Mampaso, A., Perinotto, M. 2000, A&A, 355, 713
23. Magrini, L., Corradi, R.L.M., Walton, N.A., Zijlstra, A. A., Pollacco, D.L., Walsh, J.R., Perinotto, M., Lennon, D.J., Greimel, R., 2002, A&A, 386, 869
24. Magrini, L., Corradi, R.L.M., Greimel, R., Leisy, P., Lennon, D.J., Mampaso, A., Perinotto, M., Pollacco, D.L., Walsh, J.R., Walton, N.A., Zijlstra, A.A., 2003, A&A, 407, 51
25. Magrini, L., Corradi, R.L.M., Greimel, R., Leisy, P., Lennon, D.J., Mampaso, A., Perinotto, M., Pollacco, D.L., Walsh, J.R., Walton, N.A., Zijlstra, A.A., 2004, MNRAS, in preparation
26. Magrini, L., Leisy, P., Corradi, R.L.M., Perinotto, M., Mampaso, A., Vilchez, J.M., this book
27. Marigo, P., Girardi, L., Weiss, A., Groenewegen, M.A.T., Chiosi, C., 2004, A&A, submitted
28. Mateo, M., 1998 A&A Rev 36, 435
29. Merrett, H., et al., this book
30. Parker, Q.A., Hartley, M., Russeil, D., Acker, A., Ochsenbein, F., Morgan, D.H., Beaulieu, S., Morris, R., Phillipps, S., Cohen, M. 2003, IAU Symp. 209, S. Kwok, M. Dopita, & R. Sutherland eds, ASP/IAU, p.41
31. Prada, F., et al., 2004, in preparation
32. Renzini, A., Buzzoni, A., 1986 in *Spectral Evolution of Galaxies*, eds. Chiosi C. and Renzini A., Reidel, Ap. Space Sci. Lib., Vol. 122, p.195
33. Stasinska, G., Vílchez, J.M., Pérez, E., Gonzalez Delgado, R.M., Corradi, R.L.M., Mampaso, A., Magrini, L., this book
34. van den Bergh, S., 2000, in *The Galaxies of the Local Group*, (Cambridge: Cambridge University Press), p.177 (vdB00)
35. Westerlund, B.E., IAU Symp. n. 34, Osterbrock D.E. & O'Dell C.R. eds., Dordrecht, p.23
36. Zijlstra, A.A., 1999, in Asymptotic Giant Branch Stars, IAU Symposium 191, Eds T. Le Bertre, A. Lebre & C. Waelkens, p. 551
37. Zijlstra, A.A., et al., this book

Planetary Nebulae in the Outer Disk and Halo of M31

Alexei Kniazev[1], Eva K. Grebel[2], Dan Zucker[1], Eric Bell[1], and Hugh Harris[3]

[1] Max-Planck-Institut für Astronomie, Königstuhl 17, D-69117 Heidelberg, Germany
[2] Astronomisches Institut, Universität Basel, Venusstrasse 7, CH-4102 Binningen, Switzerland
[3] U.S. Naval Observatory, Flagstaff Station, P.O. Box 1149, Flagstaff, USA

Abstract. We have developed a method to identify planetary nebula (PN) candidates in imaging data of the Sloan Digital Sky Survey (SDSS). This method exploits the SDSS's five-band sampling of emission lines in PN spectra, which results in a color signature distinct from that of other sources and which can be applied to galaxies where PNe appear as point sources. We tested this method with SDSS data of M31. In areas searched by previous authors, we were able to recover about 6/7 of the previously known PNe. All unrecovered PNe are located in overcrowded fields and were not detected by the standard SDSS pipeline. We then extended this search to the full area of M31's halo and disk sampled by the SDSS.

1 Introduction

Tidal streams from disrupting dwarf satellites provide us information both about the progenitor satellite and the dark matter halo of the parent galaxy. To constrain the orbit of the stream and of the progenitor one needs the kinematics of stream stars which requires 8–10m class telescopes at the distance of M31. At the same time planetary nebulae (PNe) provide a representative tracer of the progenitor stars and permit velocity measurements based on emission lines, which only requires 2–4m class telescopes for the Local Group. In addition, abundances from both HII regions and PNe allow us to derive an approximate enrichment history for a galaxy from intermediate ages to the present day.

2 Search for PNe Candidates in SDSS M31 Data

M31 was observed by the Sloan Digital Sky Survey (SDSS; [6,1,2]) when the primary survey areas were inaccessible. Since PNe can be used as kinematic tracers and, to a limited extent, also as tracers of the star formation history, we searched the outer disk and halo of M31 for PN candidates. We recovered a number of previously known PNe in the SDSS data from [4] and from [3] and constructed selection criteria on the basis of their SDSS colors and magnitudes (see Figure 1). Since the detected flux from PNe comes almost entirely from nebular emission lines in the optical, the range of colors characteristic of the detected PNe is defined by the ratios of these emission lines and their corresponding contributions in different SDSS passbands. Compared to regular point sources in the SDSS

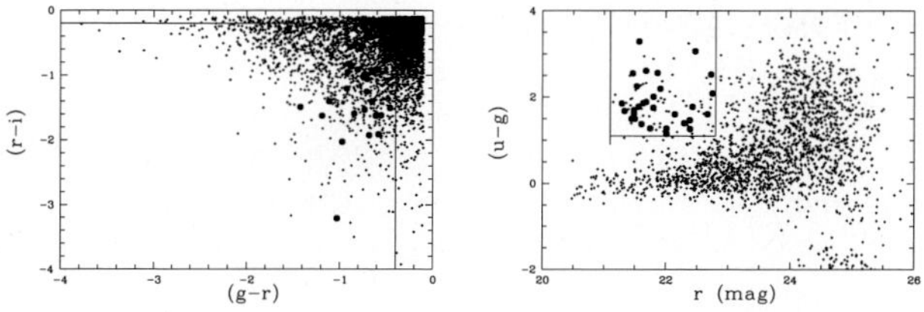

Fig. 1. Color-magnitude diagrams for stellar sources from SDSS M31 data to which we applied our selection procedure. All previously known re-identified PNe from the literature are shown with fat dots. Our color-magnitude criteria for selecting these candidates are indicated by solid lines. All previously identified PNe are located within a relatively small range of luminosities of ~ 1.5 mag that shows the depth of previously published surveys for PNe in the outer disk and halo of M31.

(mainly stars), PNe stand out as being bluer in $(g-r)$ vs. $(r-i)$, and redder in $(u-g)$ vs. $(g-r)$, due to the contribution of the very strong emission lines Hβ and [OIII] $\lambda\lambda$4959,5007 in the SDSS g-band and Hα in the SDSS r-band.

We then applied the derived criteria to the whole area of M31 (as scanned by the SDSS) and selected PN candidates. With our criteria we selected more than 85% of the previously known and spectroscopically confirmed PNe in the outer disk and halo of M31. This percentage is a lower limit, since non-detections of previously identified PNe occurred in areas where the SDSS photometry software is affected by strong crowding. Finally, we selected ~ 90 new PN candidates and 30 known PNe. Most of these are located in the outer disk of M31. We found PN candidates at projected distances as far as $\sim$40 kpc from the center of M31.

To understand possible contamination by point sources such as QSOs and stars, we carefully analysed their distribution as compared to [5]. We found that our criteria select starlike sources far from both QSO and stellar loci. There are only a few spurious data points located in our areas of interest, which would yield perhaps 1–10 contaminants per 1000 deg^2. With the SDSS M31 data we have only about 45 deg^2, and such contamination should thus be extremely small.

Acknowledgments

Funding for the creation and distribution of the SDSS Archive has been provided by the Alfred P. Sloan Foundation, the Participating Institutions, the National Aeronautics and Space Administration, the National Science Foundation, the U.S. Department of Energy, the Japanese Monbukagakusho, and the Max Planck Society. The SDSS Web site is http://www.sdss.org/.

References

1. Fukugita, M., et al. 1996, AJ, 111, 1748

2. Gunn, J.E., et al. 1998, AJ, 116, 3040
3. Jacoby, G.H., & Ford, H.C 1986, ApJ, 304, 490
4. Nolthenius, R., & Ford, H.C. 1987, ApJ, 317, 62
5. Richards, G.T., Fan, X., Newberg, H.J. et al. 2002, AJ, 123, 2945
6. York, D.G., Adelman, J., Anderson, J.E. et al. 2000, AJ, 120, 1579

Deep PN Surveys Beyond the Local Group

Roberto H. Méndez

Institute for Astronomy, University of Hawaii, 2680 Woodlawn Drive,
Honolulu, HI 96822, USA

Abstract. The introduction describes the evolution of the motivations for discovery of PNs beyond the Local Group. Then I briefly describe the on-band, off-band discovery method, and emphasize how slitless grism spectra, taken through the on-band filter, help with the PN identification and provide radial velocity information. Contamination of the PN samples by background emission-line sources is discussed. I also consider the contamination by H II regions ionized by massive hot stars in galaxies with recent star formation. I report on PN abundances in the elliptical galaxy NGC 4697. Finally, I also discuss how to understand the shape of the PN luminosity function.

1 Introduction: Motivations for Deep PN Surveys Beyond the Local Group

Historically, the first objective was distance determination using the PN luminosity function (PNLF); see Robin Ciardullo's review in these proceedings. Once the PNs have been found, however, we also want to use them for other purposes.

PN radial velocities in other galaxies are easy to measure, because essentially all the detected photons from each nebula have the same wavelength (see Section 2). The PNs can then be used as kinematic probes of the stellar population they represent, allowing us to investigate the dark matter and angular momentum distribution in early-type galaxies. The PNs detected at large angular distances from the galaxy's center are especially valuable, because they provide unique information; absorption-line studies of the integrated spectrum of the galaxy cannot reach much further than one effective radius, while PNs can be detected in the outskirts, to at least 3 or 4 effective radii, and every detected PN gives a reliable radial velocity.

The radial velocity work led to the next motivation: the discovery by Arnaboldi et al. (1996) that a few PNs in the Virgo cluster galaxy NGC 4406 have the cluster radial velocity (around + 1400 km s^{-1}), instead of the galaxy's velocity of about − 230 km s^{-1}. The indication was that there must be an intergalactic or intracluster stellar population, presumably lost to the cluster galaxies as a consequence of close but non-destructive interactions. Surveys have been undertaken to study the intracluster PNs in the Virgo and Fornax clusters, as well as in other nearby galaxy groups (see Feldmeier, these proceedings).

There is a final, most ambitious goal: we would like to determine abundances and abundance gradients in elliptical galaxies from PNs. In a similar way as with radial velocities, this would provide important information at large angular distances from the galaxy's center, where absorption-line studies from long-slit spectra are impossible due to the low surface brightness of the stellar population.

In Section 2 of this review I present a brief review of the discovery methods, and on the use of grisms for slitless radial velocity measurements. Section 3 is devoted to the problem of contamination of the PN samples by other emission-line objects at higher redshifts. Section 4 deals with the problem of searching for PNs in galaxies with recent massive star formation. Section 5 presents a preliminary report on PN abundance determinations in the elliptical galaxy NGC 4697, which is at a distance of about 11 Mpc. Section 6 is a late addition on the shape of the PN luminosity function, triggered by several comments during the workshop sessions.

2 How To Find the PNs

The basic discovery method for extragalactic PNs was introduced by Holland Ford and George Jacoby: an on-band image is taken through a narrow-band filter passing the redshifted [O III] λ5007 emission line, and an off-band image is taken through a broader filter passing no strong nebular emission lines. The PNs, which are point sources, are visible in the on-band image but absent in the off-band image.

Here I would like to describe the advantages of taking a third image, through the on-band filter and a grism. The insertion of the grism in the light path has the following effects: it transforms the images of all the continuum sources into segments of width determined by the on-band filter transmission curve; while point sources with an unresolved emission line in the spectral band passed by the filter remain as point sources. Thus the slitless grism spectrum helps in the identification of the PN candidates. In addition, and most important, the grism insertion produces a shift relative to the positions in the image taken without grism. The shift is a function of the wavelength of the emission line (the red-shifted 5007 in our case) and of position on the CCD. This shift can be measured and calibrated, giving the emission-line wavelength for all detected sources and therefore their radial velocities, irrespective of the number of sources and their distribution in the field. This method has been used to discover and measure the radial velocities of 535 PNs in the elliptical galaxy NGC 4697 (Méndez et al. 2001). The most efficient variation of the slitless technique is implemented in the PN Spectrograph, described by Nigel Douglas in these Proceedings. But I would like to emphasize that the slitless method can be used with any imaging spectrograph equipped with grisms, for example, FORS at the VLT or FOCAS at the Subaru telescope. I would strongly recommend to consider taking the grism images, if possible, whenever you are planning a survey for PNs. They increase dramatically the value of the survey.

3 Contamination of the PN Samples

The PN search method is affected by some contamination by other sources. The main contaminants are background sources with emission lines redshifted into the on-band filter; for example [O II] λ3727 at a redshift z=0.35, or Lyman α at z=3.1.

The contamination becomes more important as the surface density of PNs decreases. It is also a function of distance to the galaxy being searched for PNs. I would like to report on the contamination expected in the case of NGC 4697. The distance to this galaxy is approximately 11 Mpc. A survey for PNs produced 535 detections (Méndez et al. 2001). In addition to the PN candidates, some 20 objects of different kinds were found, namely:

- Many extended sources, presumably star forming regions at low redshift. These pose no confusion problem, because of their extended nature.
- Two point sources bright in the on-band and fainter but visible in the off-band. These could be high-redshift quasars or star forming regions, or even a PN within a globular cluster that belongs to NGC 4697 (one such globular cluster PN was reported in NGC 5128 by Minniti and Rejkuba 2002). Again there is no contamination in this case, because the off-band shows an image and we reject the source as a PN.
- Three point sources invisible in the off-band. They were recognized as non-PNs because their radial velocities were wrong for NGC 4697 (another example of the usefulness of the grism images through the radial velocity information they provide). We can expect a few more sources like this, but with velocities typical of NGC 4697; if they exist we cannot distinguish them from PNs without more information (a deeper spectrum). This level of contamination (a few in more than 500) is not important for our purposes.

The contamination becomes more troublesome as the distance increases. As we just saw, at a distance of 10 Mpc there is no trouble. But at the distance of the Virgo cluster (about 17 Mpc) the luminosity function of the contaminants imitates the luminosity function of a low-surface-density PN population. This produced some confusion in early studies of the Virgo intracluster PNs (Méndez et al. 1997; Kudritzki et al. 2000). It turns out that the surface density of Virgo intracluster PNs is not constant; in some places there are no ICPNs and all you find are the contaminants. This is frustrating but it teaches us that the distribution of ICPNs is far from relaxed in Virgo.

As a conclusion of this experience we can state that in deep searches for intracluster PNs we need spectroscopic confirmation, i.e. we need to detect at least two emission lines to confirm the redshift (e.g. [O III] $\lambda\lambda$ 4959, 5007).

The bad news is that working with PNs beyond 25 or 30 Mpc is going to be hard; the contaminants are in that case brighter than the PNs, and just a few contaminants can deform the PNLF. Spectroscopic confirmation will be essential, and that will be expensive in terms of telescope time.

4 Searching for PNs in Regions of Recent Massive Star Formation

Normally we would rather avoid galaxies showing recent massive star formation because of the confusion with compact H II regions ionized by massive hot stars. If we just want a PNLF distance, for example to compare with the Cepheid distance, then we can always restrict the search to regions without recent star formation, as done by Feldmeier et al. (1997).

However, if we are interested in finding e.g. how the PNLF shape depends on stellar population effects, then we must purposely search for PNs in difficult environments. Two methods have been proposed to disentangle PNs from unresolved massive star H II regions: (1) the PN central star is not visible, while the massive star is visible (Soffner et al. 1996). (2) massive star H II regions tend to be of low excitation, while PNs at the bright end of the PNLF are of higher excitation (e.g. Ciardullo et al. 2002).

I would like to show an image from a survey of PNs in the late spiral galaxy NGC 300, done in collaboration with Wolfgang Gieren (Univ. of Concepcion, Chile) with the ESO-MPI 2.2 m telescope and the Wide Field Imager. We used three filters: on-band, off-band, and Hα. Figure 1 shows a "false color" image, produced assigning blue color to the on-band, red to the off-band, and green to Hα (the initial assignment of red to Hα produced such an ugly result that it was unacceptable). Therefore the stellar population looks reddish, high excitation is blue and low excitation is green.

The high-excitation PNs are easily recognizable as "pale blue dots". But we also find several cases of high-excitation massive star H II regions (blue, extended). Therefore in principle there is a risk of contamination by compact, high-excitation massive star H II regions. In conclusion, it looks dangerous to depend solely on the excitation criterion. The survey should be deep enough to ensure detection of individual massive hot stars in the off-band image. If that is accomplished, then we do not need to rely on the nebular excitation level.

5 Deep Spectroscopy of PNs in the Elliptical Galaxy NGC 4697

I am collaborating with Daniel Thomas, Roberto Saglia, Claudia Maraston, Rolf Kudritzki and Ralf Bender in an attempt to obtain information about abundances and abundance gradients in elliptical galaxies from PNs. The idea is to compare with and supplement the data inferred from long-slit spectroscopy of elliptical galaxies.

As our first target we chose NGC 4697 because we had already found the PNs (Mendez et al. 2001). A long-slit spectrogram along the major axis of NGC 4697 taken with VLT+FORS shows clear indication of a metallicity gradient, going from two times solar at the center to 0.5 solar at one effective radius. An extrapolation would suggest even smaller metallicities at two or three effective radii. Do the PNs confirm this trend?

Fig. 1. A false-color image of PNs and H II regions in the spiral galaxy NGC 300. For an explanation, please refer to the text.

We selected 14 bright PNs along the major axis of NGC 4697. The multi-object spectral exposures were taken with VLT+FORS (3 nights in 2002) and Keck I + LRIS-B (1 night in 2003). Figure 2 shows the positions of the 14 objects relative to the center of the galaxy. Figures 3 and 4 show the average spectrum of the 14 PNs.

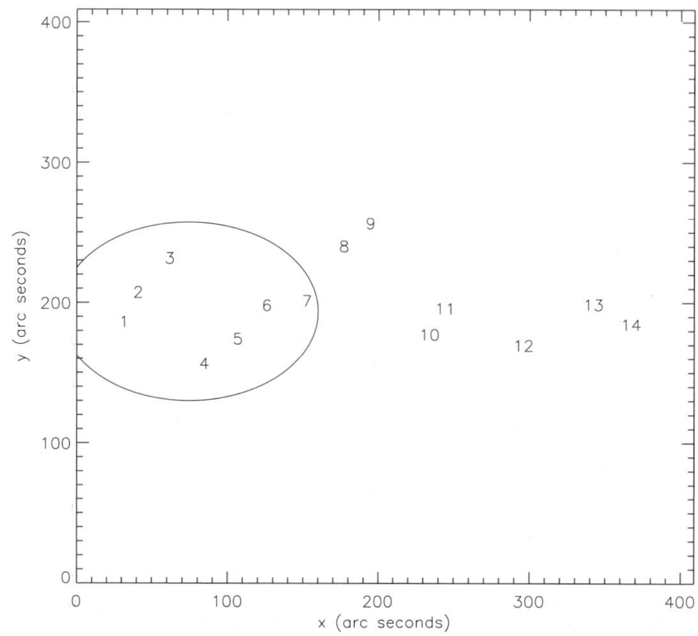

Fig. 2. Positions of the 14 bright PNs relative to the core of NGC 4697. The area and orientation of this figure are as in Fig. 7 of Méndez et al. (2001). The effective radius of NGC 4697 is 95 arc seconds

The most remarkable result is that we find many PNs with very strong 5007/Hβ ratios, near 20. Although we could not detect the diagnostic line [O III] λ4363 in individual spectra, we can argue anyway that the PNs with such a strong λ5007 (also showing strong Ne lines) must have O and Ne abundances near solar. The argument can be illustrated using Fig. 3 from Dopita et al. (1992). We have confirmed our conclusion using dozens of CLOUDY runs (Ferland et al. 1998) for a variety of nebular and central star parameters. In particular we have found near-solar abundances for 4 PNs located at more than 2 effective radii from the center of NGC 4697.

We are still discussing these results, but they are not incompatible with the existence of a metallicity gradient (the metallicity can be even higher near the core, but we cannot use our spectra to verify this, because a very high metallicity can produce a weak 5007/Hβ ratio, as shown in the same Fig. 3 of the Dopita et al. paper). The very strong 5007/Hβ ratios at large angular distances from the

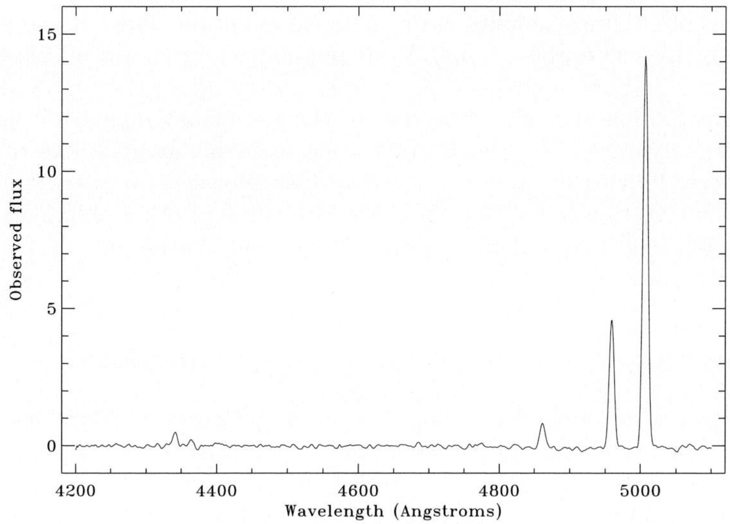

Fig. 3. Average of all spectra of PNs in NGC 4697. Visible features are Hγ λ4340, [O III] λ4363, possibly a very weak He II λ4686, Hβ λ4861, and [O III] $\lambda\lambda$4959, 5007

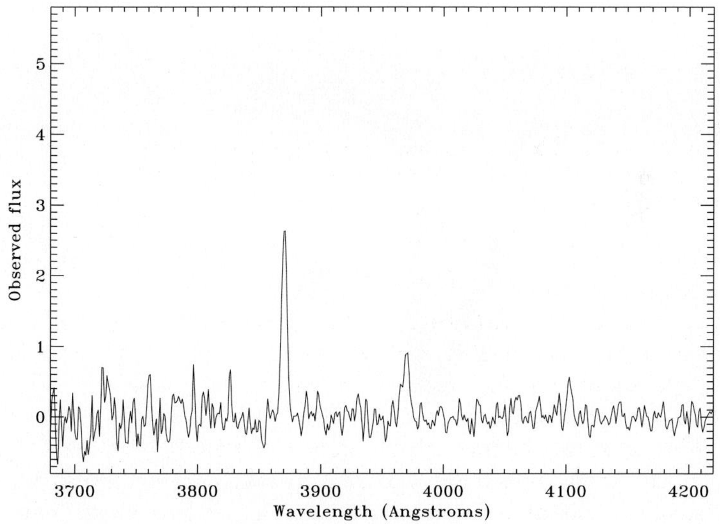

Fig. 4. Average of all Keck LRIS-B spectra of PNs in NGC 4697. Visible features are [Ne III] $\lambda\lambda$3868, 3967, and Hδ λ4101

center, together with the absorption-line data, probably mean that there must be a large spread in metallicity in the stellar population. The PNs provide direct evidence of the existence of a metal-rich population far from the center of NGC 4697.

We have found that the detection of the important diagnostic line [O III] λ4363 in individual PN spectra of elliptical galaxies at distances of 10 or 15 Mpc is very difficult for 8m or 10m telescopes. However, we also found that it would be interesting to explore the 5007/Hβ ratios in other ellipticals; this can be accomplished in reasonable exposure times and can provide useful limits on the PN abundances.

6 The Shape of the PN Luminosity Function

This is a late addition to this review, triggered by George Jacoby's description of evidence that the PN luminosity function has a "camel shape" or rather a deficit of PNs beyond the famous bright end used for distance determinations. In fact, this kind of PNLF shape is predicted naturally from Monte Carlo simulations

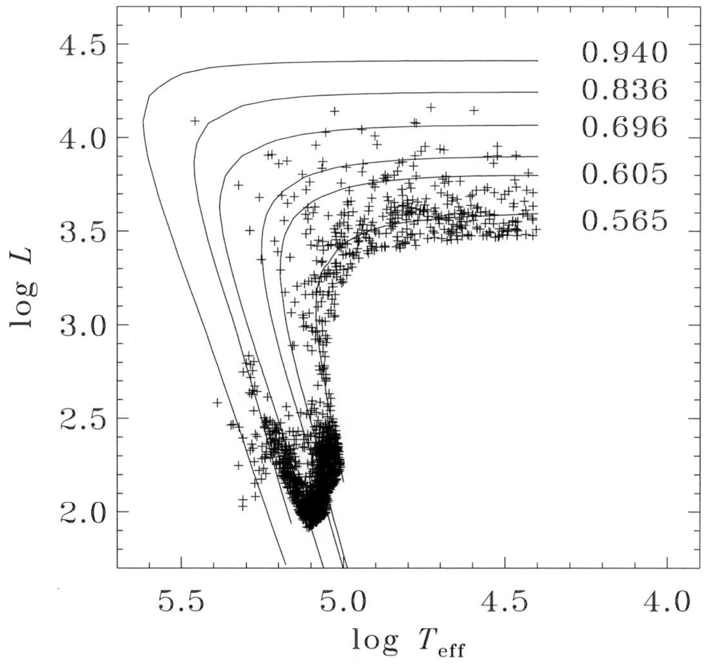

Fig. 5. The values of luminosity and surface temperature for the central stars of 1500 randomly generated PNs. The solid lines are post-AGB evolutionary tracks (H-burning) for 6 different central star masses, taken from Schönberner (1989) and Blöcker (1995). The unlabeled track is for 0.625 solar masses. See Méndez and Soffner (1997) for further details

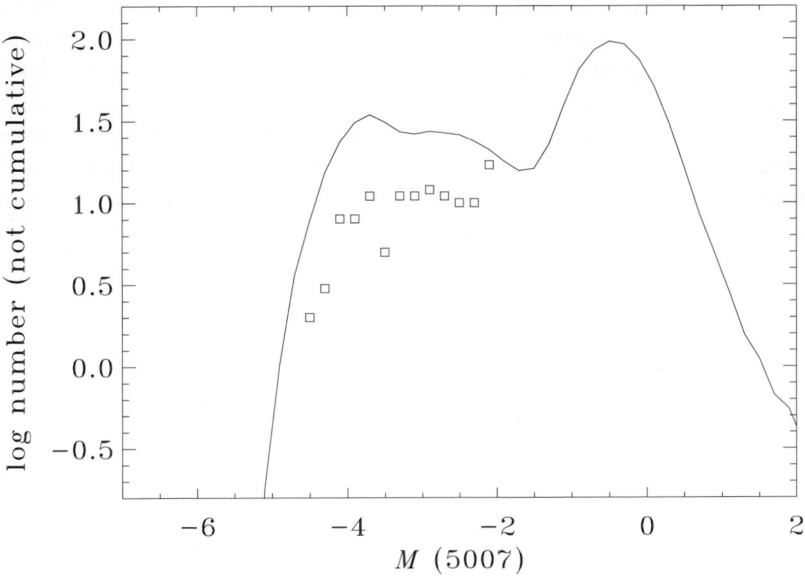

Fig. 6. A PNLF simulation assuming that all PNs are completely optically thick. The open squares represent the observed PNLF of the bulge of M 31, assumed to be at a distance of 770 Kpc. An extinction correction of 0.28 mag was applied. The solid line is the simulated PNLF, with data binned, like the observed ones, into 0.2 mag intervals. This simulated PNLF has too many bright PNs because most PNs are leaking H-ionizing photons.

if we assume that most central stars are H-burners (Méndez and Soffner 1997; Méndez 1999).

For an easier understanding of the argument, I would like to briefly summarize the procedure described in the Méndez and Soffner paper:

- Generate a set of central stars with random post-AGB ages and masses.
- Using evolutionary tracks, derive the corresponding central star luminosities and surface temperatures.
- Using recombination theory and empirical information about leaking of H-ionizing photons, calculate the nebular Hβ luminosities.
- Assuming a distribution of ratios 5007/Hβ, obtain the luminosities in 5007 and compute the PNLF.

Figure 5 shows a simulation of a collection of 1500 central stars burning H. The evolutionary tracks show a sharp drop in luminosity as the H-burning shell is extinguished and the star goes into the white dwarf cooling track. For that reason there is a lack of central stars at log L between 2.5 and 3. This lack of central stars offers a natural explanation for the lack of intermediate-brightness PNs. Figure 6 shows a PNLF simulation built under the (wrong) assumption that all the PNs are completely optically thick (i.e. no leaking of H-ionizing photons). The open squares represent the observed λ5007 PNLF of the bulge of M 31, from Ciardullo et al. (1989). The shape of the simulated PNLF shows clearly a deficit at intermediate values of the 5007 absolute magnitudes. The simulated PNLF lies everywhere above the observed points because in practice many PNs do leak a fraction of their H-ionizing photons. The final shape of the PNLF is determined mostly by the evolutionary tracks, by how many H-ionizing photons are leaking from the PNs, and by how the "leaking efficiency" evolves with time.

References

1. M. Arnaboldi, K. C. Freeman, R. H. Méndez et al.: ApJ **472**, 145 (1996)
2. T. Blöcker: A&A **299**, 755 (1995)
3. R. Ciardullo, J. J. Feldmeier, G. H. Jacoby: ApJ **577**, 31 (2002)
4. R. Ciardullo, G. H. Jacoby, H. C. Ford, J. D. Neill: ApJ **339**, 53 (1989)
5. M. A. Dopita, G. H. Jacoby, E. Vassiliadis: ApJ **389**, 27 (1992)
6. J. J. Feldmeier, R. Ciardullo, G. H. Jacoby: ApJ **479**, 231 (1997)
7. G. J. Ferland, K. T. Korista, D. A. Verner et al.: PASP **110**, 761 (1998)
8. R. P. Kudritzki, R. H. Méndez, J. J. Feldmeier et al.: ApJ **536**, 19 (2000)
9. R. H. Méndez: 'Distances from the planetary nebulae luminosity function'. In: *Post-Hipparcos Cosmic Candles*, ed. by A. Heck, F. Caputo (Kluwer, Dordrecht 1999) pp. 161–180
10. R. H. Méndez, M. A. Guerrero, K. C. Freeman et al.: ApJ **491**, L23 (1997)
11. R. H. Méndez, A. Riffeser, R. P. Kudritzki et al.: ApJ **563**, 135 (2001)
12. R. H. Méndez, T. Soffner: A&A **321**, 898 (1997)
13. D. Minniti, M. Rejkuba: ApJ **575**, L59 (2002)
14. D. Schönberner: 'Evolutionary tracks for central stars of planetary nebulae'. In: *Planetary Nebulae, IAU Symposium 131 in Mexico City, Mexico, October 5-9, 1987*, ed. by S. Torres Peimbert (Kluwer, Dordrecht 1989) pp. 463–471
15. T. Soffner, R. H. Méndez, G. H. Jacoby et al: A&A **306**, 9 (1996)

Planetary Nebulae in NGC 5128 (Centaurus A)

Eric W. Peng[1], Holland C. Ford[2], and Kenneth C. Freeman[3]

[1] Rutgers University, Piscataway, NJ 08854, USA
[2] Johns Hopkins University, Baltimore, MD 21218, USA
[3] Australian National University, Canberra, ACT, Australia

Abstract. The study of planetary nebulae (PNe) in the nearby post-merger elliptical galaxy NGC 5128 (Cen A) now has a history of nearly twenty years. As the nearest giant elliptical, it is a prime target for extragalactic PN studies. These studies have addressed many issues including the galaxy's distance, dark matter content, halo structure, merger history, and stellar populations. We review the main PN studies that have been conducted in NGC 5128, and introduce a new study where we measure the [NII]/Hα ratio for 134 PNe. We find that there are no PNe in our sample that are obviously of Type I, supporting the idea that the last major star formation event in the galaxy halo occurred over 1–2 Gyr ago.

1 Introduction

The study of planetary nebulae (PNe) beyond the Local Group has largely been conducted in luminous early-type galaxies where PNe are present in large numbers and are relatively easy to detect. The nearest, easily observable elliptical galaxy, NGC 5128 (Cen A), has thus been a recurring target for PN studies over the last twenty years. At a distance of 3.5–4 Mpc (e.g. Hui et al. 1993a; Harris et al. 1999; Rejkuba 2004), NGC 5128 is ~ 3 mag closer than the ellipticals in the Virgo cluster, which not only makes intrinsically fainter PNe more accessible, it also greatly reduces the problem of contamination from Lyα galaxies at $z = 3.14$. The galaxy is also interesting in its own right as it is clearly the product of a recent merger, exhibiting a warped central gas and dust disk, stellar and HI shells, and both radio and x-ray jets. A thorough review of this galaxy is presented by Israel (1998). Aside from its intrinsic complexity, disadvantages of studying NGC 5128 include its large extent on the sky ($1' \sim 1$ kpc), and its relatively low Galactic latitude ($b = +19$). Nevertheless, modern wide-field and multiplexing instrumentation have made studies both practical and fruitful.

2 PN Surveys

The first PN survey of NGC 5128 was conducted by Hui et al. (1993b), and covered 661 arcmin^2 over 43 CCD fields (see Figure 1). This coverage extended 20 kpc along the major axis an 10 kpc along the minor axis. At these projected distances (~ 4 and 2 r_e, respectively) PNe were detected well beyond the radii at which long-slit spectroscopy was readily obtainable. They identified 785 PN candidates using the method of on-/off-band imaging of the [OIII]λ5007 emission

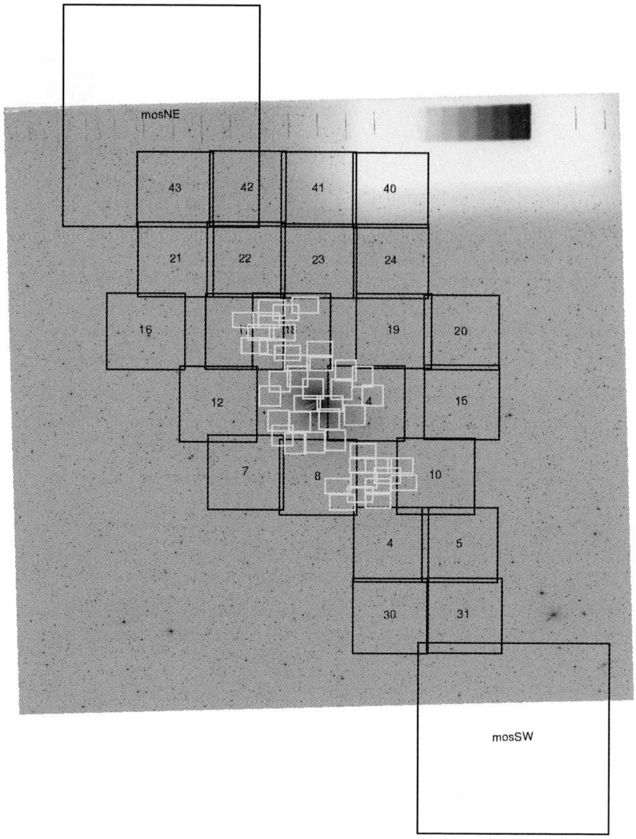

Fig. 1. NGC 5128 PN survey areas. The smaller fields toward the center are the first survey fields by Hui et al. (1993a). The larger fields that cover a much wider region of the halo are from our most recent survey (Peng et al. 2004a).

line. These PNe formed a complete sample for the brightest 1.5 mag of the PN luminosity function (PNLF). They also spectroscopically confirmed 432 PNe, on which they based their dynamical studies of dark matter and the intrinsic shape of the potential (Hui et al. 1995).

Larger CCD cameras and fiber spectrographs allowed us to conduct a second, wider PN survey of the NGC 5128 halo. This survey covered nearly 3 $\deg^2$ on the sky and extended to 100 kpc along the major axis and 40 kpc along the minor axis (Figure 1). In total, we increased the number of imaged PNe to 1141, 780 of which are spectroscopically confirmed. The spatial distribution of PNe is concentrated toward the major axis and mirrors that of the low surface brightness halo light seen in deep photographic imaging by Malin (1978); the PNe at the largest galactocentric distances in our sample are, however, ~ 30 kpc beyond the light seen in the photographic images.

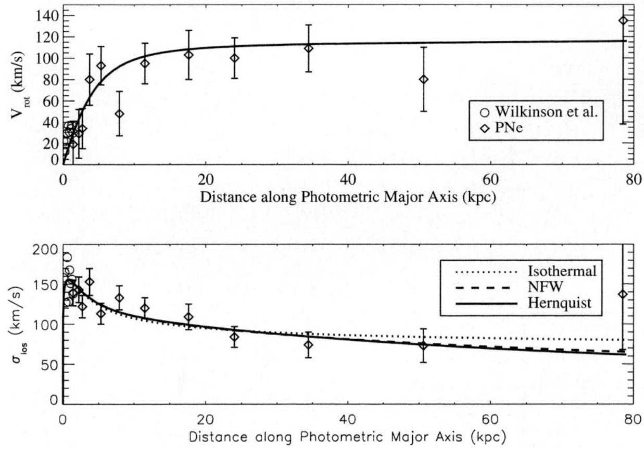

Fig. 2. (a) Major axis rotation curve. We fit a parametrized flat rotation curve. (b) Jeans Equation fit to the major axis velocity dispersion profile. In the bottom panel the best fit line-of-sight velocity dispersion profiles for three different dark halo mass models are shown.

3 Mass Estimates

PNe are some of the most effective kinematic tracers in the halos of elliptical galaxies, where their radial velocities can be used to constrain the dynamical mass of their parent galaxies (e.g. Arnaboldi et al. 1998; Romanowsky et al. 2003). Hui et al. (1993) used the original sample of 432 PN velocities to derive a total mass of $3.1 \times 10^{11} M_\odot$ and $M/L_B \sim 10$ out to a radius of 25 kpc. Our recent work, which extends out to 80 kpc, shows that the total mass increases only slowly with radius. The fitted major axis rotation curve and velocity dispersion profile are shown in Figure 2. Although the rotation curve remains flat, the dispersion profile falls with radius. Within this volume, we estimate a dynamical mass of $\sim 5 \times 10^{11} M_\odot$ and $M/L_B \sim 13$ (Peng et al. 2004a). These numbers are relatively unchanged whether we use the tracer mass estimator (Evans et al. 2003) or spherical Jeans equation under the assumption of isotropy. While it may be that this value for M/L_B can be produced from theoretical expectations (see Napolitano, this volume), the value we derive is certainly low, like the ones seen in the sample of Romanowsky et al. (2003), when compared to luminous ellipticals such as M87 and M49 (Côté et al. 2001, 2003).

4 The Two-Dimensional Velocity Field

One of the advantages of using PNe as kinematic probes is that we get full two-dimensional spatial information on the stellar velocity field. When there are hundreds of measured PNe, we can construct velocity maps that are similar

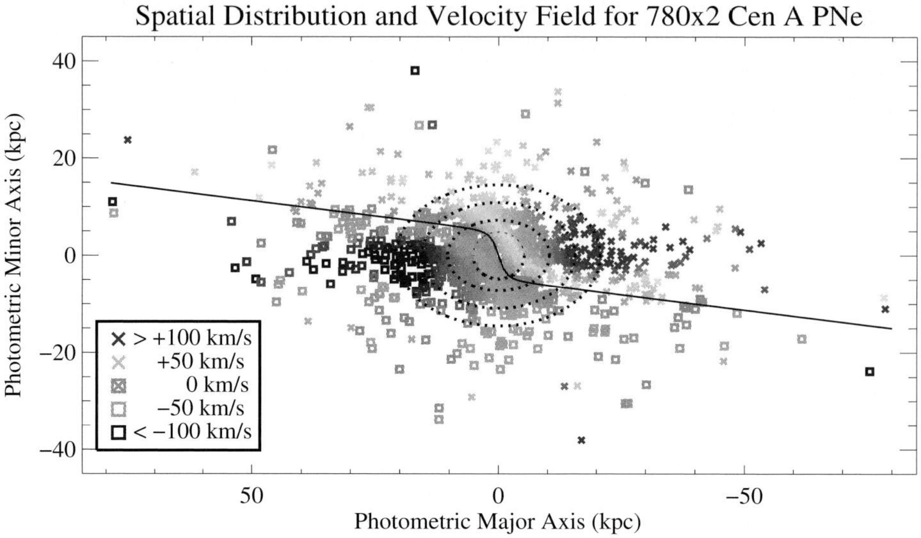

Fig. 3. Smoothed antisymmetrized velocity field with zero velocity contour (ZVC) plotted. Dotted ellipses trace the rough isophotes for NGC 5128 from 1–4 r_e. PN with velocities larger than systemic (receding) are represented by crosses, and those with velocities smaller than systemic (approaching) are represented by open boxes. The color of each point shows the magnitude of the velocity with respect to systemic. Notice how the rapid major axis rotation is only present within ± 7 degrees of the major axis, while the rest of the galaxy halo undergoes a slower minor axis rotation.

to those derived from integral field spectrograph observations, but that extend much farther out into the halo. In the case of NGC 5128, Hui et al. (1995) determined that the kinematic axis of rotation was not aligned with either of the photometric axes, suggesting that the intrinsic shape of the potential is triaxial. Since then, we have used our more extensive PN data to show that the zero-velocity surface not only does not coincide with the photometric axes, but is in fact twisted in an 'S'-shape (Figure 3). This severe twist, which asymptotes to a minor axis angle of 83 degrees, suggests that the potential could be triaxial, tending toward prolate (Statler 1991).

Separate lines of evidence, which include our survey of globular clusters in this system (Peng et al. 2004b) indicate that there may be a substantial intermediate age population of stars with an mean age of ~ 5 Gyr (Peng et al. 2004c). Bekki et al. (2003) have already produced simulations of globular cluster velocity fields that result from galaxy mergers. In the future, it will be valuable to compare the PN velocity field to those produced by merger simulations to constrain the types of mergers that could have produced this galaxy.

5 PNe as Stellar Populations

One of the useful characteristics of PNe is that, like globular clusters, it is possible in principle to obtain both their radial velocities *and* some information on their ages or abundances, producing a chemodynamical picture of the galaxy. The metallicity distribution of the field star population of NGC 5128 has been measured using Hubble Space Telescope color-magnitude diagrams of the red giants and is generally metal-rich (Harris & Harris 2002). Assuming that the PNe trace the same stellar population, this is one way to derive the overall properties of the PN system. However, one can derive abundances from PNe directly from their emission line spectra, although this requires a significant investment in observing time.

Through a heroic effort, Walsh et al. (1999) derived [O/H] for 5 PNe in NGC 5128, finding that they were generally sub-solar in abundance with a mean value of ~ -0.5. Walsh (this volume) has also recently obtained spectra of much higher quality for ~ 50 PNe, showing that these types of observations are now possible and interesting.

Another way to link PNe with their progenitor stellar populations is to find PNe in globular clusters. Minniti & Rejkuba (2002) found one such object in an NGC 5128 globular cluster. However, these objects appear to be rare, and no other PNe in extragalactic clusters have been reported.

We have investigated the PN population in NGC 5128 by looking for the young, Peimbert (1978) Type I PNe, which should have progenitors that are younger than 1–2 Gyr. In the Milky Way, these PNe are found in the disk, and are rich in N and He due to the dredge-up and hot-bottom burning processes found only in these more massive stars. Because obtaining spectra of sufficient quality to derive N or He abundances would have been prohibitively expensive, we instead chose to measure the [NII] lines $\lambda\lambda6548, 6584$ and compare their fluxes to Hα. This [NII]/Hα ratio can be used as a crude estimator of relative nitrogen abundance. Type I PNe in both Magellanic Clouds have very high [NII]/Hα ratios.

We used the 2-degree Field spectrograph on the Anglo-Australian Telescope to observe 221 PNe in NGC 5128. Of these, we detected Hα at $>$ 3-sigma in nearly all of them (219), and both Hα and [NII]λ6584 in 134 PNe. Because these were fiber observations and some of the PNe sit atop the bright galaxy continuum, it was also important that we model the galaxy and subtract it off. In order to do this, we used the models of Bruzual & Charlot (2003) to obtain a best fit simple stellar population to the blue side of the spectrum (used in Peng et al. 2004b). We then scaled this model to the appropriate flux level, and shifted it to the expected mean velocity at that position, as derived from our 2-d PNe velocity field. Only after this continuum subtraction did we measure the line fluxes.

Figure 4 shows the distribution of measured [NII]/Hα values, with incompleteness setting in for values less than ~ 0.55. In order to determine whether any PNe at the high end of the distribution could plausibly be of Type I, we compared them to the extragalactic PNe found in the Magellanic Clouds (Stanghellini et al.

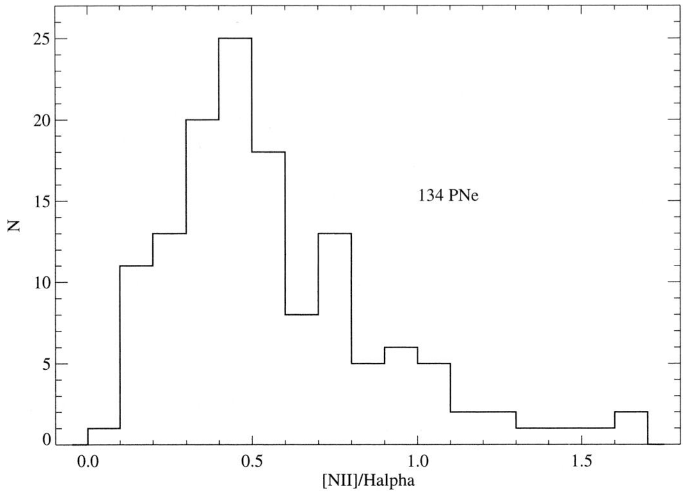

Fig. 4. Distribution of measured [NII]/Hα. Our sample is complete for all values greater than ~ 0.55.

2002, 2003), M31 (Jacoby & Ciardullo 1999), and NGC 5128 (Walsh et al. 1999). In Figure 5, we plot the value of [NII]/Hα against the brightness in [OIII]λ5007. A general trend in metallicity is evident with the SMC being most metal-poor, and NGC 5128 being most metal-rich. It is known that the Type I PNe in the Clouds are typically faint, and this is seen here, where the objects that have high [NII]/Hα (relative to the galaxy mean) are all at the faint end. These PNe can have [NII]/Hα values that are $\sim$ 10 times higher than what is typical in these galaxies. In NGC 5128, however, while the overall population has fairly high [NII]/Hα values, we see no evidence for outliers that would be good candidate Type I PNe. Even the PN from Walsh et al. (1999) that was tentatively determined to be Type I, does not stand out from the crowd. It is important to note that the cutoff values of N/O and He/H typically used to define Type I PNe are defined only for the Galaxy, and may not apply to systems of different metallicities.

6 Conclusions

Only in the Milky Way and M31 are there more known PNe than in NGC 5128. Starting with Hui et al. (1993a), extragalactic PN studies in NGC 5128 have addressed a wide array of issues from its distance, dark matter content, stellar populations, and formation history. We now know that while the halo of this galaxy does contain dark matter, it falls into a growing class of elliptical galaxies that have low to moderate mass-to-light ratios. It's 2-dimensional halo kinematics are consistent with a triaxial potential, and a galaxy that was formed in a

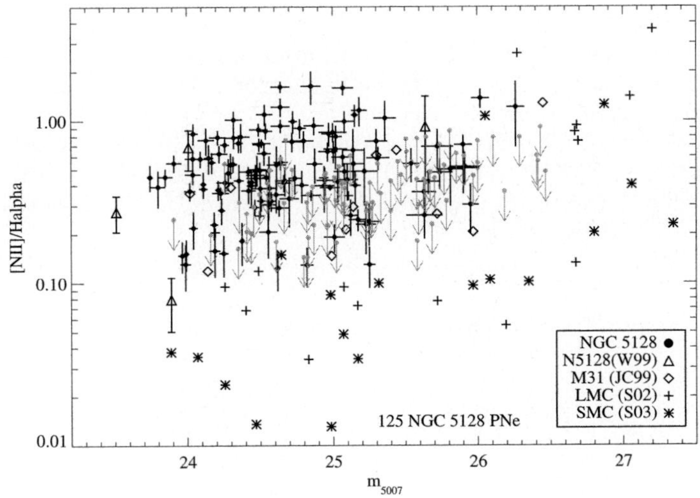

Fig. 5. [NII]/Hα versus m_{5007}. NGC 5128 values (points with error bars) show no evidence for obvious Type I PNe. Gray points are those for which we can only derive an upper limit for [NII/Hα.

recent merger. Presumably, this merger was not likely so recent as to have produced many stars younger than ~ 2 Gyr that would now be detectable as Type I PNe. Future deep, wide-field imaging and spectroscopy will be able to probe the PN abundance distribution and the star formation history of this galaxy, as well as enable studies of the PN stage of stellar evolution in an entirely different environment from that in the Local Group.

References

1. Arnaboldi, M., Freeman, K. C., Gerhard, O., Matthias, M., Kudritzki, R. P., Méndez,
2. Bekki, K., Harris, W. E., & Harris, G. L. H. 2003, MNRAS, 338, 587
3. Bruzual, G. & Charlot, S. 2003, MNRAS, 344, 1000
4. Côté, P. et al. 2001, ApJ, 559, 828
5. Côté, P., McLaughlin, D. E., Cohen, J. G., & Blakeslee, J. P. 2003, ApJ, 591, 850
6. Harris, G. L. H., Harris, W. E., & Poole, G. B. 1999, AJ, 117, 855
7. Harris, W. E. & Harris, G. L. H. 2002, AJ, 123, 3108
8. Hui, X., Ford, H. C., Ciardullo, R., & Jacoby, G. H. 1993a, ApJ, 414, 463
9. Hui, X., Ford, H. C., Ciardullo, R., & Jacoby, G. H. 1993b, ApJs, 88, 423
10. Hui, X., Ford, H. C., Freeman, K. C., & Dopita, M. A. 1995, ApJ, 449, 592
11. Israel, F. P. 1998, A&A Rev., 8, 237
12. Jacoby, G. H. & Ciardullo, R. 1999, ApJ, 515, 169
13. Malin, D. F. 1978, Nature, 276, 591
14. Minniti, D. & Rejkuba, M. 2002, ApJL, 575, L59

15. Peimbert, M. 1978, IAU Symp. 76: Planetary Nebulae, 76, 215
16. Peng, E. W., Ford, H. C., & Freeman, K. C. 2004, ApJ, 602, 685
17. Peng, E. W., Ford, H. C., & Freeman, K. C. 2004b, ApJs, 150, 367
18. Peng, E. W., Ford, H. C., & Freeman, K. C. 2004c, ApJ, 602, 705
19. Rejkuba, M. 2004, A&A, 413, 903
20. Romanowsky, A. J., Douglas, N. G., Arnaboldi, M., Kuijken, K., Merrifield, M. R., Napolitano, N. R., Capaccioli, M., Freeman, K. C. 2003, Science, 301, 1696
21. Stanghellini, L., Shaw, R. A., Mutchler, M., Palen, S., Balick, B., & Blades, J. C. 2002, ApJ, 575, 178
22. Stanghellini, L., Shaw, R. A., Balick, B., Mutchler, M., Blades, J. C., & Villaver, E. 2003, ApJ, 596, 997
23. Statler, T. S. 1991, AJ, 102, 882
24. Walsh, J. R., Walton, N. A., Jacoby, G. H., & Peletier, R. F. 1999, A&A, 346, 753

Extragalactic Planetary Nebula: Methods of Discovery Using Infrared Photometry

Pedro García-Lario, Amelia Bayo, and Carlos Alfageme

ESAC/ISO Data Centre, Apartado de Correos 50727, E-28080 Madrid, Spain

Abstract. Methods for discovering new planetary nebulae based on their characteristic near-, mid- and far-infrared photometric colours are presented. We describe the methods and review the results obtained for Galactic sources. Their possible application to extragalactic searches is discussed.

1 Introduction

Planetary Nebulae (PNe, hereafter) are well known strong infrared emitters since the first detection in the 1960's of a strong infrared excess at 10 μm in NGC 7027 (Gillett et al. 1967). Following this discovery, dedicated observations of a limited number of PNe were carried out from ground-based observatories in the 8–20 μm range (Cohen & Barlow 1974; 1980; Aitken et al. 1979; Aitken & Roche 1982). Simultaneously, systematic infrared surveys carried out at various wavelengths from airborne stratospheric observatories showed that a strong infrared emission from 1 to 100 μm was a characteristic feature of most PNe (specially the youngest ones). Their immediate precursors, strongly obscured Mira variable stars at the tip of the Asymptotic Giant Branch (AGB, hereafter), were also found to show very similar properties, as well as a small number of stars located at high Galactic latitudes with bright optical counterparts, later identified as post-AGB stars, also in their way to becoming PNe.

2 IRAS Colours

An important step forward was achieved with the all-sky survey performed by the Infrared Astronomical Satellite (IRAS) at 4 photometric bands centred at 12, 25, 60 and 100 μm. It soon became clear that most PNe and post-AGB stars were reasonably well separated from other objects in the IRAS [12]−[25] *vs.* [25]−[60] two-colour diagram. Only a small overlap was found with some young stellar objects and a few Seyfert galaxies. About 20% of the most evolved OH/IR stars (oxygen rich AGB stars with thick circumstellar envelopes) had also similar colours to PNe and were also identified as such in subsequent OH maser surveys at radio wavelengths.

The conclusion that was drawn from the systematic analysis of the IRAS colours of well known PNe and evolutionary related sources (Pottasch et al. 1988; García-Lario 1992) was that they usually showed the following colours:

$$F_\nu(12)/F_\nu(25) \leq 0.50; \qquad F_\nu(25)/F_\nu(60) \leq 0.35 \tag{1}$$

Searches for new candidate PNe satisfying the above (or very similar) conditions were carried out by Preite-Martinez (1988) or García-Lario (1992) among others, resulting in the discovery of many new young PNe and transition objects in the last two decades, confirmed as such with the help of optical spectroscopy and/or near-infrared photometry.

3 Near-Infrared JHK Colours

Whitelock (1985) and Persi et al. (1987) showed that another efficient method of identification of PNe can be based on the characteristic colours of these sources in the near-infrared two-colour diagram J−H *vs.* H−K. The majority of PNe have JHK colours completely dominated by the emission from hydrogen and helium plasma. In particular, many PNe show a characteristic J band excess produced by the large contribution from the He I triplet line at 1.083 μm to the emission detected in this filter.

This means that sources with the following colours:

$$J - H \leq 0.2; \qquad H - K \geq 0.4 \tag{2}$$

can be identified with a high degree of confidence as PNe. The method was found to be especially successful when used to determine the nature of those sources previously selected as candidate PNe because of their characteristic IRAS colours (García-Lario et al. 1997). However, it is important to notice that the use of this method to search for PNe introduces a bias towards PNe of intermediate excitation class, since they are expected to show the largest strength of the He I triplet and, thus, a stronger J-band excess.

4 MSX Colours

Photometric data from the Midcourse Space Experiment (MSX) Point Source Catalogue (Egan et al. 2003) can also be used, just as we did for IRAS, to provide a separation between different types of source based on the observed MSX colours. The infrared instrument on MSX was named SPIRIT III and provided high resolution, high sensitivity photometry of more than 400,000 sources in 4 main bands named A (centred at 8.28 μm and covering the range 6.8–10.8 μm), C (12.13 μm; 11.1–13.2 μm), D (14.65 μm; 13.5–15.9 μm) and E (21.34 μm; 18.2–25.1 μm). MSX covered the Galactic Plane region (±6 degrees) with a much better resolution than IRAS and a similar sensitivity.

In Fig. 1 we show the location of the 17,657 sources included in the MSX5C Point Source Catalogue with good quality photometry in the A, C and D bands in a two-colour diagram. Around 20% of these sources were identified with the help of SIMBAD either as normal stars, AGB stars, young stellar objects, compact H II regions and, of course, as planetary and proto-planetary nebulae or post-AGB stars. Normal stars appear well segregated from the rest of the sources as expected, but there is a strong overlap between young and evolved sources that we need to avoid. Thus, in order to reduce as much as possible the contamination from young stellar objects and compact H II regions in our sample we restricted our analysis to the regions of the Galactic Plane located above or below 2 degrees in Galactic latitude, where star forming regions are not abundant, with very successful results, as we show in Fig. 2.

We can see that there is an excellent segregation of sources of different nature in the [A]−[C] *vs.* [C]−[D] two-colour diagram, and a very little contribution from young stellar objects. Most of the identified sources are normal stars, AGB and post-AGB stars or PNe and they occupy clearly distinguished regions of the diagram, so that we can establish the following colours:

$$[A] - [C] \geq 0.70; \qquad [C] - [D] \geq 0.70 \tag{3}$$

as excellent selection criteria to search for new PNe among the unidentified sources in the sample (small dots in Figure 2).

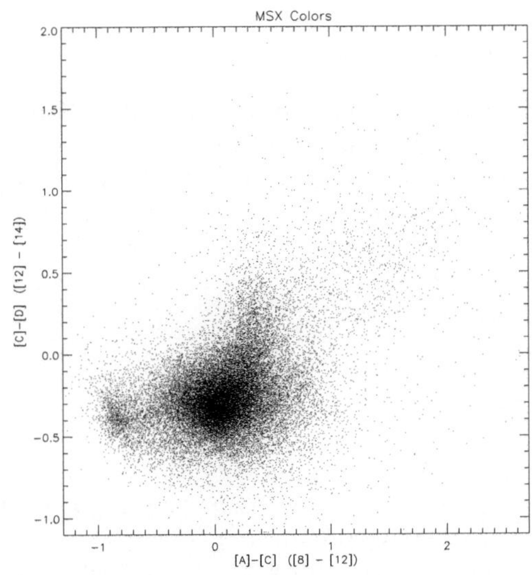

Fig. 1. [A]−[C] *vs.* [C]−[D] two-colour diagram showing all the sources with good quality MSX photometry in the A, C and D bands.

5 2MASS Colours

The combination of mid-infrared data from the MSX satellite mission and ground-based near-infrared photometry can be used, as we have already mentioned, to confirm the nature of potential candidate sources and identify new PNe among the unidentified sources detected by MSX, even at low Galactic latitudes. This

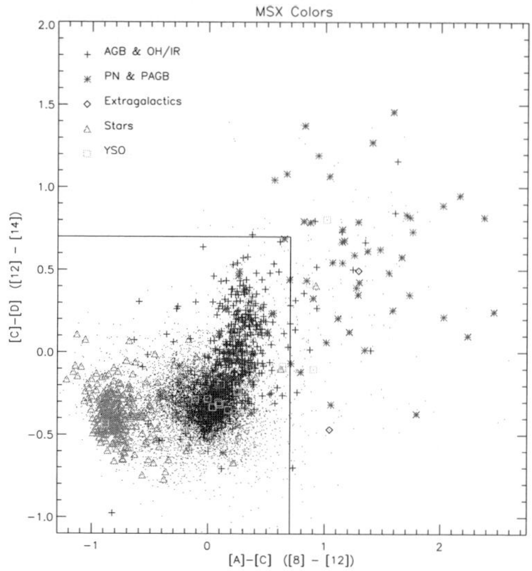

Fig. 2. [A]−[C] *vs.* [C]−[D] two-colour diagram showing all the sources with a good identification in SIMBAD (see legend), together with those for which their nature is unknown (small dots). Note the characteristic colours shown by different types of source. The solid lines indicate the colour criteria defined in the text to search for new PNe among the unidentified sources.

can now be done for all the sources satisfying the above colour criteria with the availability of the 2MASS survey data (Cutri et al. 2003) and the cross-identification of sources is at this moment on-going.

6 Application to Extragalactic PNe

In addition to the Galactic Plane, Version 6C of the MSX Point Source Catalogue includes data from a few small areas located at high Galactic latitude, the Large and the Small Magellanic Clouds and from eight nearby galaxies that have been used as a test to check whether our selection criteria can be extended to search for extragalactic PNe using MSX data.

In Fig. 3 we show as an example the results obtained for the Large Magellanic Cloud. We can see that there is only a very small number of sources with good quality photometry in the three bands (A, C and D) because of the limited sensitivity of MSX. Those sources satisfying our selection criteria are being considered for further analysis. Two of them are well known AGB stars belonging to the LMC, while the rest are not yet identified.

In the near future we would strongly encourage the same kind of analysis to be made with more sensitive instrumentation and higher spatial resolution. This would result in an increase in the number of sources detected in other galaxies

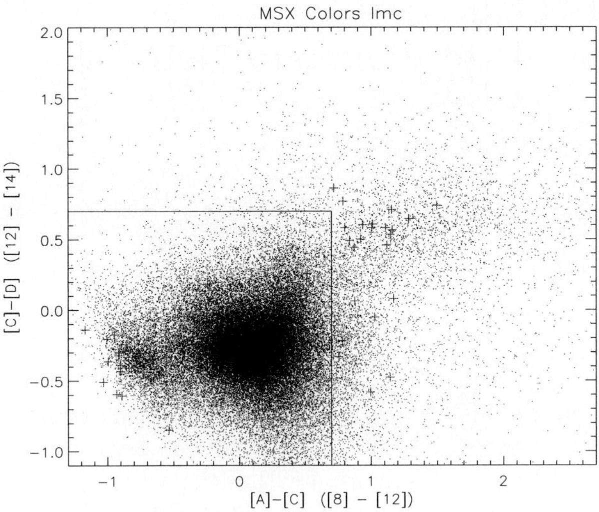

Fig. 3. [A]−[C] *vs.* [C]−[D] two-colour diagram showing all the sources with a good quality MSX photometry in the A, C and D bands detected in the direction of the Large Magellanic Cloud.

of our Local Group (something that can be done now using the Spitzer Space Telescope) and an extension of the analysis toward longer wavelengths in order to avoid the confusion with young stellar objects for those sources located at low Galactic latitudes. In this regard ASTRO-F will soon provide an all-sky survey at several photometric bands covering a wide range from 9 to 170 μm.

References

1. D.K. Aitken & P.F. Roche: MNRAS **200**, 217 (1982)
2. D.K. Aitken, P.F. Roche, P.M. Spenser & B. Jones: ApJ **233**, 925 (1979)
3. M. Cohen & M.J. Barlow: ApJ **238**, 585 (1974)
4. M. Cohen & M.J. Barlow: ApJ **238**, 585 (1980)
5. R.M. Cutri, M.F. Strutskie, S. van Dyk et al.: '2MASS All-Sky Catalogue of Point Sources' (2003)
6. M. Egan, S.D. Price, K.E. Kraemer et al.: 'MSX6C Infrared Point Source Catalogue' (Version 2.3)
7. P. García-Lario: Ph.D. Thesis, Univ. de La Laguna, Spain (1992)
8. P. García-Lario, A. Manchado, W. Pych & S.R. Pottasch: A&AS. **126**, 479
9. F.C. Gillett, F.J. Low & W.A. Stein: ApJ **149**, L97 (1967)
10. P. Persi, A. Preite-Martinez, M. Ferrari-Toniolo, L. Spinoglio. In: *'Planetary and Proto-Planetary Nebulae: from IRAS to ISO'*, ed. by A. Preite-Martinez (Reidel-Dordrecht) p. 221 (1987)
11. A. Preite-Martinez: A&AS **76**, 317 (1988)
12. S.R. Pottasch, C. Bignell, R. Olling, A.A. Zijlstra: A&A **205**, 248 (1988)
13. P.A. Whitelock: MNRAS, **213**, 59 (1985)

Proposed Nomenclature for Extragalactic Planetary Nebulae

Quentin A. Parker[1,2] and Agnes Acker[2]

[1] Macquarie University/Anglo-Australian Observatory, Sydney, Australia, qap@ics.mq.edu.au
[2] Observatoire de Strasbourg, France, acker@astro-u-strasbg.fr

1 Introduction

The ability to identify and distinguish between the wide variety of celestial objects benefits from application of a systematic and logical nomenclature. This often includes value-added information within the naming convention which can aid in placing the object positionally either via an RA/DEC or l,b concatenation. All new nomenclatures should be created following IAU guidelines. However as the number density of specific object types on the sky increases, as in the case of PN in external galaxies, a useful positional identifier becomes problematic. This brief but timely paper attempts to progress the debate on this vexing issue for the case of extragalactic planetary nebulae (PNe). There is a clear need to rationalise the current ad-hoc system now that many thousands of Extragalactic PN are being discovered.

2 Existing Structures and Conventions

A comprehensive web site exists at the Centre de Données Astronomiques de Strasbourg (CDS) to assist the process of astronomical nomenclature and a registration form is available at http://www.vizier.u-strasbg.fr/viz-bin/DicForm. The current convention for Galactic PN is to use 'PN Glll.l+/-bb.b'. However, the positional information included is no longer sufficient to discriminate amongst all the known galactic PNe with the advent of MASH (Macquarie/AAO/Strasbourg H-alpha Planetary Nebula project) which has added $\sim$ 1000 new Galactic PNe including several PNe 'pairs' within an arcminute or so of each other (Parker et al. 2003). PN Glll.l+/-bb.b a,b,c etc is used in the 'unlikely' event of PNe being found in such close proximity. This untidy fix demonstrates the need to try to future proof such designations against use of too crude positional identifiers if they are to be used to any effect in the designation at all.

2.1 Application of Current Convention

A first pass proposal has been suggested: PN X JHHMMSSs+DDMMSS. This includes a PN prefix 'PN' followed by a galaxy identifier 'X' (e.g. NGC7793) then 'J' to indicate equinox J2000 and then an RA/DEC positional identifier.

Lets look at the initial proposal applied to a PN in galaxy NGC7793:
We would have: PN NGC7793 J2357499-323520. This is quite a mouthful even if we replace NGC with N. Thousands of PNe are being found in galaxies at greater distance which are of small angular extent. An individual PN would require an extremely long identifier if positional concatenation is to be used to any effective purpose. Hence PN NGC7793 J23574998-3235206 at least may be required. Furthermore, any identifier should be future proofed to work when PNe are found a hundredth of an arcsecond from another in a distant galaxy. Perhaps an RA/DEC concatenation is not the best way forward.

3 What's in a Name?

The unwieldy nature of a simple logical extension of the current system and its potential ambiguity has motivated us to investigate a simple alternative. Clearly the host galaxy is a key parameter immediately placing the PNe in context. Then we need to identify the object as an extragalactic PN (or PN candidate) and we need to discriminate it from all the other (perhaps hundreds or even thousands) of PN that may be discovered in an individual galaxy or its environs.

We propose a simple 'EPN' prefix to denote extragalactic PN, then a galaxy identifier and then a running number, perhaps in order of PN discovery epoch. The greater the number the more recent the discovery. Also the largest number immediately tells you approximately how many PN have been found in the particular galaxy to date. The name is also of manageable length and can be easily cross-referenced to all relevant data and accurate positions which may be updated as astrometry improves. Crucially the ID will remain the same.

3.1 Proposed New Nomenclature

EPN-NGC7793-1, EPN-N7793-1157 or EPN-IC5175-32 etc and that's it! The key 'EPN' prefix can be used to search entire CDS databases and will naturally pick up all extragalactic PN regardless of galaxy affiliation. We assume that any galaxy will always have a suitable identifier and that for brevity NGC can be concatenated to 'N'. The hyphens serve as useful delimiters between the key nomenclature components and assist in readability.

4 Problems and Issues

We have successfully registered the EPN acronym with the relevant IAU commission but there are of course some problems with this simple system and no scheme will ever satisfy the whole community. However there is strong community support for the essence of the new proposal together with a keen desire to urgently sort out the naming convention. Any adopted system cannot function by self-regulation so each group/discovery team working on extragalactic PN can present their own identifiers for each object as their 'usual names' and will not be expected to assign any adopted standardised nomenclature themselves. We propose that a small committee be tasked to assign nomenclatures to new extragalactic PNe. This would convene once or twice a year to incorporate all results published in the interim. An electronic only catalogue of such EPN will be made available. The preliminary committee membership is suggested as Acker, Jacoby and Parker.

4.1 Intracluster PN

How do we name them? The most supported suggestion is to attach them to the closest named galaxy. Many felt we can simply retain EPN-N7793-2415 for example. This is especially germane as the question of where does a galaxy end and the intergalactic medium begin? An alternative is to simply divide the sky into a few hundred cells and allocate PN accordingly. This was seen less favourably. PN associated with free-floating globular clusters or whatever would still work IF we assume such objects will have a name.

4.2 Contaminants

In many extragalactic PN projects many 'new' PN are candidates, being selected on the basis of a single line (usually [OIII] 5007Å). When such candidates are subsequently identified as contaminants in a numbered scheme '1-n' a removed object 'm' will leave a gap in what may have been a nice ordered sequence. This does not matter. Even in commercial data-bases such gaps are quite acceptable and common and their number can be easily determined. Furthermore we need only accept a PN and assign it the new nomenclature once it has been confirmed. Also it does not matter if pet designations for specific projects are also retained, after all any SIMBAD query for a Galactic PN can turn up a plethora of individual identifiers for the same object.

5 Conclusions

There is an urgent need to adopt a working extragalactic PN nomenclature that can then be implemented and adhered to. An extension to existing conventions could easily yield an identifier containing nearly 30 characters. This was felt by most to be too unwieldy and anonymous for general use. We propose a simple alternative whilst not perfect is tractable and easy to understand. It will require careful management. We commend it to the community.

Acknowledgements

The authors are grateful for the positive feedback from the community and the useful comments on an earlier version of this paper by George Jacoby.

References

1. Q. Parker, et al. (2003), IAU Symp.209, pg25, eds. S.Kwok, M.Dopita & R.Sutherland, Kluwer

Part III

The Planetary Nebula Luminosity Function

The Planetary Nebula Luminosity Function

Robin Ciardullo

The Pennsylvania State University, Department of Astronomy and Astrophysics, 525 Davey Lab, University Park, PA 16803, USA

Abstract. The [O III] λ5007 planetary nebula luminosity function (PNLF) occupies an important place on the extragalactic distance ladder: it is the only standard candle that can be applied to all the large galaxies of the Local Supercluster. We review the method's precision, and use it to show that the distance scale defined by Cepheids and the Surface Brightness Fluctuation method is likely too large by $\sim$ 7%. We also discuss some of the physics underlying the phenomenon, and present clues as why the technique is so resilient.

1 Introduction

Planetary nebulae (PNe) are unique probes of galactic chemical and dynamical evolution. But before they can be used, they must be found, and the best way of doing this involves searching for PN candidates via their bright [O III] λ5007 emission. As a result, the first piece of information obtained about a set of extragalactic PNe is their [O III] λ5007 luminosity function. Since this luminosity function contains information about both distance and parent population, one should make every effort to understand its properties and features.

2 The PN Luminosity Function as a Distance Indicator

Although the idea of using PNe as extragalactic distance indicators was first suggested the early 1960's [1,2], it was not until the late 1970's that a PN-based distance estimate was made [3], and the first study of the [O III] λ5007 planetary nebula luminosity function (PNLF) was not performed until 1989 [4,5]. In fact, it is an irony of the subject that PN distance measurements inside the Local Group [6] were made only after the technique had been applied to more distant systems [3,7], and the first application of the PNLF inside the Milky Way [8] occurred after the method had been used to estimate the Hubble Constant [9]. This odd chronology, of course, was due to the fact that individual PNe are definitely not standard candles [10–12], and that distance estimates to Galactic PNe are extremely poor. Nevertheless, today the [O III] λ5007 PNLF is one of the most important standard candles in extragalactic astronomy, and the only method that can be applied to all the large galaxies of the Local Supercluster, regardless of environment or Hubble type (see Fig. 1).

The use of the PN luminosity function is extremely simple. One surveys a distant galaxy or cluster and identifies point sources that are present in [O III]

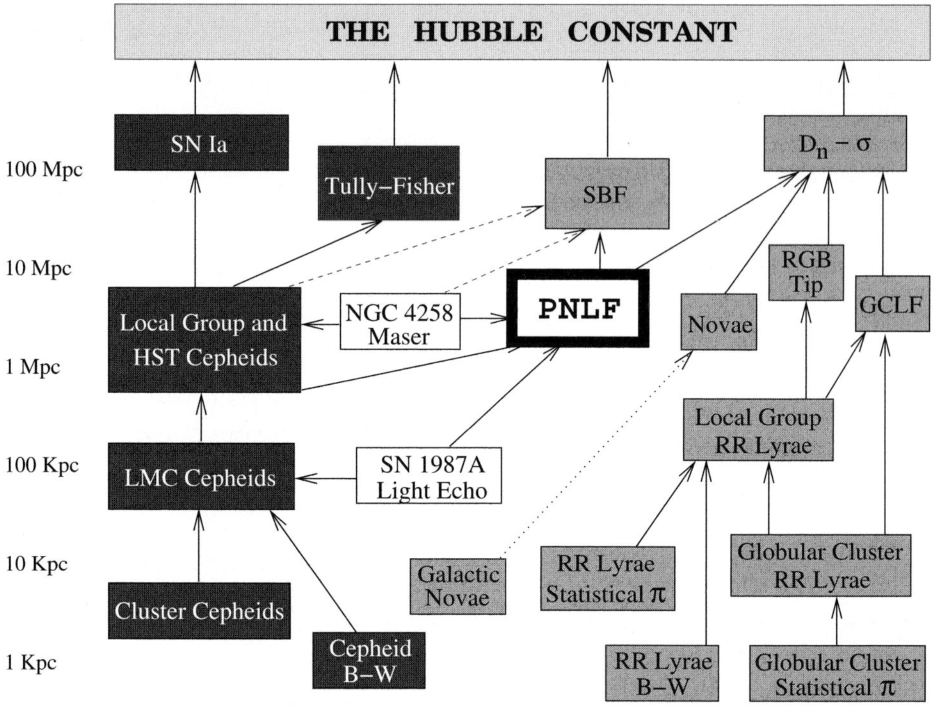

Fig. 1. The extragalactic distance ladder. The dark boxes show techniques useful in star-forming galaxies, the lightly-filled boxes give methods that work in Pop II systems, and the open boxes represent geometric distance determinations. Uncertain calibrations are noted as dashed lines. The PNLF is the only method that is equally effective in all the populations of the Local Supercluster.

λ5007 but completely invisible in the continuum. These are the planetary nebula candidates. Unfortunately, mixed in with the PNe are three types of contaminants:

H II Regions: Star formation is not an issue in most elliptical and lenticular galaxies, but in late-type systems, the H II regions around O and B stars will far outnumber the PNe. Fortunately, most H II regions are resolvable under good seeing ($\lesssim 1''$) conditions, and their exciting stars may be visible on deep continuum images. This contrasts with PNe which (in galaxies more distant than the Magellanic Clouds) are always stellar and have invisible central stars. Moreover, observations in M31's bulge, M33's disk, and the LMC have shown that all planetaries in the top $\sim$ 1 mag of the PNLF have [O III] λ5007 to Hα+[N II] line ratios greater than $\sim$ 1.6 [13]. Most H II regions have Hα brighter than [O III] [14].

Supernova Remnants: Unresolved high-excitation supernova remnants can masquerade as planetary nebulae, especially in galaxies that have a cold, high-density interstellar medium. Because compact supernova remnants are rare, their effect on the PNLF is minimal. Nevertheless, one needs to be aware of this source of contamination, especially at the bright limit of the luminosity function. Any single object that appears overluminous in [O III] $\lambda 5007$ could be the result of a supernova explosion.

Lyα Galaxies: At $z = 3.12$, Lyα is redshifted to 5007 Å, and at fluxes below $\sim 10^{-16}$ ergs cm^{-2} s^{-1}, unresolved and marginally resolved high-redshift galaxies with extremely strong Lyα emission (equivalent widths $\gtrsim$ 300 Å in the observers frame) can mimic planetary nebulae [15,16]. Since the surface density of these starbursting objects is relatively low, ~ 1 arcmin^{-2} per unit redshift interval brighter than 5×10^{-17} ergs cm^{-2} s^{-1} [17], PN surveys in galaxies are not strongly affected by this contaminant. However, in the intracluster environment of systems such as Virgo and Fornax, Lyα galaxies are the predominant source of error, and they limit the effectiveness of any luminosity function analysis. To identify these interlopers, one must either obtain deep broadband images (to detect the underlying continuum of the galaxies) or perform spectroscopy (to test for the existence of [O III] $\lambda 4959$ or resolve the ~ 400 km s^{-1} width of the Lyα line).

Once the PNe are found, all one needs to do to determine a distance is to measure their monochromatic [O III] $\lambda 5007$ fluxes, define a statistical sample of objects, and fit the observed luminosity function to some standard law. For simplicity, Ciardullo *et al.* [5] have described the PNLF via a truncated exponential

$$N(M) \propto e^{0.307M}\{1 - e^{3(M^* - M)}\} \tag{1}$$

where

$$m_{5007} = -2.5 \log F_{5007} - 13.74 \tag{2}$$

though other forms of the relation are possible [18]. In the above equation, the key parameter is M^*, the absolute magnitude of the brightest planetary. Despite some efforts at Galactic calibrations [8,18], the PNLF remains a secondary standard candle. The original value for the zero point, $M^* = -4.48$, was based on an M31 infrared Cepheid distance of 710 kpc [19] and a foreground extinction of $E(B-V) = 0.11$ [20]. Since then, M31's distance has increased [21], its reddening has decreased [23], and, most importantly, the Cepheid distances to 12 additional galaxies have been included in the calibration [13]. Somewhat fortuitously, the current value of M^* is only 0.01 mag fainter than the original value, $M^* = -4.47 \pm 0.05$.

Note that all PNLF distances require some estimate of the foreground interstellar extinction. There are two sources to consider. The first, extinction in the Milky Way, is readily available from reddening maps derived from H I measurements and galaxy counts [22] and/or from the DIRBE and IRAS satellite experiments [23]. The second, internal extinction in the host galaxy, is only a problem in late-type spiral and irregular galaxies. Unfortunately, it is difficult to quantify. In the Galaxy, the scale height of PNe is significantly larger than that of the dust [24]. If the same is true in other galaxies, then we would expect the bright end of the PNLF to always be dominated by objects foreground to the dust layer. This conclusion seems to be supported by observational data [25,13] and numerical models [25], both of which suggest that the internal extinction which affects a galaxy's PN population is $\lesssim 0.05$ mag. We will revisit this issue in Section 3.

3 Tests of the Technique

In the past decade, the PNLF has been subjected to a number of rigorous tests. These tests fall into four categories.

3.1 Internal Tests Within Galaxies

Five galaxies have large enough PN samples to test for radial gradients in the location of the PNLF cutoff. Two are large-bulge Sb spirals (M31 [26] and M81 [27]), one is a pure-disk Sc spiral with a strong metallicity gradient (M33 [28]), one is large elliptical galaxy (NGC 4494 [29]), and one is a blue, interacting peculiar elliptical (NGC 5128 [30]). No significant change in the PNLF has been seen in *any* of these systems. Given the diversity of the stellar populations sampled in these systems, this result, in itself, is impressive proof of the robustness of the method.

3.2 Internal Tests Within Galaxy Groups

To date, six galaxy clusters have multiple PNLF measurements: the M81 Group (M81 and NGC 2403 [31,25]), the NGC 1023 Group (NGC 891 and 1023 [32]), the NGC 5128 Group (NGC 5102, 5128, and 5253 [33,30,34]), the Fornax Cluster (NGC 1316, 1380, 1399, and 1404 [35]), the Leo I Group (NGC 3351, 3368, 3377, 3379, and 3384 [13,25,36]), and the Virgo Cluster (NGC 4374, 4382, 4406, 4472, 4486, and 4649 [9]). In each system, the observed galaxies have a range of color, absolute magnitude, and Hubble type. Despite these differences, no discrepant distances have been found, as the galaxies of each group always fall within ~ 1 Mpc of each other. Indeed, the PNLF measurements in Virgo easily resolve the infalling M84/M86 Group, which is background to the main body of the cluster [37].

3.3 External Tests with Different Methods

Because planetary nebulae are found in all types of galaxies, it is possible to compare the results of the PNLF method directly with distances derived from all of the other techniques at the top of the distance ladder. The most instructive of these comparisons involve Cepheid variables and the Surface Brightness Fluctuation (SBF) Method.

Figure 2 compares the PNLF distances of 13 galaxies (derived using the foreground extinction estimates from DIRBE/IRAS [23]) with the final Cepheid distances produced by the *HST Key Project* [21]. Neither dataset has been corrected for the effects of metallicity. Since the absolute magnitude of the PNLF

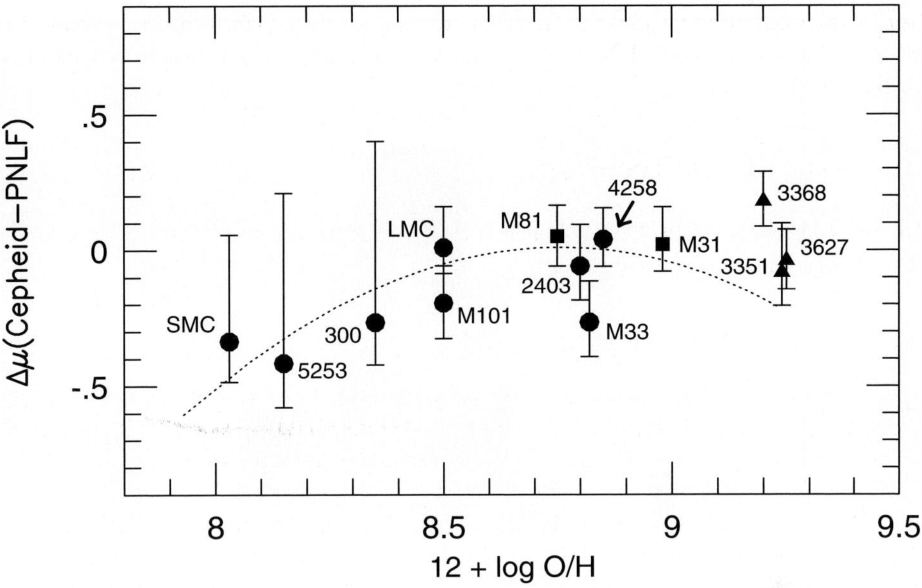

Fig. 2. A comparison of the PNLF and Cepheid distance moduli as function of galactic oxygen abundance, as estimated from the systems' H II regions [39]. No metallicity correction has been applied to either distance indicator. Circles represent PNLF measurements in galactic inner disks, squares in galactic bulges, and triangles in outer disks/halos. The error bars display the formal uncertainties of the methods added in quadrature; small galaxies with few PNe have generally larger errors. The curve shows the expected reaction of the PNLF to metallicity [38]. Note that metal-rich galaxies should not follow this relation, since these objects always contain enough lower metallicity stars to populate the PNLF's bright-end cutoff. The scatter between the measurements is consistent with the internal errors of the methods.

cutoff, M^*, is calibrated via these Cepheid distances, the weighted mean of the distribution must, by definition, be zero. However, the scatter about this mean, and the systematic trends in the data, are valid indicators of the reliability of the technique.

As Fig. 2 illustrates, the scatter between the Cepheid and PNLF distances is impressively small. Except for the most metal-poor systems, the residuals are perfectly consistent with the internal uncertainties of the methods. Moreover, the systematic shift seen at low-metallicity is exactly that predicted by models of post-AGB evolution [38]. There is little room for any additional metallicity term, either in the Cepheid or PNLF relations.

Figure 3 compares the PNLF distances of 29 galaxies with distances derived from SBF measurements. As is illustrated, the relative distances of the two methods are in excellent agreement: the scatter in the data is exactly that predicted from the internal errors of the measurements. However, the absolute distance scale derived from the SBF data is ~ 0.3 mag larger than that inferred

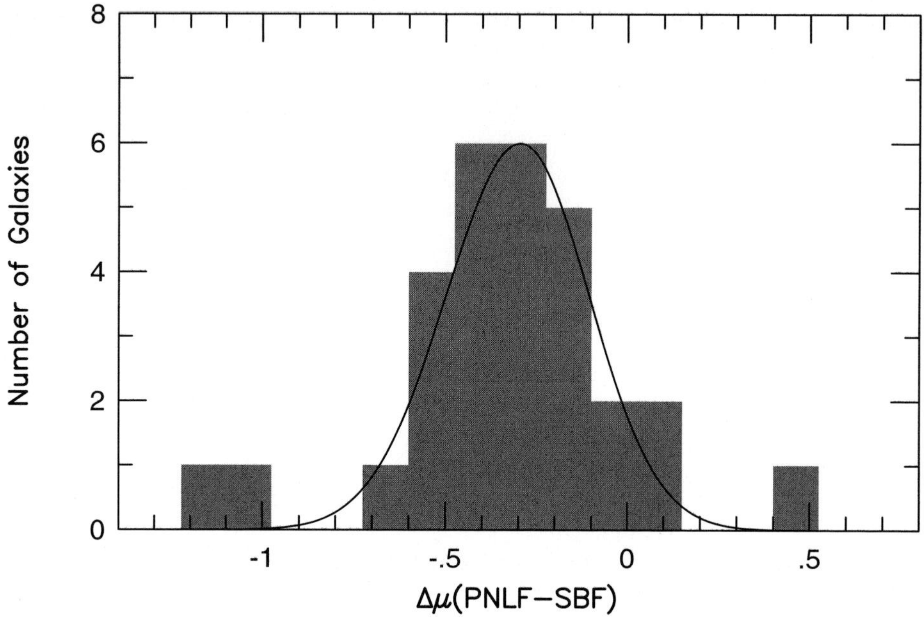

Fig. 3. A histogram of the difference between the PNLF and SBF distance moduli for 29 galaxies measured by both methods. The two worst outliers are the edge-on galaxies NGC 4565 ($\Delta\mu = -0.80$) and NGC 891 ($\Delta\mu = +0.71$). The curve represents the expected dispersion of the data. The figure demonstrates that the absolute scales of the two techniques are discrepant by ~ 0.3 mag, but the internal and external errors of the methods agree.

from the PNLF. Since both techniques are calibrated via Cepheid measurements, and both have formal zero-point uncertainties of ~ 0.05 mag, this type of offset should not exist.

The most likely cause of the discrepancy is a small amount of internal extinction in the Cepheid calibration galaxies. For most methods (including the PNLF), an underestimate of extinction produces an underestimate of luminosity. As a result, undetected internal extinction within a Cepheid calibration galaxy propagates into an underestimate of the standard candle's brightness, and an extragalactic distance scale that is too small. For the SBF technique, however, the opposite is true. The SBF standard candle, $\bar{M}_I$, is extremely sensitive to color, i.e.,

$$\bar{M}_I = -1.74 + 4.5[(V - I)_0 - 1.15] \tag{3}$$

[40]. If internal extinction exists in an SBF calibrator, the galaxy's underlying population will appear too red, and the inferred value of $\bar{M}_I$ will be too bright. The result will be a set of distances that are systematically too large.

If internal extinction really is responsible for the offset displayed in Fig. 3, then the zero points of both systems must be adjusted. These corrections propagate all the way up the distance ladder. For example, according to the *HST Key Project,* the SBF-based Hubble Constant is 69 ± 4 (random) ± 6 (systematic) km s^{-1} Mpc^{-1} [21]. However, if we assume that the calibration galaxies are internally reddened by $E(B-V) \sim 0.04$, then the zero point of the SBF system fades by 0.17 mag, and the SBF Hubble Constant increases to 75 km s^{-1} Mpc^{-1}. This one correction is as large as the technique's entire systematic error budget. Such an error could not have been found without the cross-check provided by PNLF measurements.

3.4 External Comparisons with Geometric Distances

Two galaxies have distance measurements that do not depend on the distance ladder. The first is NGC 4258, which has a resolved disk of cold gas orbiting its central black hole. The proper motions and radial accelerations of water masers associated with this gas yield an unambiguous geometric distance of 7.2 ± 0.3 Mpc [41]. The second benchmark comes from the light echo of SN 1987A in the Large Magellanic Cloud. Although the geometry of the light echo is still somewhat controversial, the most detailed and complete analysis of the object to date gives a distance of $D < 47.2 \pm 0.1$ kpc [42]. In Table 1 we compare these values with the distances determined from the PNLF [13] and from Cepheids [21].

Table 1. Benchmark Galaxy Distances

Method	LMC	NGC 4258	$\Delta\mu$ (mag)
Geometry	$< 18.37 \pm 0.04$	29.29 ± 0.09	10.92 ± 0.10
Cepheids	18.50	29.44 ± 0.07	10.94 ± 0.07
PNLF	18.47 ± 0.11	29.43 ± 0.09	10.96 ± 0.14

As the table demonstrates, the Cepheid and PNLF methods both overestimate the distance to NGC 4258 by ~ 0.14 mag, i.e., by $\sim 1.3\,\sigma$ and $1.1\,\sigma$, respectively. In the absence of some systematic error affecting both methods, the probability of this happening is $\lesssim 5\%$. On the other hand, there is no disagreement concerning NGC 4258's distance *relative to that of the LMC:* the Cepheids, PNLF, and geometric techniques all agree to within ±2%! Such a small error is probably fortuitous, but it does suggest the presence of a systematic error that affects the entire extragalactic distance ladder.

In fact, the *HST Key Project* distances are all based on an LMC distance modulus of $(m-M)_0 = 18.50$ [21], and, via the data of Fig. 2, the PNLF distance scale is tied to that of the Cepheids. If the zero point of the Cepheid scale were shifted to $(m-M)_0 = 18.37$, then all the measurements would be in agreement. This consistency supports a shorter distance to the LMC, and argues for a 7% increase in the *HST Key Project* Hubble Constant to 77 km s^{-1} Mpc^{-1}.

4 The Physics of the Luminosity Function

The fact that the PNLF method is insensitive to metallicity is not a surprise. Since oxygen is a primary nebular coolant, any decrease in its abundance raises the plasma's electron temperature and increases the rate of collisional excitations per ion. This mitigates the effect of having fewer emitting ions in the nebula, so that a decrease in oxygen abundance only lessens the emergent [O III] λ5007 line flux by (roughly) the square root of the abundance difference [4]. Meanwhile, the PN's core reacts to metallicity in the opposite manner. If the metallicity of the progenitor star is decreased, then the PN's central star will be slightly more massive and its emergent UV flux will be slightly greater [43,44]. This additional energy almost exactly compensates for the decreased emissivity of the nebula. As a result, the total [O III] λ5007 flux that is generated by a planetary is virtually independent of metallicity. This fortuitous cancellation has been confirmed by more detailed models of PN evolution: only when the metallicity drops to sub-LMC levels does the the brightness of the PNLF cutoff fade by more than a few percent [38].

The insensitivity of the PNLF cutoff to population age is more difficult to understand. A PN's [O III] λ5007 flux is directly proportional to the luminosity of its central star, and this luminosity, in turn, is extremely sensitive to central star mass [45,46]. Since central star mass is directly proportional to progenitor mass via the initial mass-final mass relation [47], one would think that the PNLF cutoff would fade dramatically with time [48]. Yet observations demonstrate that this fading does not occur.

Although we do not as yet have a complete theory for the age-invariance of the PNLF, a few clues are beginning to emerge. The first concerns the faint end of the luminosity function. Although the location of the PNLF cutoff does not change with stellar population, the same is not true for the function's overall shape. As Fig. 4 demonstrates, old populations, such as M31's bulge, have a PNLF that increases exponentially following $N(m) \propto e^{0.307m}$. This type of behavior is expected if the PNe of these systems have low-mass cores, so that the timescale for central star evolution is much longer than that for nebular expansion [1,28]. In contrast, star-forming galaxies have a "dip" in the PNLF at magnitudes between ~ 2 and ~ 4 mag down from M^*. This deficit is a natural characteristic of models in which the evolution of [O III] λ5007 is governed by the rapid evolution of a high-mass core. In such systems, the non-monotonic PNLF is simply a reflection of the bimodal luminosity function expected from post-AGB stars [45].

A second clue to the PNLF comes from the fortuitous correlation between the maximum UV emission achieved by a PN central star and the mass of its envelope. High luminosity PN cores have large circumstellar envelopes, and, consequently, large amounts of circumstellar dust. The extinction caused by this dust can act to regulate the amount of [O III] λ5007 flux generated by a planetary, so that the luminosity of high core-mass objects never exceeds M^*. Support for this scenario comes from the fact that nine Magellanic Cloud planetaries have intrinsic [O III] λ5007 magnitudes brighter than M^*, but *all* are self-extincted below the PNLF cutoff [49–51]. In addition, there is observational evidence for a

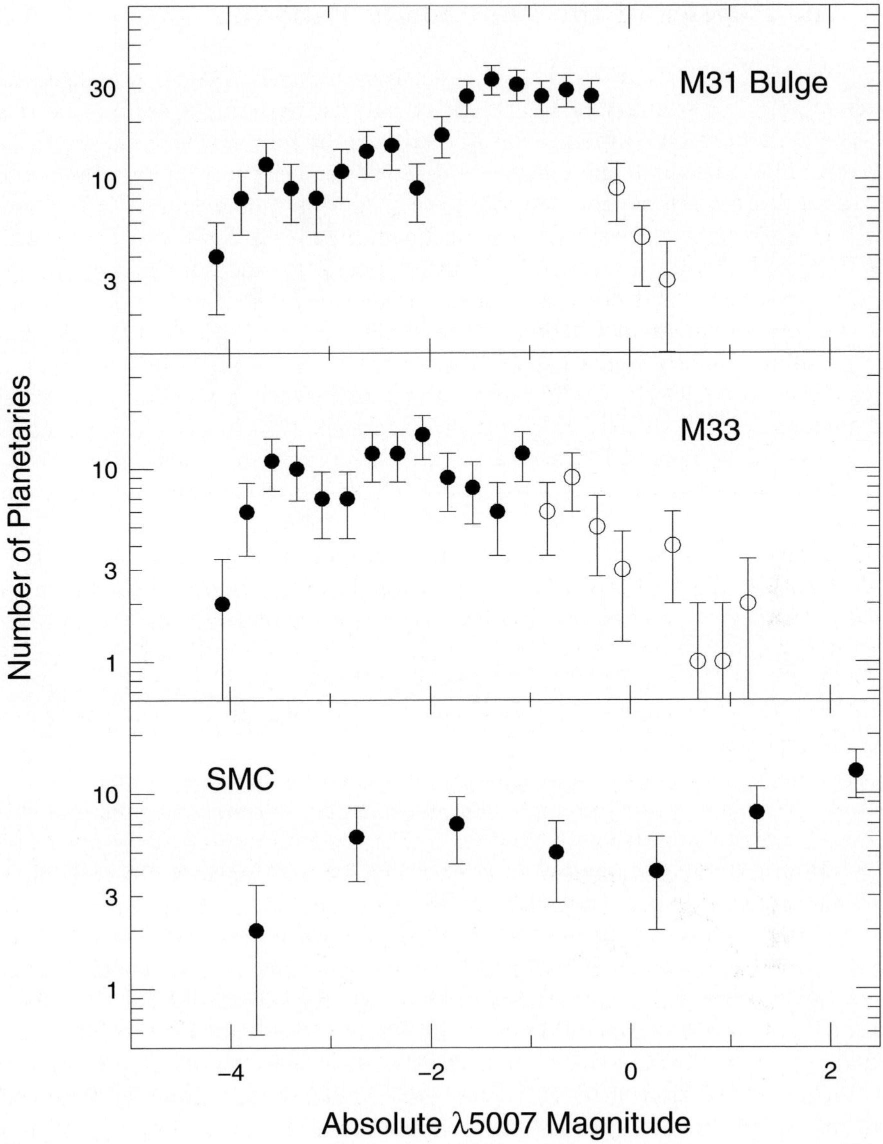

Fig. 4. The [O III] λ5007 planetary nebula luminosity functions of of three Local Group stellar populations. Data past the completeness limit are represented by open circles. Old populations, such as that found in the bulge of M31, have an exponentially increasing PNLF. In contrast, the PNLFs of star-forming systems are non-monotonic, with a "dip" that begins $\sim$ 2 mag below M^*. The location of this dip is consistent with models in which the luminosity evolution of planetary nebulae is governed by the rapid evolution of high-mass central stars.

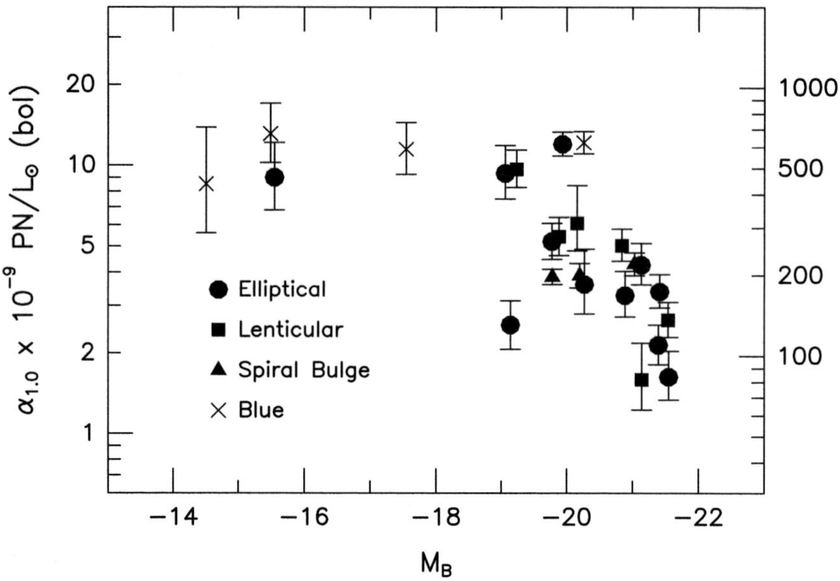

Fig. 5. The bolometric luminosity-specific number of PNe in the top 1 mag of the [O III] λ5007 PNLF plotted as a function of galactic absolute B magnitude. The right side of the y-axis translates this number into a constraint on the PN lifetime using a luminosity-specific stellar evolutionary flux of 2×10^{-11} stars yr−1 $L_\odot$. Either the [O III]-bright PNe of giant ellipticals live very short lives, or a substantial fraction of the stars in these galaxies do not contribute to the bright-end of the PNLF.

correlation between the amount of PN circumstellar extinction and central star mass for young, [O III]-bright objects [52].

The final constraint on the physics of the PNLF comes from the number of PN produced by different stellar populations. According to the theory of stellar energy generation [53], all populations older than ~ 1 Gyr create ~ 1 PN per year per $5 \times 10^{12} L_\odot$ (bolometric). If this is true, then measurements of the luminosity-specific number of PNe present in a galaxy can be immediately translated into constraints on the lifetime of an [O III]-bright nebula and on the fraction of stars which go through the bright planetary nebula phase.

This constraint is given in Fig. 5. For small and medium-sized galaxies, PNe typically spend ~ 500 yr within ~ 1 mag of M^*. However, the apparent PN timescale decreases dramatically for objects in giant elliptical and lenticular galaxies. Either the [O III]-bright planetaries in these systems have much shorter lives than their small-galaxy counterparts, or a substantial fraction of the stars of giant ellipticals do not evolve into these types of objects.

5 The Future

Of course, attempting to derive the evolution of ensembles of planetary nebulae from [O III] data alone is like trying to infer the evolution of star clusters from photometry performed through only one filter. In order to gain a fuller understanding of the global properties of planetary nebulae, one must study the luminosity function of PNe in a multi-dimensional emission-line space. The data for such a study are just now beginning to be available. Narrow-band photometric surveys are yielding information on the absolute line strengths of PNe in [O III] λ5007 and Hα+[N II] for a variety of stellar populations. Follow-up radial velocity surveys are then producing data on Hβ relative to [O III] λ5007 and, in some cases, [N II] and [S II] relative to Hα. By comparing these measurements to models, we should be able to gain a much more complete understanding of planetary nebula phenomenon.

This work was supported in part by NSF grant AST 00-71238.

References

1. K.G. Henize, B.E. Westerlund: Ap. J. **137**, 747 (1963)
2. P.W. Hodge: *Galaxies and Cosmology.* (McGraw-Hill, New York 1966)
3. H.C. Ford, D.C. Jenner: Bull. A.A.S. **10**, 665 (1978)
4. G.H. Jacoby: Ap. J. **339**, 39 (1989)
5. R. Ciardullo, G.H. Jacoby, H.C. Ford, J.D. Neill: Ap. J. **339**, 53 (1989)
6. G.H. Jacoby, M.P. Lesser: A. J. **86**, 185 (1981)
7. D.G. Lawrie, J.A. Graham: Bull. A.A.S. **15**, 907 (1983)
8. S.R. Pottasch: Astr. Ap. **236**, 231 (1990)
9. G.H. Jacoby, R. Ciardullo, H.C. Ford: Ap. J. **356**, 332 (1990)
10. L. Berman: Lick Obs. Bull. **18,** 57 (1937)
11. R. Minkowski: Pub. Obs. Univ. Mich. **10,** 25 (1951)
12. I.S. Shkolvski: Astron. Zh. **33,** 222 (1956)
13. R. Ciardullo, J.J. Feldmeier, G.H. Jacoby, R.K. de Naray, M.B. Laychak, P.R. Durrell: Ap. J. **577**, 31 (2002)
14. P.A. Shaver, R.X. McGee, L.M. Newton, A.C. Danks, S.R. Pottasch: M.N.R.A.S. **204**, 53 (1983)
15. R.-P. Kudritzki, R.H. Méndez, J.J. Feldmeier, R. Ciardullo, G.H. Jacoby, K.C. Freeman, M. Arnaboldi, M. Capaccioli, O. Gerhard, H.C. Ford: Ap. J. **536**, 19 (2000)
16. K.C. Freeman, M. Arnaboldi, M. Capaccioli, R. Ciardullo, J. Feldmeier, H. Ford, O. Gerhard, R. Kudritzki, G. Jacoby, R.H. Méndez, R. Sharples: 'Intracluster Planetary Nebulae in the Virgo Cluster'. In: *Dynamics of Galaxies: From the Early Universe to the Present, ASP Conference 197*, ed. by F. Combes, G.A Mamom, V. Charmandaris (Astronomical Society of the Pacific, San Francisco 2000) pp. 389-392
17. R. Ciardullo, J.J. Feldmeier, K. Krelove, G.H. Jacoby, C. Gronwall: Ap. J. **566**, 784 (2002)
18. R.H. Méndez, R.P. Kudritzki, R. Ciardullo, G.H. Jacoby: Astr. Ap. **275**, 534 (1993)

19. D.L. Welch, C.W. McAlary, R.A. McLaren, B.F. Madore: Ap. J. **305**, 583 (1986)
20. R.D. McClure, R. Racine: A. J. **74**, 1000 (1969)
21. W.L. Freedman, B.F. Madore, B.K. Gibson, L. Ferrarese, D.D. Kelson, S. Sakai, J.R. Mould, R.C. Kennicutt, Jr., H.C. Ford, J.A. Graham, J.P. Huchra, S.M.G. Hughes, G.D. Illingworth, L.M. Macri, P.B. Stetson: Ap. J. **553**, 47 (2001)
22. D. Burstein, C. Heiles: Ap. J. Suppl. **54**, 33 (1984)
23. D.J. Schlegel, D.P. Finkbeiner, M. Davis: Ap. J. **500**, 525 (1998)
24. D. Mihalas, J. Binney: *Galactic Astronomy*. (W.H. Freeman, New York 1981)
25. J.J. Feldmeier, R. Ciardullo, G.H. Jacoby: Ap. J. **479**, 231 (1997)
26. X. Hui, H. Ford, G. Jacoby: Bull.A.A.S. **26,** 938 (1994)
27. L. Magrini, M. Perinotto, R.L.M. Corradi, A. Mampaso: Astr. Ap. **379**, 90 (2001)
28. R. Ciardullo, P.R. Durrell, M.B. Laychak, K.A. Herrmann, K. Moody, G.H. Jacoby, J.J. Feldmeier: Ap .J. in press
29. G.H. Jacoby, R. Ciardullo, W.E. Harris: Ap. J. **462**, 1 (1996)
30. X. Hui, H.C. Ford, R. Ciardullo, G.H. Jacoby: Ap. J. **414**, 463 (1993)
31. G.H. Jacoby, R. Ciardullo, H.C. Ford, J. Booth: Ap. J. **344**, 704 (1989)
32. R. Ciardullo, G.H. Jacoby, W.E. Harris: Ap. J. **383**, 487 (1991)
33. R. McMillan, R. Ciardullo, G.H. Jacoby: A. J. **108**, 1610 (1994)
34. M.M. Phillips, G.H. Jacoby, A.R. Walker, J.L. Tonry, R. Ciardullo: Bull. A.A.S. **24**, 749 (1992)
35. R. McMillan, R. Ciardullo, G.H. Jacoby: Ap. J. **416**, 62 (1993)
36. R. Ciardullo, G.H. Jacoby, H.C. Ford: Ap. J. **344**, 715 (1989)
37. H. Böhringer, U.G. Briel, R.A. Schwarz, W. Voges, G. Hartner, J. Trümper: Nature **368**, 828 (1994)
38. M.A. Dopita, G.H. Jacoby, E. Vassiliadis: Ap. J. **389**, 27 (1992)
39. L. Ferrarese, H.C. Ford, J. Huchra, R.C. Kennicutt, J.R. Mould, S. Sakai, W.L. Freedman, P.B. Stetson, B.F. Madore, B.K. Gibson, J.A. Graham, S.M. Hughes, G.D. Illingworth, D.D. Kelson, L. Macri, K. Sebo, N.A. Silbermann: Ap. J. Suppl. **128**, 431 (2000)
40. J.L. Tonry, A. Dressler, J.P. Blakeslee, E.A. Ajhar, A.B. Fletcher, G.A. Luppino, M.R. Metzger, C.B. Moore: Ap. J. **546**, 681 (2001)
41. J.R. Herrnstein, J.M. Moran, L.J. Greenhill, P.J. Diamond, M. Inoue, N. Nakai, M. Miyoshi, C. Henkel, A. Riess: Nature **400**, 539 (1999)
42. A. Gould, O. Uza: Ap. J. **494**, 118 (1998)
43. J.C. Lattanzio: Ap. J. **311**, 708 (1986)
44. E. Brocato, F. Matteucci, I. Mazzitelli, A. Tornabé: Ap. J. **349**, 458 (1990)
45. E. Vassiliadis, P.R. Wood: Ap. J. Suppl. **92**, 125 (1994)
46. T. Blöcker: Astr. Ap. **299**, 755 (1995)
47. V. Weidemann: Astr. Ap. **363**, 647 (2000)
48. P. Marigo, L. Girardi, A. Weiss, M.A.T. Groenewegen, C. Chiosi: Astr. Ap., in press
49. S.J. Meatheringham, M.A. Dopita: Ap. J. Suppl. **75**, 407 (1991)
50. S.J. Meatheringham, M.A. Dopita: Ap. J. Suppl. **76**, 1085 (1991)
51. G.H. Jacoby, A.R. Walker, R. Ciardullo: Ap. J. **365**, 471 (1990)
52. R. Ciardullo, G.H. Jacoby: Ap. J. **515**, 191 (1999)
53. A. Renzini, A. Buzzoni: In: *Spectral Evolution of Galaxies*, ed. by C. Chiosi, A. Renzini (Reidel, Dordrecht), p. 195 (1986)

The PNLF Distance to the Sculptor Group Galaxy NGC 55

G.C. Van de Steene[1], G.H. Jacoby[2], C. Praet[3], R. Ciardullo[4], and H. Dejonghe[3]

[1] Royal Observatory of Belgium, Ringlaan 3, 1180 Brussels, Belgium
[2] WIYN Observatory, PO Box 26732, Tucson, AZ, 85726, USA
[3] Universiteit Gent, Krijgslaan 281 (S9), 9000 Gent, Belgium
[4] Department of Astronomy and Astrophysics, Penn State University, 525 Davey Lab University Park, PA 16802, USA

Abstract. We have identified 21 new Planetary Nebula (PN) candidates in the Sculptor Group galaxy NGC 55. We determined a most likely distance of 2.00 ± 0.2 Mpc using the Planetary Nebula Luminosity Function (PNLF) method. The distance to NGC 55 is larger than previously determined distances, which means that the Sculptor Group is a little farther away from the Local Group than previously thought. The distance to NGC 55 is again similar to the distance of NGC 300, adding support to the suggestion that these galaxies form a bound pair.

1 NGC 55

NGC 55 is a highly inclined (i=85±5°) late type galaxy (SB(s)m) in the Sculptor Group. Its Holmberg radius is 20ʹ.2 and blue magnitude m_B=8.42. Sculptor is a very loosely concentrated and almost freely expanding aggregate of galaxies of prolate shape which we view pole-on. It stretches from the outskirts of the Local Group at 1.5 Mpc out to 5 Mpc. Besides the 5 main spiral galaxies, the Sculptor Group contains several dwarf galaxies (Jerjen et al. 1998).

NGC 55 is one of the 5 bright spiral galaxies of the Sculptor Group, which could be the one nearest to the Milky Way. Distance estimates for NGC 55 range from 1.34 Mpc based on carbon stars (Pritchet et al. 1987) to 1.8 Mpc based on the Tully-Fisher relation (Karachentsev et al. 2003). The two bright spiral galaxies NGC 55 and NGC 300 are both at the near side of the Sculptor group and it has been suggested that NGC 55 and NGC 300 form a bound pair (Graham 1982; Pritchet et al. 1987; Whiting 1999). The distance to NGC 300 is well determined via Cepheids (Freedman et al. 2001: 2.07 ± 0.07 Mpc) and via the PNLF method (Soffner et al. 1996: 2.4 ± 0.4 Mpc). NGC 55 has remained without a reliable distance estimate.

2 Observations

The observations were done with the Wide Field Imager (WFI) on the 2.2-m telescope at the La Silla Observatory of the European Southern Observatory on the 4 and 5 July 1999. The weather was photometric. WFI consists of a 4 times

2 mosaic of 2k x 4k CCDs. The pixel scale is 0.″238 and the total field of view is 34′ x 33′, which covers the entire galaxy. NGC 55 was observed through the [O III]/8 filter (FWHM 80.34 Å), the off-band filter 518/16 and the Hα/7 filter for a total of 3120 sec, 1620 sec, and 3600 sec respectively. On the first night the galaxy was offset to the east and south by 3′ to fill in the CCD gaps. The seeing was 1.″4.

3 Data Reduction and Results

The ESOWFI external package in IRAF was used to convert the ESO headers to suit the MSCRED package and to set the instrument files and astrometry solution. Next the reductions were done using the MSCRED mosaic reduction package in IRAF according to the guide by Valdes (1998).

We followed the survey technique as described in e.g. Jacoby et al. (1989b) and Ciardullo et al. (1989a). The on- and off-band images were aligned. The off-band images were then scaled to the level of the on-band images and subtracted from them to produce the difference image. We identified the PNe by blinking this difference image with the off-band image, and the Hα+[N II] images. In order to discriminate PNe from other emission-line sources, we used the following criteria (Feldmeier et al. 1997): (1) PN candidates had to have a point-spread function (PSF) consistent with that of a point source as all PNe are expected to be unresolved at the distance of NGC 55; (2) PN candidates had to be invisible on the off-band image to exclude bright OB stars exciting H II regions; and (3) PN candidates had to be significantly fainter in Hα+[N II] image than [O III], in order to reduce further the possibility of contamination from H II regions. With all these constraints, only compact, H II regions that have high nebula excitation and faint central OB associations will have been mistaken for PNe. However, even these will be faint and hence cannot significantly affect the PNLF. In total we identified 21 candidate planetaries in NGC 55.

The PN candidates were measured photometrically in the off-band image using PHOT in IRAF and flux calibrated using standard stars and the procedures outlined in Jacoby et al. (1987). In order to determine the filter transmission for the PNe we need to take the redshift into account and the bandpass shift of the interference filter to the blue. The systemic velocity of NGC 55 was taken to be 116 km/s (Puche & Carignan 1991). The percentage of transmission is about 88% across the filter for the [O III] line and about 87% for the Hα filter at the location of our objects. The difference in transmission across the field has negligible effect on the derived flux values and magnitudes.

The resulting monochromatic [O III] flux values F_{5007} were converted to [O III]λ5007 magnitudes using Jacoby (1989a):

$$m_{5007} = -2.5\ log(F_{5007}) - 13.74. \tag{1}$$

4 The PNLF Distance

[O III] magnitudes were corrected for the interstellar extinction. We consider only the foreground Galactic extinction towards NGC 55 and adopt E(B-V) = 0.013 mag (Schlegel et al. 1998). Identifications become incomplete beyond m_{5007} = 23.5 mag. Besides missing faint PNe, the probability of overlap with a star or with an H II region increases towards the centre, particularly in the Hα+[N II] image. Hence brighter PNe than the ones recovered in the outskirts may have been missed towards the center.

From these data the PNLF distance to the galaxy can normally be derived by convolving the empirical model for the PNLF given by Ciardullo et al. (1989):

$$N(m) \propto e^{0.307M}[1 - e^{3(M^*-M)}] \tag{2}$$

with the photometric error function and fitting the data to the resultant curve via the method of maximum likelihood. This takes into account that the probability of observing PNe near the cutoff magnitude M* decreases for small sample sizes. Fig. 1 plots the PNLF for the sample. Assuming M*= −4.47, based on the calibration to M 31 (Ciardullo et al. 2002), we obtain a most likely distance modulus (m − M*) = 26.95. The PNLF cut-off is fainter in small, low metallicity galaxies, but well modeled by the theoretical relation of Dopita et al. (1992). The metallicity correction, as determined from the oxygen abundance, is needed only for galaxies with metallicities smaller than the LMC (12+log(O/H) < 8.5) (Ciardullo et al. 2002). The oxygen abundance of NGC 55 as determined based

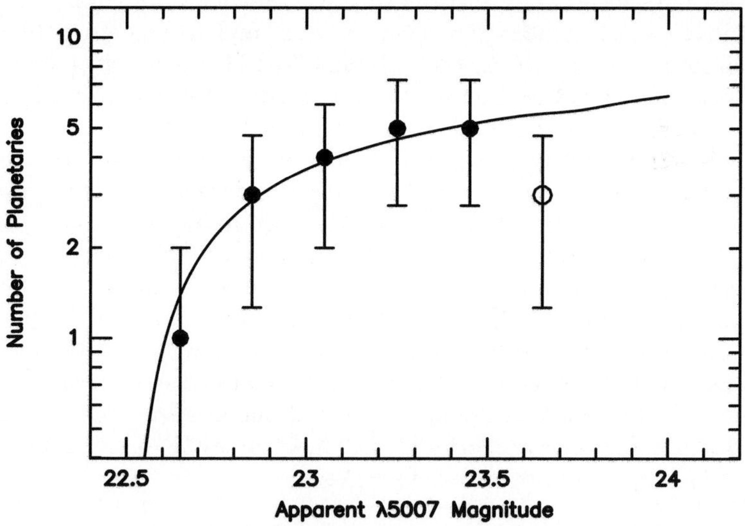

Fig. 1. Observed [O III]λ5007 PNLF of NGC 55. The curve represents the best-fitting empirical PNLF convolved with the photometric error function and shifted to the most likely distance. The open circle represents a point past the completeness limit.

on H II regions is 12+log(O/H)=8.05 (Tüllmann et al. 2003), which is much lower than this value and similar to the SMC metallicity. Assuming a solar abundance of oxygen of 12+log(O/H) = 8.87 (Grevesse et al. 1996), the metallicity corrected value of the distance modulus is $(m - M^*) = 26.50 \pm 0.2$, which corresponds to a distance of 2.00 ± 0.2 Mpc. This value is a slightly farther than previously determined distances to NGC 55, which would mean that the Sculptor Group would be more distant than previously thought.

NGC 300 also has a low metallicity, though not as low as NGC 55. The metallicity correction to be applied is 0.15 mag (Ciardullo et al. 2002). The distance modulus to NGC 300 determined via the PNLF maximum likelihood method is 26.8 mag (Soffner et al. 1996) and after correction for metallicity 26.65 mag. This corresponds to a distance of 2.14 $\pm$0.4 Mpc , which is in very good agreement with the Cepheid distance determination of 2.02 ± 0.07 Mpc by Freedman et al. (2001).

The PNLF distances to NGC 55 and NGC 300 are again similar, which adds support to the fact that they form a bound pair and illustrates the consistency of the PNLF method for distance determination.

5 Conclusions

We identified 21 new PNe candidates in the Sculptor Group galaxy NGC 55. The PNLF method gives us a most likely distance of 2.00 $\pm$0.2 Mpc, which would mean that the Sculptor Group is a little farther away from the Local Group than previously thought. The distance to NGC 55 turns out to be similar to the distance of NGC 300, adding support to the suggestion that these galaxies form a bound pair and illustrates the consistency of the PNLF method for distance determination.

References

1. Ciardullo, R., Jacoby, G. H, Ford, H.C., 1989, ApJ, 344, 715
2. Ciardullo, R., Jacoby, G. H., Ford, H.C., Neill, J. D., 1989b, ApJ, 339, 53
3. Ciardullo, R. Feldmeier, J. J., Jacoby, G. H., Kuzio de Naray, R., Laychak, M. B., Durell, P. R., ApJ, 577, 31
4. Dopita, M. A., Jacoby, G. H., Vassiliadis, E., 1992, ApJ, 389, 27
5. Feldmeier, J. J., Ciardullo, R., Jacoby, G.H., 1997, ApJ, 479, 231
6. Freedman, W. L., Madore, B. F., Gibson, B. K., Ferrarese, L., Kelson, D. D., Sakai, S., Mould, J. R., Kennicutt, R. C., Jr., Ford, H. C., Graham, J. A., Huchra, J. P., Hughes, S. M. G., Illingworth, G. D., Macri, L. M., Stetson, P. B., ApJ, 553, 47
7. Graham, J. A. 1982, ApJ, 252, 474
8. Grevess, N., Noels, A., Sauval, A.J., 1996, in ASP Conf. Ser. 99, Cosmic Abundances, ed. S. S. Holt & G. Sonneborn (San Francisco: ASP), 117
9. Jacoby, G. H., Quigley, R. J., Africano, J. L., 1987, PASP, 99, 672
10. Jacoby, G. H, 1989, ApJ, 339. 39.
11. Jacoby, G.H., Ciardullo, R., Ford, H.C., Booth, J., 1989, ApJ, 344, 70
12. Jerjen, H., Freeman, K. C., Binggeli, B., 1998, AJ, 116, 2873

13. Karachentsev, I. D., Grebel, E. K., Sharina, M. E., Dolphin, A. E., Geisler, D., Guhathakurta, P., Hodge, P. W., Karachentseva, V. E., Sarajedini, A., Seitzer, P., 2003, å, 404, 93
14. Méndez, R. H., Kudritzki, R. P., Ciardullo, R., Jacoby, G. H., 1993, A&A, 275, 534
15. Pritchet, C. J., Schade, D., Richer, H. B., Crabtree, D., Yee, H. K. C. 1987, ApJ, 323, 79
16. Puche, D., Carignan, C. 1988, AJ, 95, 1025
17. Puche, D., Carignan, C., Wainscoat, R. J., 1991, AJ, 101, 447
18. Schlegel, D.J., Finkbeiner, D.P., Davis, M., 1998, ApJ, 500 525
19. Soffner, T., Méndez, R. H., Jacoby, G. H., Ciardullo, R., Roth, M. M., Kudritzki, R. P. 1996, A&A, 306, 9
20. Tüllmann, R., Rosa, M. R., Elwert, T., Bomans, D. J., Ferguson, A. M. N., Dettmar, R.-J., 2003, A&A, 412, 69
21. Valdes, F., 1998, "Guide to the NOAO mosaic data handling software"
22. Whiting, A. B., 1999, AJ, 117 202

Part IV

PN Stellar Progenitors

Mass Loss At the Tip of the AGB: Essentials for Population Synthesis

Lee Anne Willson[1,2]

[1] Iowa State University, Ames IA 50011, USA
[2] Uppsala University, Uppsala, Sweden

Abstract. It is not yet possible to construct models that include details of all processes that play a role in Asymptotic Giant Branch tip mass loss. However, the very sensitivity of mass loss rates to stellar parameters, that causes problems for modellers, also provides the means for synthesizing populations, because we can interpret observed mass loss rates as telling us which stars have entered the "death zone" near the edge of the "cliff" in mass vs. luminosity. This understanding gives us a first estimate of the effect of metallicity on the Asymptotic Giant Branch (AGB) terminal luminosity, as well. With this understanding of the observations, we can test mass loss laws based on existing models: most fail to reproduce the death-line. The cliff formulation cannot yet predict the masses of the central stars of planetary nebulae, however, because Miras are demonstrably overluminous with respect to their most likely core masses. Observers are encouraged to combine planetary nebula surveys with searches for the Mira variables, that mark the tip of the AGB, to shed light on this problem.

1 Introduction

That the luminosity function of planetary nebulae is a powerful tool for learning about the stellar populations of external galaxies, and that it can be a useful means of establishing the distance scale out to tens of megaparsecs, emerges clearly from other presentations made at this conference. However, before this can be relied on to give accurate information for populations very different from those that we can study in other ways, we must be able to explain those observations and extrapolate reliably to new situations. For this, we look to models of stellar evolution that incorporate an understanding of mass loss, because the mass loss process is what selects the luminosity, L_{final}, (and related quantities radius, R_{final}, or effective temperature, T_{final}) at which stars leave the AGB.

I shall argue that we are far from having mass loss models that may be used for all purposes - for predicting mass loss rates for individual stars, for providing the pattern of mass loss for a population of stars, for providing a structure that may be used to match observations of spectral lines or spectral energy distribution, and for modelling the details of planetary nebula formation. However, we have quite a number of observational studies of mass loss for AGB stars and models that allow us to interpret these observations correctly, as indicating which stars are now losing mass at an interesting rate.

2 Mass Loss Modelling

Four distinct processes must be adequately modelled in order to get mass loss rates that match observational constraints. Additional physics is needed to get a reasonable match of models with observed spectra. Further details are needed if a truly self-consistent model is desired.

2.1 Essential Elements of Mass Loss Models

The essential elements of mass loss models for AGB tip stars are pulsation, non-LTE, chemistry, and radiative transfer.

Pulsation gives rise to atmospheric shocks and levitation of the material in the atmosphere. Shocks compress the material to temperatures well above the radiative equilibrium temperature and expansion between shocks dominates cooling in parts of the atmosphere. Pressure gradients resulting from this cyclic process help drive mass loss, along with wave pressure, in the cases where dust grains do not form.

A comment regarding the mode of pulsation: 25 years ago we pointed out [12], [31] that the velocity jump across the shocks observed in Miras could not be accommodated in stars with low gravity, including any pulsating in overtone modes. Recently, two additional arguments have been added to this one: (a) The Period – Luminosity ($P - L$) sequences obtained for the LMC and SMC by surveys like MACHO show clearly fundamental and overtone sequences [33]; and (b) discovery of warm molecular circumstellar material around Miras, explaining the earlier apparently large angular diameters and allowing new stellar angular diameters to be measured ([28], [29], [17]; [23]; [25]), show that the stars are small and warm. There should be little doubt left about the mode of pulsation, and the new results fit the fundamental mode period-mass-radius relations from [21], [6], and [7].

Non-LTE. At photospheric densities, LTE is a good approximation to the excitation of atoms and molecules, and the resulting coupling to the radiation field is strong. Shocks are effectively isothermal and the temperature of the gas between shocks is approximately the radiative equilibrium temperature. However, at densities encountered relatively close to the photosphere, the excitation begins to depart from LTE, and as a result the gas temperature begins to depart from the radiative equilibrium temperature. This produces both additional temperature and pressure gradients, to help drive winds, and a refrigerated zone near the star where molecules and grains can form efficiently.

Chemistry. Molecules and grains need to be included in order to reproduce the highest mass loss rates and the observed qualities of mass losing stars. The chemistry is complex and grain formation and growth an area where much remains to be done. Even the composition of the grains may differ substantially from the predictions of near-equilibrium theory (e.g. [20], [19], [18]).

Radiative Transfer. The temperature structure of the atmosphere and wind result from the interaction of radiation with the gas, as well as the expansion or compression of it. Höfner and Loidl ([10], [11], [15]) have demonstrated that models with non-grey opacities differ substantially from ones with grey opacity, as well as that spectra from dynamical models are quite different from those from static ones, even for LTE.

2.2 Types of Models

Most models available today fall in either of two categories: (a) those computed with very detailed physics, whose principal value is as a laboratory for studying those processes in a stellar context; and (b) those constructed for use in interpreting observations of individual stars. A large group of authors with connections to Berlin, including Sedlmayr, Winters, Gauger, Fleischer, Woitke, and Gail, have focused on trying to understand the physics of dust-driven winds from carbon stars. They have found, as have others before, that purely dust-driven winds occur only for stars with very extreme parameters - very high L/M, low T_{eff}, and high Carbon/Oxygen for making carbon-based grains grow efficiently. Their models use chemical networks and detailed nucleation theory, but tend to rely on equilibrium formulations for key rates and balances. These models allow us to develop our understanding of some of the processes, and may tell us something about the final mass loss process that happens after a star has passed the Mira stage. However, the high L/M and high C/O required by these models to get substantial mass loss are probably not reached by most stars until after they have entered their final superwind stage, if ever, and thus these models do not provide us with useful mass loss rules for terminating AGB evolution - see below.

Another group of investigators with ties to Vienna, Copenhagen, and Uppsala, including Dorfi and Höfner, have concentrated on making pulsation models that can be used for spectrum synthesis to match observations, such as those from the ISO satellite, and also visibility functions for interferometry. These models are very careful to treat the heating/cooling and radiative transfer consistently in the deeper layers, where LTE is not a bad approximation, and thus have been reasonably successful in matching observed spectral energy distributions and other features. However, as they do not include the non-LTE, density-dependent cooling that is important in the wind driving region, they fail to produce the high mass loss rates found for oxygen as well as carbon-rich stars at the tip of the AGB.

The models that I have used for interpreting observations, and that inform what follows, the Bowen models, differ from the above models in that: (a) they include all the essential processes, but all of them in very simple approximations; (b) many models have been constructed to cover parameter space, and most of these have been constrained by some consistency conditions to provide insight into the pattern of mass loss that ends the AGB.

In 2000 I reviewed the theory and observations of mass loss from cool stars, emphasizing the AGB stars [30]. In that article, I demonstrated how to test a mass loss law for use in stellar evolution codes: assume a consistent set of evolutionary tracks to replace (L, R, M) with (L, M) vs time; compute the resulting $\dot{M}(L)$; find where $\dot{M} \approx M\dot{L}/L = \dot{M}_{crit}$; this locus is the "deathline" or cliff and stars with mass loss rates between 0.1 and $10 \times \dot{M}_{crit}$ are in the "deathzone". This same approach, applied to the mass loss law of [27], yields Fig. 1. Their mass loss law would allow stars to evolve to much higher luminosities than are observed, if it applied to stars during their ascent of the AGB. While the resulting L_{final} vs. M can be shifted slightly by assuming different evolutionary tracks, it cannot be shifted enough to match observations. This should perhaps not be surprising since the models they are using correspond to stars that are already "over the cliff" in (for example) Fig. 7 of [30] or Fig. 2 of [2]. The models may well describe what happens to stars that have already lost quite a bit of mass and, perhaps in consequence, become carbon stars, but they do not help us identify the "deathline" for stars.

3 Identifying the Stars That Are Dying Now

Observations of mass loss tell us which stars are losing mass at an interesting rate. If the observed mass loss rate is $M\dot{L}/L$ then the mass loss time scale is approximately the nuclear evolution time scale, and the star is being torn apart

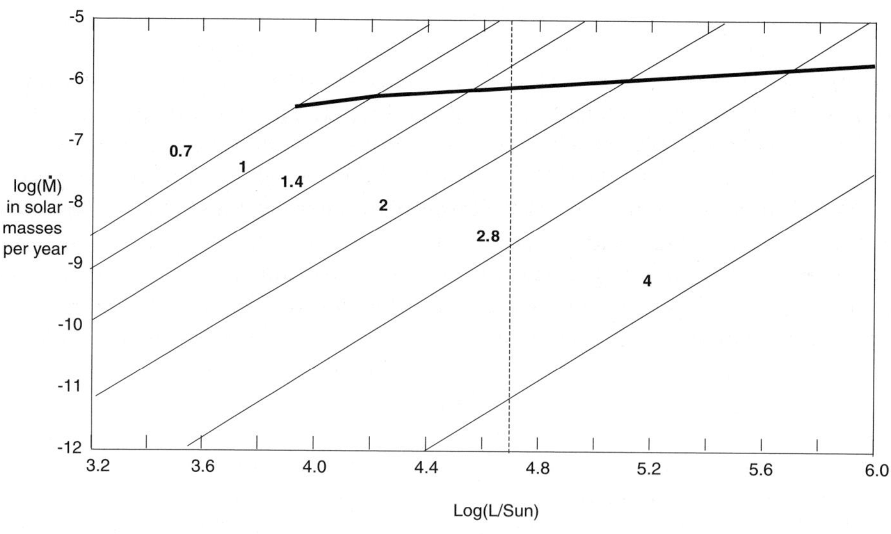

Fig. 1. The mass loss law of [27], plus evolutionary tracks from [14], imply termination of the AGB at much too high L. Shifting the evolutionary tracks cannot solve this problem. Therefore, this mass loss law should not be used for finding the luminosity at which stars leave the AGB.

by its wind. Because models show that mass loss rates are very sensitive to stellar parameters, therefore a star does not evolve very far from this condition before its envelope is entirely gone. Note that this statement is *not* true about empirical correlations, such as Reimers' [24] relation, as those do *not* show how the mass loss evolves for a single star but rather *which stars have reached an interesting level of mass loss.*

Miras are stars at the edge of the "cliff" and therefore mark the luminosity of the tip of the AGB [30]. Should the Mira luminosities match the luminosities of the central stars of the planetary nebulae? Below, I demonstrate that the Miras are almost certainly overluminous compared with standard core mass luminosity relations. When this overluminosity is understood, this will tell us whether it continues beyond the AGB; however, it may be that observations will provide the answer to this question before the mechanism is fully understood.

The steep dependence of the mass loss rate on stellar parameters means that the models we are using could easily be producing mass loss rates substantially different from the correct values. How sensitive are the results (the predicted cliff location) to errors in $\dot{M}(L, R, M)$? If the mass loss rates are in error by 1 dex, then the shift in L for the cliff edge is found from $d \log \dot{M}/d \log L$ and $d \log \dot{M}/d \log M$. These are both large (> 10), the resulting offset in the position of the cliff edge, or deathline, is small even for a relatively large change in $\dot{M}(L, R, M)$.

4 Extrapolating to Lower Metallicity

Once the pattern of the death zone is understood, it can be used to make some predictions about the effects of changing metallicity. In Fig. 9 of [30] the Bowen cliff lines for five metallicities are shown. At a given mass, $d \log L_{final}/d \log Z$ deduced from this figure is 0.1 to 0.2. In the Bowen models two factors contribute to this shift. For the larger shift, from $Z =$ Solar and $M > 1$ to $Z = 0.03$ Solar, $M > 1$, these two factors both contribute: (a) the lower metallicity models are less able to make dust and therefore dust plays a minor or no role in their mass loss process; (b) the evolutionary tracks shift to higher T_{eff} for lower Z (due to lower opacity in the stellar envelope, presumably). Note that *any* models in which the mass loss rate is sensitive to the radius of the star will encounter this second effect, and will therefore predict higher L_{final} for lower Z populations.

Can this logic be applied to AGB-manqué stars, i.e. stars that leave the RGB with enough mass loss to populate the blue horizontal branches of their populations and become WD without ever ascending the AGB to its tip? There are two mechanisms for mass loss near the tip of the RGB: (a) the same mechanism that works at the tip of the AGB; and (b) ejection associated with He core flash. We do not know enough about (b) to make any predictions, but for (a), the answer is almost certainly that mass loss is usually an all-or-nothing affair. Either the star enters the death-zone and is very unlikely to emerge with much of an envelope, or it never gets to the death-zone and does not lose mass in a steady wind (e.g. [38] and [37]).

5 Why the Core Mass – Luminosity Relations Must Be Wrong

According to the period distribution N(P) of Miras from Wood and Cahn's pivotal paper ([32], Fig. 2), the majority of Miras have periods > 330 days and the median period is 380 days. In Table 1 the luminosities and core masses associated with these two periods are shown for three core mass – luminosity relations: Paczynski's [22] formula, Boothroyd and Sackmann's [1] relation, and Iben's mass sensitive relation [14] for $M = 1$ and $M = 2$ $M_\odot$. Most white dwarfs have masses between 0.5 and 0.6 $M_\odot$ [16] whereas these three core mass – luminosity relations and the Mira $N(P)$, translated to luminosity distribution ($N(L)$), suggest the typical core mass should be more like 0.6 to 0.7 $M_\odot$ for stars leaving the AGB.

Table 1. Core masses ($M_\odot$) corresponding to Mira periods

$L - M_{core}$ relation	330 days	380 days
Paczynski [22]	0.62	0.73
Boothroyd & Sackmann [1]	0.56	0.70
Iben [14] $M = 1$	0.63	0.87
Iben [14] $M = 2$	0.59	0.77

There are a number of factors that could create this apparent contradiction between the observed and theoretical core masses.

1. Perhaps only high mass stars become Miras, or perhaps lower mass stars become Miras only during the peak L of the shell flash cycle. The counter-argument to these two options is that there are too many Miras to be accounted for only by higher mass progenitors, or only by a short-lived part of the shell flash cycle. Also, the mapping of progenitor masses to period groups supports the idea that the full range of < 1 to > 4 $M_\odot$ are represented in the Mira sample – see [13], [2].
2. The white dwarf mass function is established for nearby stars, while the period – luminosity relation, $L(P)$, is determined from the LMC, so perhaps LMC stars have larger cores at the same P, or perhaps the distance to the LMC is wrong. A counter-argument is that the Hipparcos distances to the Miras in our vicinity average to about the same $P - L$ relationship, and that is also what we obtain from the $P - L$ relationship for the Bowen models, that is not particularly sensitive to metallicity.
3. Perhaps the core loses mass in the final transition to the PN locus. A counter-argument is that the core is in a very deep potential well from which it is difficult to remove mass by most mechanisms; also, this would produce very interesting abundances in the last ~0.1 $M_\odot$ of material that goes into forming the PN.

4. The luminosity of the Miras could be greater than that predicted by the core mass – luminosity relation, for example through either deep dredge-up or hot bottom burning (e.g. [8], [9]). This seems to me to be the most likely explanation, although there are not yet standard models that have higher L for core masses in the needed range of 0.5–0.6 $M_\odot$.

6 Conclusions

From the modelling of mass loss at the tip of the AGB there is good news and bad news. The bad news is that the problem is sufficiently complex that detailed models of, for example, the time leading up to the final envelope loss and evolution to a PN will not be available for some time yet. The good news is that for some purposes we don't need those detailed models. We have learned enough from the current round of modeling to interpret the observations of mass loss and of AGB stars and PN nuclei in a consistent way and to learn from them some of the details that are needed for including mass loss in stellar evolution.

The pattern of evolution at the tip of the AGB is quite clear and these results are robust. Stars evolve up the AGB with little or no mass loss until they reach the "death zone", where $\log \dot{M} = \log(M\dot{L}/L) \pm 1\,dex$. They evolve through the death zone in about 105 years, and then from there lose the rest of their envelopes as mostly obscured OH-IR sources and/or carbon stars over another 10^4 to 10^5 years. The details of the final "ejection", if there is such an event, are still obscure; in many cases, this event may depend on the involvement of a close companion star or a massive planet. Also, for stars that become carbon stars, the picture may change in several ways. At a given L the star may be larger, increasing the mass loss rate. At the same time if enough carbon dust or carbon-rich molecules are formed they assist the mass loss process. For stars with $O > C$ this assistance will be reduced. So an entirely different set of models may be needed for the carbon star mass loss process. There is still, also, no consensus on which stars become carbon stars and how. If prior mass loss is necessary, then stars may only become carbon stars during the post-cliff "free fall", as is suggested by the family of models that show mass loss as modeled e.g. by [27].

The overwhelming impression with which I left the meeting was the power of survey techniques now in use to probe the AGB and PN populations of a much wider variety of systems than just our Milky Way, with the added benefit of all the targets then being at a known and uniform distance. I would encourage the simultaneous gathering of data on AGB tip stars and PNe. Several epoch observations of each field should make it possible to select the AGB tip Miras, since they both vary in brightness by a considerable amount over timescales of hundreds of days. They also show strong emission in hydrogen lines, particularly Hβ, Hγ, and Hδ (Hα and Hϵ are normally heavily over-absorbed and not strongly in emission), during the descending branch of the light curve, or overall during about half the pulsation cycle. They also stand out as a large fraction of the very luminous stellar sources at 2 microns.

Acknowledgements

Thanks to the organizers for a very interesting meeting and the invitation to come. Thanks to Uppsala University for hosting me for a prolonged visit before and after the conference, with time to write this contribution. Thanks also to Iowa State University for a "Faculty Professional Development Assignment" for 2003-4 during which I have been able to carry forward a number of related projects.

References

1. A. J. Boothroyd; I.-J. Sackmann: Ap.J., **328**, 641 (1988)
2. G. H. Bowen, G., L. A. Willson: Ap.J. Letters, **375**, L53 (1991)
3. G. H. Bowen: Ap.J., **329**, 299 (1988)
4. G. H. Bowen, in *Numerical Modelling of Nonlinear Stellar Pulsations Problems and Prospects*, 155 (1990)
5. J. P. Cox, R. T. Giuli: *Principles of Stellar Structure* , (New York, Gordon and Breach 1968)
6. A. N. Cox, A. N. & D. A. Ostlie: 1993, ApSS, **210**, 311 (1993)
7. M. W. Fox, P. R. Wood, Ap.J.: **259**, 198 (1982)
8. F. Herwig: A&A: **360**, 952 (2000)
9. F. Herwig, D. Schoenberner, D., T. Bloecker: A&A, **340**, L43 (1998)
10. S. Höfner, U. G. Jørgensen, & R. Loidl: ApSS, **255**, 281 (1997)
11. Höfner, S., Gautschy-Loidl, R., Aringer, B., & Jørgensen, U. G. 2003, A&A, **399**, 589 (2003)
12. S. J. Hill, L. A. Willson: Ap.J.: **229**, 1029 (1979)
13. S. M. G. Hughes, P. R. Wood: AJ, **99**, 784 (1990)
14. I. Iben Jr.: Ap.J., **277**, 333 (1984)
15. R. Loidl, S. Höfner, U. G. Jørgensen, B. Aringer, A&A, **342**, 531 (1999)
16. J. Madej, M. Należyty, & L. G. Althaus: A&A, **419**, L5 (2004)
17. B. Mennesson, G. Perrin, G. Chagnon, V. Coudé du Foresto, S. Ridgway, A. Merand, P. Salome, P. Borde, W. Cotton, S. Morel, P. Kervella, W. Traub, M. Lacasse : Ap.J., **579**, 446 (2002)
18. J. A. Nuth, F. J. M. Rietmeijer, H. G. M. Hill, Meteoritics and Planetary Science, 37, 1579 (2002)
19. F. J. M. Rietmeijer, S. L. Hallenbeck, J. A. Nuth, J. M. Karner: Icarus, 156, 269 (2002)
20. F. J. M. Rietmeijer, J. A. Nuth, J. M. Karner: Ap.J. 527, **395** (1999)
21. D. A. Ostlie, A. N. Cox: Ap.J., **311**, 864 (1986)
22. B. Paczyński, Acta Astronomica, **20**, 47 (1970)
23. G. Perrin, V. Coudé du Foresto, S. T. Ridgway, B. Mennesson, C. Ruilier, J.-M. Mariotti, W. A. Traub, M. G. Lacasse: A&A, **345**, 221 (1999)
24. D. Reimers, D., *Memoires of the Societé Royale des Sciences de Liege*, **8**, 369 (1975)
25. T. Tsuji, K. Ohnaka, W. Aoki, I. Yamamura: A&A **320**, L1 (1997)
26. Vassiliadis, E. & Wood, P. R. 1993, Ap.J., **413**, 641
27. A. Wachter, K.-P. Schröder, J. M. Winters, T. U. Arndt, E. Sedlmayr: A&A, **384**, 452 (2002)

28. J. Weiner, D. D. S. Hale, & C. H. Townes: Ap.J., **589**, 976 (2003)
29. J. Weiner, D. D. S. Hale, & C. H. Townes: Ap.J., **588**, 1064 (2003)
30. L. A. Willson: ARAA **38**, 573 (2000)
31. L. A. Willson, S. J. Hill: Ap.J., **228**, 854 (1979)
32. P. R. Wood, J. H. Cahn: Ap.J., **211**, 499 (1977)
33. P. R. Wood, K. M. Sebo: M.N.R.A.S., **282**, 958 (1996)
34. P. R.Wood: **1999**, IAU Symp. 191: Asymptotic Giant Branch Stars, 191, 151 (1999)
35. P. R. Wood: Ap.J., **190**, 609 (1974)
36. P. R Wood: Ap.J., **227**, 220 (1979)
37. S. Yi, P. Demarque, A. Oemler, A: ApJ, **492**, 480 (1998)
38. S. Yi, P. Demarque, A. Oemler, A: , ApJ, **486**, 201 (1997)

AGB Stars in the Local Group, and Beyond

M.A.T. Groenewegen

Instituut voor Sterrenkunde, KU Leuven, Celestijnenlaan 200B,
B-3001 Leuven, Belgium

1 Introduction

Carbon stars are tracers of the intermediate age population in galaxies. Either they are currently undergoing third dredge-up on the (Thermal-Pulsing) Asymptotic Giant Branch (TP-AGB) – the cool and luminous N-type carbon stars (or the "infrared carbon stars", when they are significantly reddened by circumstellar dust through mass-loss) –, or have been enriched with carbon-rich material in a binary system when the present-day white dwarf was on the AGB (the carbon dwarfs and CH-stars. The R-stars may be the result of a coalescing binary [McClure 1997]).

Similarly, there are "intrinsic" and "extrinsic" S-stars which represent, respectively, the stars currently on the AGB, and the stars polluted by a present-day white dwarf (e.g. Jorissen 2003).

Next to the classical oxygen-rich giants (of spectral type M and MS), there are the OH/IR stars (which are obscured by circumstellar dust) and the Barium-dwarfs and Barium-giants which are polluted by a present-day WD (e.g. Jorissen 2003).

Since their spectral signature is very different from oxygen-rich and S-type stars, it is relatively easy to identify carbon stars even at large distances. In Sect. 2 the main technique to identify carbon stars is briefly discussed, and various recent surveys for carbon stars in external galaxies are summarised. That section is an update of reviews I gave at IAU Symposium 191 (Groenewegen 1999; hereafter G99), and a Ringberg conference (Groenewegen 2002; hereafter G02), and here I will only refer to new results published since then. Please refer to G99, G02 and the similar review by Azzopardi (1999) for the full story. Sections 3 and 4 discuss aspects related to NIR work and variability properties of AGB stars in LG galaxies.

2 Optical Narrow-Band Imaging

This has been discussed in some detail in G99. Briefly, the most effective method for large scale surveys of late-type M- and C-stars uses typically two broad-band filters from the set V, R, I, and two narrow-band filters near 7800 and 8100 Å, which are centred on a CN-band in carbon stars, and a TiO band in oxygen-rich stars, respectively. In a narrow band [78-81] versus $[V-I]$ (or $[R-I]$) colour-colour plot, carbon stars and late-type oxygen-rich stars clearly separate redwards of $(V-I) \approx 1.6$. For an illustration of this, see Cook & Aaronson (1989) or Nowotny & Kerschbaum (2002).

A caveat is that, unfortunately, not all groups adopt the same "boxes" in these diagrams to select M- and C-stars (e.g. compare Nowotny et al. 2003 and Battinelli & Demers et al. 2004b). Furthermore, in many cases, only the photometry is published for the stars the authors consider to be the M- and C-stars, so it is not possible to apply ones own selection criteria a-posteriori. Furthermore, one usually applies the same lower-limit on the broad band colour ($(V-I)$, or $(R-I)$) to select M-stars (usually chosen to correspond to M0 and later for solar metallicity) from C-stars, while it is known (e.g. Nowotny & Kerschbaum 2002)) that the hottest C-stars (spectral type C0) are bluer than this limit. So, applying the same lower-limit on the broad-band colour will bias against the hottest C-stars.

2.1 Surveys

In this section the (recent, post-2002) surveys for carbon stars in external galaxies are described.

M31

Battinelli et al. (2003) present results on a 2240 arcmin^2 area in the SW disk of M31, identifying 945 carbon stars and estimating a C/M ratio of 0.084. Previous work has been mentioned in G99 and G02.

NGC 205

Demers et al. (2003a) detect 289 carbon stars inside an 10 arcmin ellipse around NGC 205, and estimate that the actual total number is between 500 and 550, the difference being due to a lack of detections near the centre (their figure 6) due to crowding and increased photometric error. A C/M ratio of 0.09 is estimated.

Davidge (2003) present JHK results on a 3.6 arcmin^2 centred on NGC 205, and he kindly provided the K-band fits image and object list. I derived the WCS information by identifying objects on a 2MASS K-image resulting in astrometry with an rms of 0.21 and 0.30 arcsec in RA and DEC using 18 objects over the entire field. Figure 1 shows the K-image with the known C-stars from Demers et al., and candidate AGB stars ($(J-K) > 1.5$ and $K < 17.2$) from Davidge overplotted.

Even in the outer annulus, where Demers et al. suggest that their survey is not hampered by crowding, there are many candidate AGB stars based on JK colours and magnitudes. Second, as both papers quote a seeing of 0.7 arcsec, it illustrates the power of infrared observations.

Of the 25 known C-stars, 23 have a counterpart in the Davidge dataset (although not all have $(J-K) > 1.5$ and $K < 17.2$). The remaining two are both visible on the K-image; one is on the edge of the frame, and the second was missed in the source detection/extraction.

One interesting exercise is to compare the bolometric magnitudes derived from JK, with those derived from RI photometry. This comparison is shown in Figure 2 both on a star-by-star basis and in the form of luminosity functions. Although these stars are expected to be LPVs (hence variable) the data suggests possible systematic effects, which warrants further investigation.

Table 1. The carbon star census

Name	D (kpc)	M_V (mag)	[Fe/H]	N_C [(b)]	Area[(c)] (kpc^2)	N_M [(d)]
M31	770	–21.2	0.0	243	12.3	789 (5+)
Galaxy[(a)]		–20.9	0.0	81	1.00	C/M ≈ 0.2
M33	840	–19.0	–0.6	15	0.20	5 (5+), 60 (0+)
LMC	50	–18.5	–0.6	1045	4.8	1300 (5+)
				7750	220.	
SMC	63	–17.1	–1.2	789	5.4	180 (5+)
				1707	12.2	
NGC 205	830	–16.4	–0.8	525	12.3	5830 (0+)
NGC 6822	490	–16.0	–1.2	904	4.5	941 (0+)
NGC 3109	1360	–15.7	–1.7	446	33	250 (0+)
NGC 185	620	–15.6	–0.8	145	4.5	850 (0+)
IC 1613	715	–15.3	–1.4	195	7.8	35 (5+), 300 (0+)
NGC 147	755	–15.1	–1.1	288	25.	1200 (0+)
SagDSph	28	–15.0	–0.5	26	7.2	
WLM	930	–14.4	–1.5	149	17.0	12 (0+)
Fornax	138	–13.1	–1.2	104	1.35	4 (5+), 25 (2+)
Pegasus	760	–12.9	–2.0	40	1.04	77 (0+)
SagDIG	1060	–12.0	–2.3	16	0.58	
Leo I	250	–11.9	–1.4	23	0.45	1 (5+), 15 (0+)
And I	790	–11.8	–1.4	0	0.33	
And II	680	–11.8	–1.5	8	0.35	1 (0+)
And III	760	–10.2	–1.7	0	0.66	
And V	810	–9.1	–1.9	0	0.66	
And VI	775	–11.3	–1.7	1	0.41	
And VII	760	–12.0	–1.5	3	0.24	
DDO210	950	–10.9	–1.9	3	0.18	1 (0+)
Leo II	205	–10.1	–1.6	8	0.47	
Cetus	775	–10.1	–1.7	1	1.1	
Sculptor	88	–9.8	–1.5	8	0.65	40 (2+), 0 (5+)
Phoenix	405	–9.8	–1.9	2	0.40	
Tucana	870	–9.6	–1.7	0	0.22	
Sextans	86	–9.5	–1.9	(0)		
Draco	79	–9.4	–2.0	6	0.50	
Carina	94	–9.4	–1.8	11	0.31	
Ursa Minor	69	–8.9	–1.9	7	0.58	
NGC 2403	3390	–20.4	0.0	4	2.0	7 (0+)
NGC 300	2170	–18.7	–0.4	16	3.2	23 (0+), 6 (5+)
NGC 55	1480	–18.0	–0.6	14	2.8	6 (5+)

[a]. In the Solar neighbourhood per kpc^2 (Groenewegen et al. 1992)
[b]. Number of known carbon stars.
[c]. Survey area at the assumed distance of the galaxy.
[d]. Number of M-stars. 0+ indicates M0 and later, etc.

It might be recalled in this respect that one advantage of using VI instead of RI as broad-band colours is the fact the bolometric correction to I based on

Fig. 1. NGC 205 in the K-band (Davidge 2003, FOV of 3.6 arcmin^2): known carbon stars from Demers et al. (triangles) and objects with $(J-K) > 1.5$ and $K < 17.2$ from Davidge (squares).

$(V-I)$ has much less scatter than the one based on $(R-I)$ (Bessell & Wood 1984).

NGC 3109

Demers et al. (2003b) present the first survey of this galaxy using the narrow-band filter technique and detect 446 C-stars, and derive a C/M ratio of about 1.8

NGC 147 & NGC 185

Nowotny et al. (2003) and Battinelli & Demers (2004b,c) independently present data on these two companions to M31, while Harbeck et al. (2004) observed NGC 147. Nowotny et al. find 146 C-stars and a C/M ratio of 0.15 in NGC 147, and 154 and 0.089 in NGC 185, respectively, over an un vignetted FOV of approximately 33 arcmin^2. Battinelli & Demers (2004c) find 288 C-stars and a C/M ratio of 0.24 for NGC 147 and Battinieli & Demers (2004b) find 145 C-

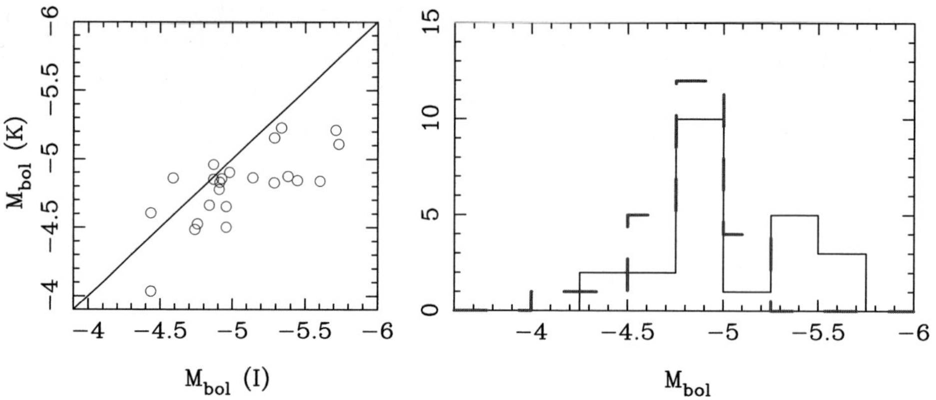

Fig. 2. Bolometric magnitudes for the 23 C-stars in common between Demers et al. (solid histogram) and Davidge (dashed histogram) for NGC 205, based on *JK* and *RI* photometry and standard bolometric correction formulae.

stars and a C/M ratio of 0.17 for NGC 185, both over an area of 1180 arcmin^2. Figure 3 in Battinieli & Demers (2004b) shows that most of their C-stars are located in the main body of NGC 185, which is still well covered by the smaller FOV of the Nowotny et al. observations. This is consistent with the very similar numbers of C-stars discovered in both surveys. On the contrary, the difference in number of C-stars detected between the two datasets in the case of NGC 147 is–at least in part–due to the difference in areal coverage of this galaxy and the more extended distribution of the C-stars. Harbeck et al. (2004) find 155 C-stars in their 92 arcmin^2 FOV.

And III, V, VI, VII

Harbeck et al. (2004) also observed And III, V, VI, VII to derive, respectively, 1, 0, 2, 5 C-stars, of which, respectively 0, 0, 1, 3 are believed to be genuine AGB C-stars, while the others are suggested to be low-luminosity CH-stars.

WLM

Battinelli & Demers (2004a) observe a 42 x 28 arcmin field centred on WLM to find 149 C-stars, and to derive a C/M ratio of 12.4 ±3.7.

Cetus

Harbeck et al. (2004) also observed the Cetus dwarf spheroidal to find 1 genuine AGB C-star, and 2 CH-stars.

Table 1 summarises the number of known carbon stars in external galaxies. Local Group members not explicitly mentioned have no published information on their C-star population. The last three entries are galaxies outside the Local Group. Listed are the adopted distance, absolute visual magnitude, metallicity (these three parameters come from Grebel et al. (2003), Mateo (1998) and van den Bergh (2000)), number of known carbon stars, the surface area on the sky of the respective survey, and the number of late-type M-stars, when known

In the near future one may expect new results from the optical narrow-band imaging technique for Carina, Sculptor, Phoenix (Groenewegen et al.), IC 10

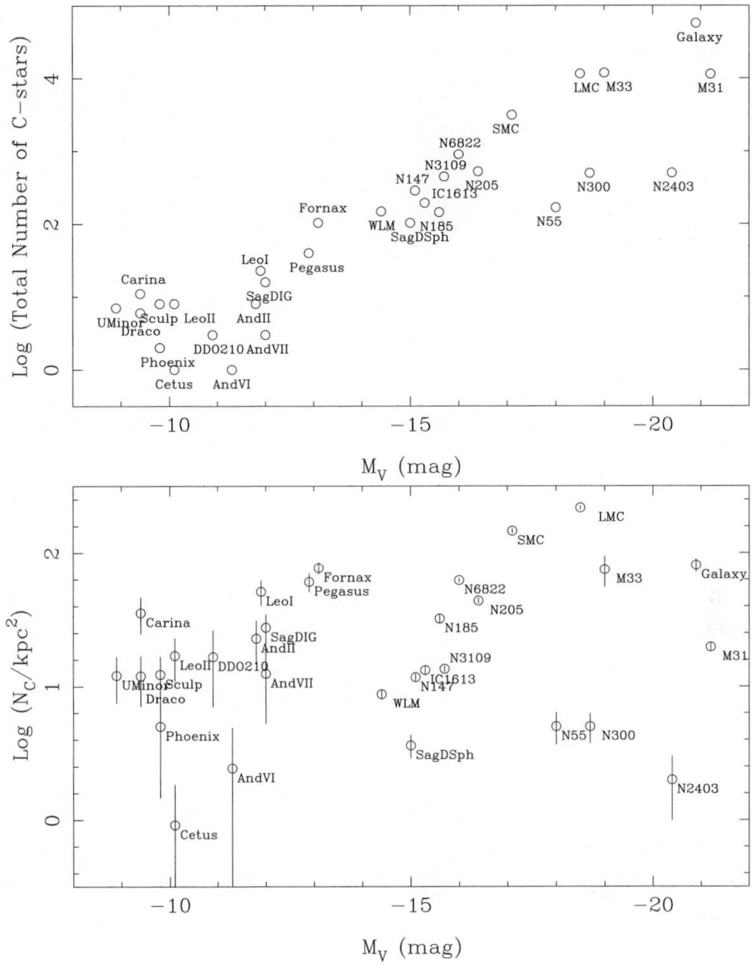

Fig. 3. Total number of carbon stars and surface density of carbon stars versus M_V. No correction for inclination effects has been made.

(Demers, Battinelli), and And II, M32, Leo I, Leo II, Draco, Ursa Minor (Kerschbaum, Nowotny, Olofsson, Schwarz). This last group also acquired funds to put these filters on the 4.1m SOAR (Southern Astrophysical Research) telescope.

3 Near-Infrared Results

The comparison between the optical narrow-band imaging and near-infrared observation of NGC 205 in the previous section already demonstrated the power of the latter. A disadvantage of broad-band NIR observations is that no distinction between M- and C-stars can be made, although one often sees in the literature the assumption that stars redder than $(J - K)_0 \sim 1.4 - 1.6$ are carbon stars. This is an oversimplification, as dust-obscured (mass-losing) M-stars can have

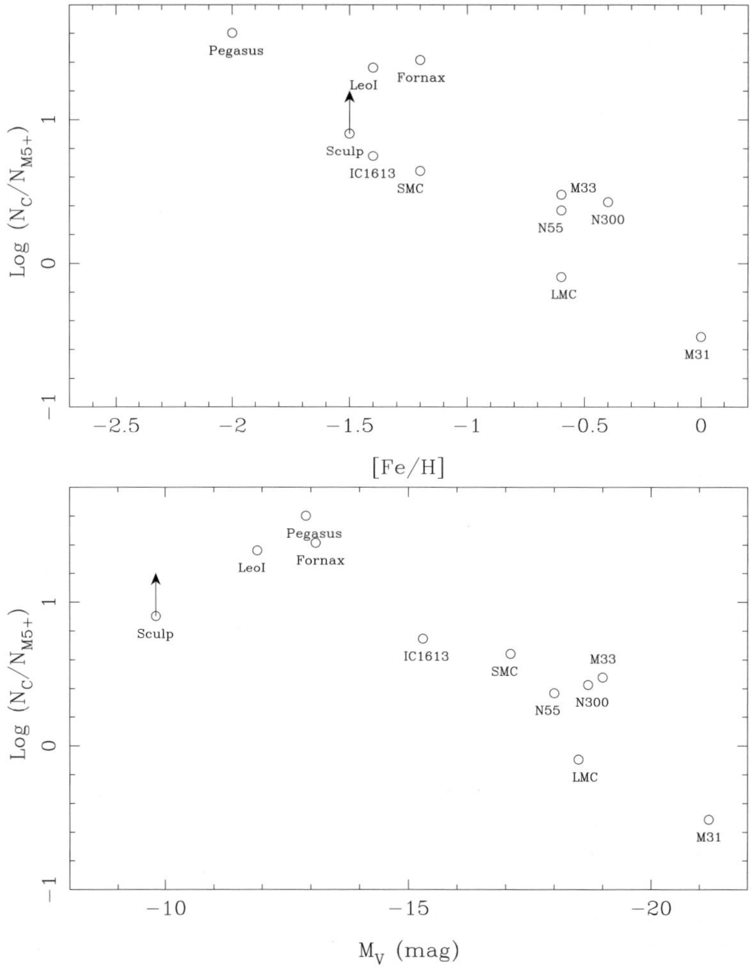

Fig. 4. Log (number carbon stars / number late M-stars) versus metallicity and M_V.

red colours, and more importantly, hotter C-stars have NIR colours that overlap with the photospheric colours of M-giants (e.g. figure 2 in G02).

The release of the 2MASS database (as well as DENIS data) has had a large impact on the study of AGB stars in the Magellanic Clouds (MCs), e.g. Cioni & Habing (2003), but results have also been presented for Fornax (Demers et al. 2002, G02), and the counterparts of optically known C-stars at the time of the 2MASS *second data release*. More recently, an undergraduate has investigated the 2MASS data of the *all-sky release* for all LG galaxies within 1 Mpc (excluding MCs, M31, M32), retaining objects with $(J-K)_0 > 1.22$, appropriate M_K-range for AGB stars, and excluding known objects using the SIMBAD database (Marescaux 2003). Table 2 compares the number of resulting candidate AGB stars (C-stars and dust-obscured M-stars) to the number of known C-stars. For galaxies beyond ∼300 kpc the number of candidates becomes very

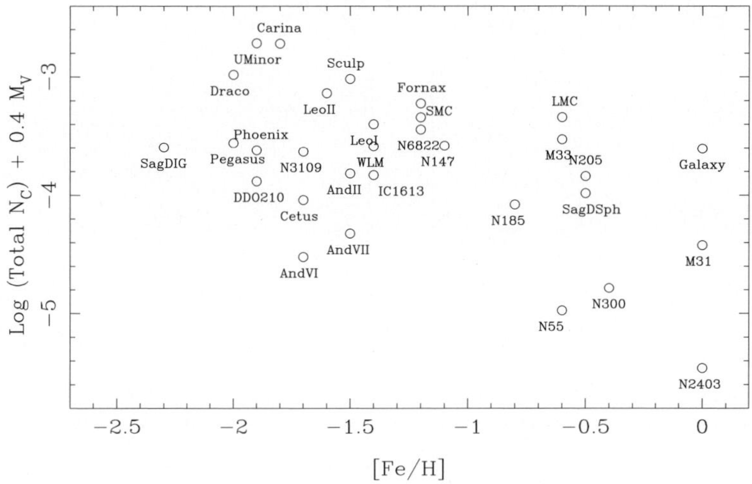

Fig. 5. Log (number of carbon stars / total visual luminosity) versus metallicity.

small and is limited by the survey limit of 2MASS, restricting the candidates to the intrinsically most luminous ones. For the more nearby galaxies, the limiting magnitude of 2MASS is not a limiting factor and one observes that the number of candidates is a non-negligible fraction of the number of known C-stars, again indicating the power of IR observations to provide a complementary picture to the optical narrow-band imaging.

In the near future deeper ground-based IR observations will become available. Several groups are observing or even monitoring LG galaxies: e.g. Fornax, Leo I, Leo II, Carina, Sculptor, Phoenix, Sextans (E. Held using SOFI); Leo I, Fornax, (all smaller LG) using the SIRIUS camera on the IRSF (e.g. Menzies et al. 2002); NGC 6822 using the new CPAPIR camera on the CTIO 1.5m (Demers, Battinelli); Leo A, Leo I, Leo II, Sex B, NGC 6822, Draco, NGC 147, NGC 185 using INGRID on the WHT (Cioni & Habing, e.g. these proceedings).

Another exciting possibility is to use narrow-band filters in the NIR. A first example was recently presented by Östlin & Mouhcine (2004) who observed a 14 x 14 arcsec field in the metal-poor ([Fe/H]= -1.7) galaxy IZw 18 (at a distance modulus of 30.5 !) using the NICMOS F171M and F180M filters. For a carbon star, the F171M filter is located in the continuum, while the F180M filter is centred on a deep CN band at 1.77 μm. Combining this with a broad-band F160W they were able to identify 5 C-stars, 1 M-type AGB and 20 supergiants.

4 Variability

One of the characteristics of AGB stars is that they are variable on different timescales and amplitudes, as has been clearly revealed by the mircolensing surveys of the Magellanic Clouds and Galactic Bulge (e.g. Wood et al. 1999, Wood 2000). It has been found that variable AGB stars occupy different sequences

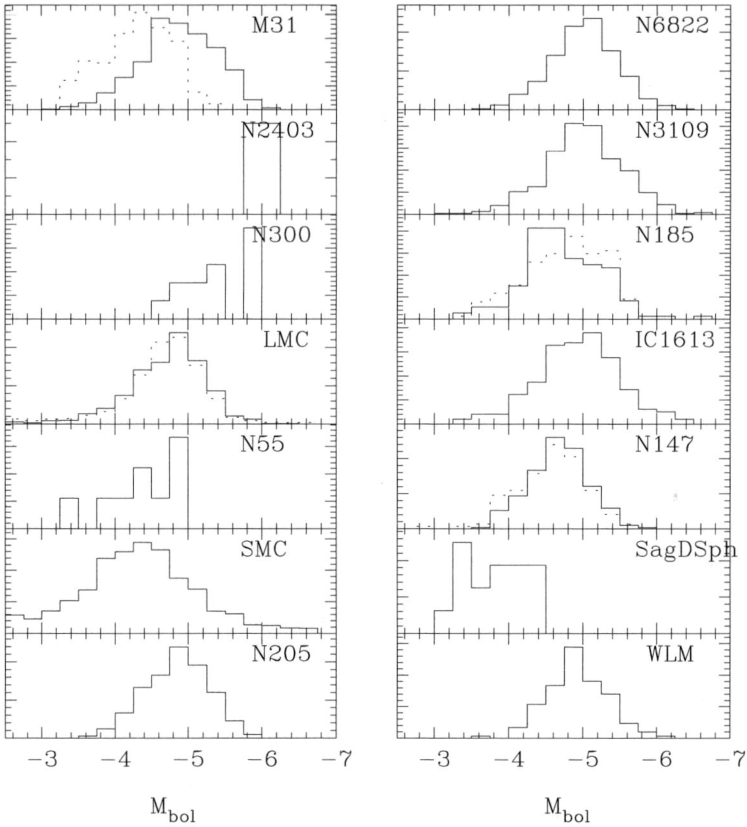

Fig. 6. Normalised carbon star luminosity functions, ordered by decreasing M_V of the galaxies, ordered top to bottom, left to right. Number of C-stars used to calculate the LFs varies from galaxy to galaxy (see Table 1. In the case of the SMC and LMC, the lowest luminosity bin is cumulative. For M31 data plotted is from Battinelli et al. (2003, solid) and Brewer et al. (1995, dashed); for the LMC from Costa & Frogel (1996, solid) and Kontizas et al. (2001, dashed); for NGC 185 from Battinieli & Demers (2004b, solid) and Nowotny et al. (2003, dashed); for NGC 147 from Battinelli & Demers (2004c, solid) and Nowotny et al. (2003, dashed) For the LMC, NGC 147 and NGC 185 these independent LFs are in good agreement. The difference for M31 appears to be real and related to the fact that these LFs have been derived for field(s) that have different locations along the major axis of M31.

(usually labelled ABCD) in period-luminosity diagrams, with the large amplitude Mira variables on sequence C.

In G02, some earlier work on (candidate) variable AGB stars in the SagDSph, Fornax and IC 1613 was reported. Since then, Snigula et al. (2004) mention 16 candidate LPVs (Long Period Variables) in Leo A, and 5 in GR 8, Gallart et al. (2004) propose 6 LPV candidates in Phoenix, and Rejkuba (2004) present a K-band PL-relation for 240 well defined Miras in NGC 5128.

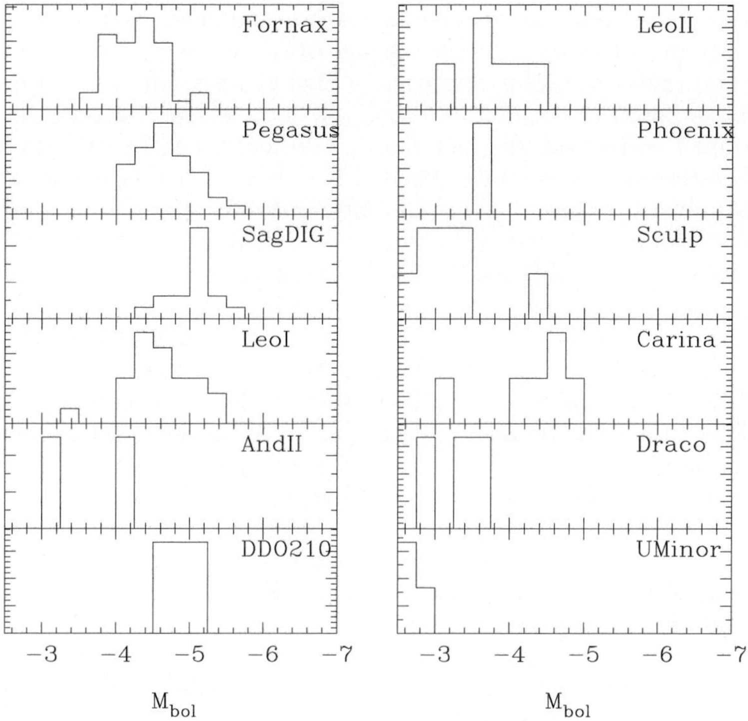

Fig. 7. Continued.

Table 2. Candidate IR AGB stars based on 2MASS compared to the number of known C-stars

Name	D (kpc)	N_{2mass}	N_C	Name	D (kpc)	N_{2mass}	N_C
Ursa Minor	69	8	7	NGC 6822	490	6	904
Sculptor	79	8	8	LGS 3	620	1	?
Draco	82	6	6	IC 10	660	9	?
Sextans	86	10	(0)	IC 1613	720	3	195
Carina	100	6	11	NGC 147	755	1	288
Fornax	138	34	104	And I	790	1	0
Leo II	205	1	8	Leo A	800	2	?
Leo I	250	2	23	Tucana	870	1	0

5 Discussion

Figures 3, 4, 5, 6, 7 are updates of those in G02, and for lack of space I refer to the last section in G02 for additional details.

Figure 3 shows the number of carbon stars in a galaxy versus M_V represented in two ways. First, the total number of C-stars in a galaxy was estimated by simply multiplying the known number by the ratio of total surface area of a galaxy to the survey area. As for some galaxies the survey area is less than a few percent, this correction factor can be quite large (and uncertain). To circumvent this, the bottom panel shows the surface density of carbon stars in the particular survey. The drawback of this approach is that it does not take into account the spatial variation of the density of carbon stars within a galaxy. In neither approach did I correct for the fact that we do not see these galaxies face-on. Some interesting things can be noticed. There is a clear relation between the (estimated) total number of C-stars and M_V, and there seems to be a maximum surface density of about 200 kpc^{-2} averaged over a galaxy. In both plots NGC 55, 300 and 2403 are clear outliers. These are the most distant galaxies surveyed, and one might suppose that the surveys have been incomplete. For NGC 55 the explanation probably lies as well in the fact that we see this galaxy almost disk-on, and so both the total number as well as the surface density have been underestimated. Reddening within the disk of the galaxy can also play a role. For NGC 2403 the small number of carbon stars lies in the fact that the survey has been incomplete. All 4 known C-stars have luminosities that are much higher than the average in galaxies for which we know the LF in more detail. The same is true for NGC 300.

Figure 4 shows the ratio of C- to late M-stars. The interpretation of the well known trend is that a star with a lower metallicity needs fewer thermal pulses to turn from an oxygen-rich star into a carbon star.

Figure 4 shows the ratio of the total estimated number of carbon stars over the visual luminosity of the galaxy. Most of the galaxies scatter between a value of –3 and –4, with a few outliers which are the same as noticed in Fig. 3.

Figures 6 and 7 shows the C-stars bolometric LF for the galaxies for which it could be constructed. The data show that in well populated LFs, the mean $M_{\rm bol}$ is between –4 and –5. It also shows that the mean in NGC 300 and NGC 2403 is much higher. Unless one would invoke a large uncertainty in the distance or a burst of recent star formation, the most natural explanation lies in the incompleteness of the surveys in these distant galaxies. Finally, the data shows that in the fainter galaxies the mean magnitude increases and that a fair number of C-stars are of low-luminosity type.

In a recent paper Mouhcine & Lançon (2003) present evolutionary population synthesis models, including chemical evolution, with special focus on intermediate age populations. Their models are the first that are able to account for the observed trend in Fig. 4 adopting 'typical' SFH for Sa, Sb, Sc and Irr Hubble type galaxies. The AGB phase is included through a semi-analytical treatment of the third dredge-up, with efficiency parameters set to values that have been determined in other studies to fit the LMC carbon star LF and C/M ratio.

6 Conclusion

In principle, the overall carbon star LF and C/M ratio contains information about the star-formation rate history from, say, 10 Gyr ago (the low-luminosity C-stars in binaries) to a few-hundred Myr ago (the high luminosity tail of the LF). It is a challenge to theoretical models to use these constraints together with other data to derive the chemical evolution and star formation history of these galaxies. The models of Mouhcine & Lançon represent a first successful step in this direction.

Acknowledgements

I would like to thank Tim Davidge for providing the K-band fits image and his list of detections for NGC 205, and Pierre Royer (KUL) for helping in deriving the WCS parameters.

References

1. Azzopardi M., 1999, Ap&SS 265, 291
2. Battinelli P., Demers S., Letarte B., 2003, AJ 125, 1298
3. Battinelli P., Demers S., 2004a, A&A 416, 111
4. Battinelli P., Demers S., 2004b, A&A 417, 479
5. Battinelli P., Demers S., 2004c, A&A 418, 33
6. Bessell M.S., Wood P.R., 1984, PASP 96, 247
7. Brewer J., Richer H.B., Crabtree D.R., 1995, AJ 109, 2480
8. Cioni M.-R. L., Habing H.J., 2003, A&A 402, 133
9. Cook K.H., Aaronson M., 1989, AJ 97, 923
10. Costa E., Frogel J.A., 1996, AJ 112, 2607
11. Davidge T.J., 2003, ApJ 597, 289
12. Demers S., Dallaire M., Battinelli P., 2002, AJ 123, 3428
13. Demers S., Battinelli P., Letarte B., 2003a, AJ 125, 3037
14. Demers S., Battinelli P., Letarte B., 2003b, A&A 410, 795
15. Gallart C., Aparicio A., Freedman W.L., et al., 2004, AJ 127, 1486
16. Grebel E.K., Gallagher J.S III, Harbeck D., 2003, AJ 125, 1926
17. Groenewegen M.A.T., de Jong T., van der Bliek N.S., Slijkhuis S., Willems F.J., 1992, A&A 253, 150
18. Groenewegen M.A.T., 1999, in: "IAU symposium 191: Asymptotic Giant Branch Stars", eds. T. Le Bertre, A. Lèbre and C. Waelkens, Kluwer, p. 535 (G99)
19. Groenewegen M.A.T., 2002, in: "The Chemical Evolution of Dwarf Galaxies", astro-ph/0208449 (G02)
20. Harbeck D., Gallagher J.S III, Grebel E.K., 2004, AJ 127, 2711
21. Jorissen A., 2003, in: "Asymptotic Giant Branch Stars", eds. H. Habing and H. Olofsson, Springer Verlag N.Y., p. 461
22. Marescaux M., 2003, undergraduate thesis available in Dutch at http://www.ster.kuleuven.ac.be/~groen/MatthiasM_thesis.ps.gz
23. Mateo M., 1998, ARAA 36, 435
24. McClure R.D., 1997, PASP 109, 256
25. Menzies J., Feast M., Tanabe T., Whitelock P., Nakada Y., 2002, MNRAS 335, 923
26. Mouhcine M., Lançon A, 2003, MNRAS 338, 572
27. Nowotny W., Kerschbaum F., 2002, Hvar Obs. Bulletin 26, 63
28. Nowotny W., Kerschbaum F., Olofsson H., Schwarz H.E., 2003, A&A 403, 93
29. Östlin G., Mouhcine M., 2004, preprint
30. Rejkuba M., 2004, A&A 413, 903

31. Snigula J., Gössl C., Hopp U., Barwig H., 2004, in: "IAU Colloquium 193: Variable Stars in the Local Group", eds. D.W. Kurtz & Karen Pollard, ASP Conf. Ser. 310, p. 70 (astro-ph/0310704)
32. van den Berg S., 2000, PASP 112, 529
33. Wood P.R., Alcock C., Allsman R.A., et al., 1999, in: "IAU Symposium 191: AGB stars", eds. T. Le Bertre, A. Lèbre and C. Waelkens, Kluwer Publishers, p. 151
34. Wood P.R., 2000, PASA 17, 18

Abundance Gradients in Local Group Galaxies Using Asymptotic Giant Branch Stars

Maria–Rosa Cioni

ESO, Karl–Schwarzschild-Str. 2, D–85748 Garching bei München, Germany

Abstract. Simultaneous IJK_s observations allow us to statistically select AGB stars, both M and C type, from RGB and younger or foreground stars in nearby galaxies. Regions of different metallicity, identified from the distribution of the C/M ratio, show a considerable variation across the surface of the Magellanic Clouds and NGC 6822.

1 Introduction

Asymptotic Giant Branch (AGB) stars are useful indicators of the properties of a galaxy. They trace the intermediate–age population (between one and several Gyr), are isolated and often the brightest objects, and can be observed beyond 1 Mpc. Furthermore, because there are two kinds of AGB stars: oxygen-rich (O-rich or M-type) and carbon-rich (C-rich or C-type), their ratio is a powerful indicator of metallicity. AGB stars are found mostly in Irregular, Spiral and Elliptical galaxies in the Local Group. In particular there are about 30000 AGB stars in the Large Magellanic Cloud (LMC), about 8000 in the Small Magellanic Cloud (SMC) and about 3000 in the central region of NGC 6822. A handful of AGB stars are also found in some Spheroidal and Dwarf Spheroidal galaxies. It is important, for the purpose of this paper to remember that there are more C-rich AGB stars in metal poor systems.

In 1980 Pagel[17] wrote that the metal abundance in external galaxies exhibits radial gradients. Searches for AGB stars in the Solar Neighborhood [16], the Magellanic Clouds [2] and Baade's Window [12] have shown that the C/M ratio correlates with metallicity in the sense that a higher ratio corresponds to a lower metallicity. Theoretically the C/M ratio indicates the metallicity of a system because in a metal poor environment the location of the giant branch shifts to higher temperatures and fewer carbon atoms are needed to form C stars [18,15].

Studies of AGB stars in the Magellanic Clouds (MCs) have strongly improved thanks to optical and near-infrared (near-IR) large scale surveys: DENIS, 2MASS, MACHO, EROS, OGLE, MCPS. Similar surveys of other Local Group galaxies, especially in near-IR, and monitoring programmes are taking place.

2 [Fe/H] Abundance

2.1 The Large Magellanic Cloud

Using the DENIS catalogue towards the MCs (DCMC) [5] defined a photometric criteria to select stars of a different type and age. In the $(I-J, I)$ diagram AGB

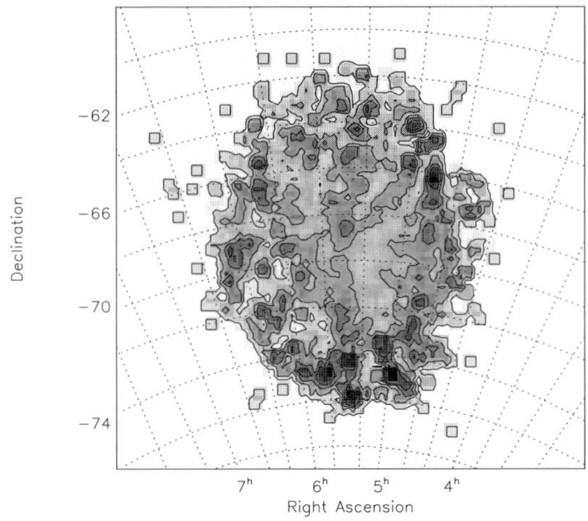

Fig. 1. Distribution of the C/M ratio in the LMC.

stars, irrespectively of type, are easily distinguished as a plume of objects above the tip of the red giant branch (TRGB – [7]), from RGB stars that are fainter and from younger stars or foreground sources, that have much bluer colours ([4]). The distribution of AGB stars across the LMC describes a smooth and regular elliptical structure without clear signs of spiral features. The same AGB stars, because of different molecule that dominate their atmosphere, are separated into O-rich and C-rich at $J - K_s = 1.4$: C-rich stars are redder than O-rich stars. The separation is a function of metallicity. Our IJK_s selection criteria includes also O-rich AGB stars of early M spectral sub-type (M0+) located below the K_s-TRGB. In addition most AGB stars with a thick circumstellar envelope ($J - K_s > 2.2$) are excluded because they are located below the I-TRGB; these can be O-rich as well as C-rich.

The distribution of the C/M ratio (Fig. 1) is rather clumpy. Regions with a high ratio are progressively located in the outer part of the galaxy suggesting that the metallicity decreases from the center to the outer galaxy. This gradient has been recently confirmed by [1] fitting the RGB in the ($J - K_s$, K_s) 2MASS diagram in different locations. In agreement with [8], the bar is more metal rich than the inner disk region and towards the bridge connecting the MCs, the metallicity is low (Arm B region), as suggested by [19]. Calibrating the C/M in terms of [Fe/H], as discussed in Sect. 2.4, we obtain a variation of 0.75 dex. Comparing with the extinction map [21], the C/M0+ ratio is not the result of differential extinction.

2.2 The Small Magellanic Cloud

A similar study, as presented above, has been done also for the SMC using DCMC data. Figure 2 shows the distribution of the C/M ratio. Here a clear gradient is

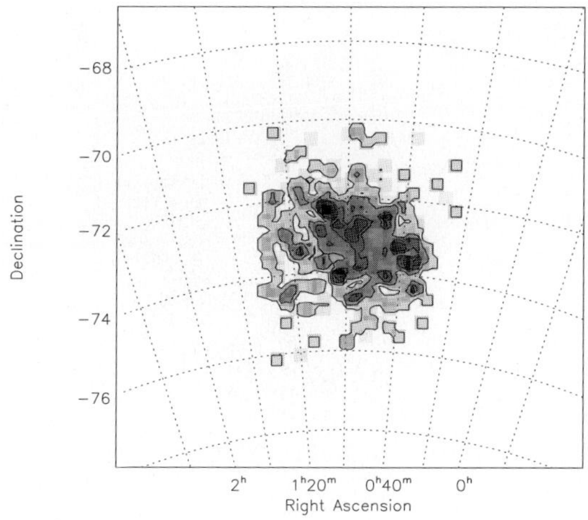

Fig. 2. Distribution of the C/M ratio in the SMC.

not present, however, there are clumps of high ratio (or low metallicity) located in the outer borders of an inner region, while the outermost parts of the galaxy have a much lower ratio. The variation of [Fe/H] also corresponds to 0.75 dex. An indication of a similar metallicity distribution has been found [14]. The authors combined $UBVI$ data from the Magellanic Cloud Photometric Survey (MCPS) with theoretical isochrones [11]. They conclude that a population about 2.5 Gyr old with $Z = 0.008$ is associated with an outer, perhaps uncertain, ring, which encloses clumps of objects about 1–1.5 Gyr old with $Z = 0.001$–0.004.

2.3 NGC 6822

NGC6822 is an isolated Irregular galaxy in the Local Group in many ways similar to the MCs (i.e. of intermediate [O/H] abundance). On account of its low latitude, it is affected by a moderate extinction ($E(B-V) = 0.25 - 0.45$) and contamination by foreground stars. It started to form stars about 10 Gyr ago with a rate that increased in the past 3 Gyr. Using IJK_s data from the William Herschel telescope in La Palma, we have surveyed the inner $20' \times 20'$ of the galaxy down to about 1 mag below the TRGB [3]. These are the first near-IR observations that cover the whole galaxy; previously it was observed in J & K_s only three very small regions [9]. Using a similar selection technique as for the MCs we have studied the distribution of AGB stars and of the C/M ratio (Fig. 3). The latter corresponds to a variation of [Fe/H] of 1.89 dex, much larger than that found within the MCs. This agrees with the spread derived by [20] from RGB stars. Clumps of high ratio (or low metallicity) are distributed in two NS semi-circles around the central bar, that has on the other hand a much lower ratio. At least some clumps correspond to regions of high HI column density ([10]).

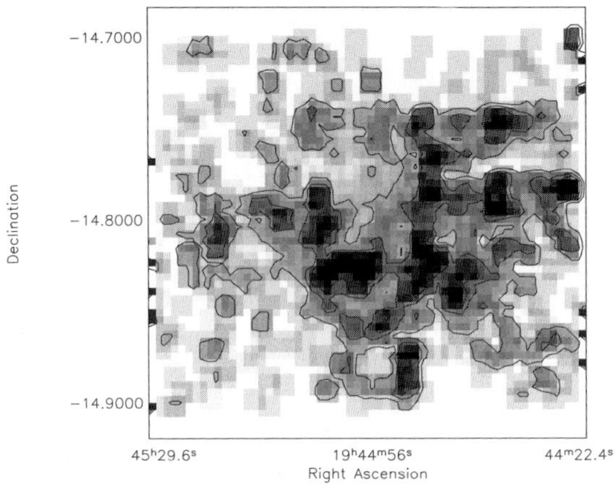

Fig. 3. Distribution of the C/M ratio in NGC 6822.

2.4 Calibration of C/M versus [Fe/H]

In order to relate the C/M to [Fe/H] we have used values available in the literature from the compilation by [13] – see also these proceedings. Though a correlation is present (Fig. 4) the fit is rather uncertain: measurements of [Fe/H] might be in error by 0.2 dex while the uncertainty on the number of AGB stars is up to 50%. In fact the [Fe/H] relies on just a few stars or HII regions while the C/M ratio depends on extrapolation to the whole galaxy of numbers of objects detected in small survey areas, often from incomplete and inhomogeneous samples. In order to improve this relation we have recently obtained spectra of about 300 AGB stars in NGC 6822 to measure the Ca II triplet and derive, using the most up-to-date calibration between these features and [Fe/H], an estimate of the metallicity. These values will be averaged in bins that contain a significant number of AGB stars and hopefully we will be able to put further constraints on the C/M versus [Fe/H] relation.

It is possible that the C/M ratio also depends on the local star formation history. In collaboration with Girardi, Marigo & Habing we are comparing the luminosity function of AGB stars in the MCs, in different locations, with a theoretical luminosity function to address this aspect.

3 Other Abundances

Work in progress on the analysis of FLAMES spectra of a sample of AGB stars in the LMC will allow us to derive the metallicity and the abundance of other elements that play a key role in the evolution of AGB stars. Preliminary results from data obtained during the science verification of the instrument are very encouraging and were followed by new observations in order to reach a necessary

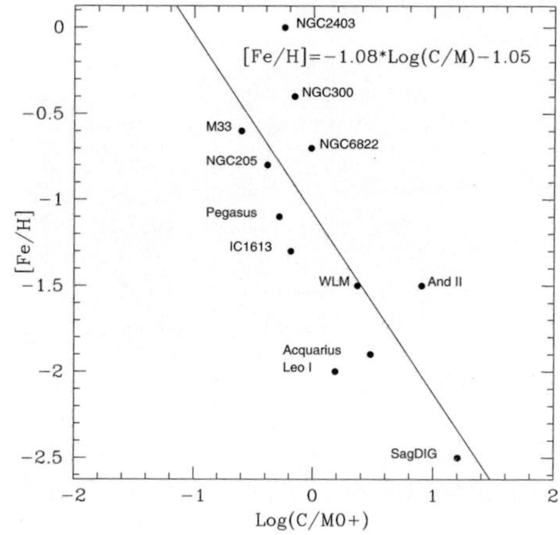

Fig. 4. Relation between the C/M ratio and [Fe/H] in the Local Group.

statistical sample to study abundances as a function of magnitude, colour and period.

References

1. D. Alves, 'Structure of the Periphery of the Large Magellanic Cloud Further Revealed by 2MASS'. Poster (2004)
2. V.M. Blanco, M.F. McCarthy, B.M. Blanco, ApJ 242, 938 (1980)
3. M.-R.L. Cioni, H.J. Habing, A&A *submitted* (2004)
4. M.-R.L. Cioni, H.J. Habing, A&A 402, 133 (2003)
5. M.-R.L. Cioni, C. Loup, H.J. Habing, et al., A&AS 144, 235 (2000a)
6. M.-R.L. Cioni, H.J. Habing, F.P. Israel, A&A 358, L9 (2000b)
7. M.-R.L. Cioni, R.P. van der Marel, C. Loup, H.J. Habing, A&A 359, 601 (2000c)
8. A.A. Cole, T.A. Smecker-Hane, E. Tolstoy, et al., MNRAS 347, 367 (2000)
9. T.J. Davidge, PASP 115, 635 (2003)
10. W.J.G. de Block, F. Walter, ApJ 537, L95 (2000)
11. L. Girardi, A. Bressan, G. Bertelli, C. Chiosi, A&AS 141, 371 (2000)
12. I.S. Glass, S. Ganesh, C. Alard, et al., MNRAS 308, 127 (1999)
13. M.A.T. Groenewegen, 'Carbon Stars in the Local Group'. In: *The Chemical Evolution of Dwarf Galaxies, Ringberg Castle, Germany, 28 July 2002*
14. J. Harris, D. Zaritsky, AJ 127, 1531 (2004)
15. I. Iben, A. Renzini, ARA&A 21, 271 (1983)
16. T. Mikami, Tokyo Astronomical Observatory, Annals, Second Ser., 17, 1, 1 (1978)
17. B. E. J. Pagel, M. G. Edmunds, G. Smith, MNRAS 193, 219 (1980)
18. J.M. Scalo, G.E. Miller, ApJ 248, L65 (1981)
19. L. Staveley-Smith, S. Kim, M.R. Calabretta, et al., MNRAS 339, 87 (2003)
20. E. Tolstoy, M.J. Irwin, A. Cole, et al., MNRAS 327, 918 (2001)
21. D. Zaritsky, J. Harris, I.B. Thompson, E.K. Grebel, AJ *in press* (2004)

Red Giant Stars in NGC 5128

Marina Rejkuba

European Southern Observatory, Karl-Schwarzschild-Str. 2, D-85748 Garching, Germany

Abstract. I present a selection of results obtained from VLT FORS1 and ISAAC photometric monitoring of late-type giants in NGC 5128 (=Centaurus A). The combination of optical and near-IR photometry allows the full metallicity range of the stars on the upper red giant branch to be probed, owing to the combined low-metallicity sensitivity of the optical and high-metallicity sensitivity of the near-IR bands. The metallicity distribution covers a wide range with the mean value around [M/H] ~ -0.45 dex. The near-IR monitoring of the variable AGB stars allows insight to be gained into the age distribution. The period distribution of these long period variables indicates only about 10% contribution of the intermediate-age component (age $\lesssim$ 5 Gyr) to the predominantly old stellar halo. Among the brightest, large amplitude and long period variables, only very few have near-IR and optical colors consistent with carbon-rich giants.

1 Introduction and Observations

The red giant stars of NGC 5128 (=Centaurus A), the nearest, easily observable giant elliptical galaxy, were resolved first time by Soria et al. (1996), who used WFPC2 camera on board HST to image this galaxy's halo. Today, with the availability of 8-10m class telescopes in excellent astronomical sites it is possible to obtain similar or even better results with imaging from the ground.

The data presented here have been obtained between April 1999 and July 2002 with FORS1 optical imager and spectrograph and ISAAC near-IR instrument at UT1 Very Large Telescope at ESO Paranal Observatory. Two halo fields were observed once in U, V, J_s and H bands and monitored with 20–24 K_s observations spread over 3 years. The observations, data reductions and photometric catalogues are presented by Rejkuba et al. (2001, 2003a). Some results concerning the metallicity and ages of red giant branch (RGB) and asymptotic giant branch (AGB) stars in the halo of NGC 5128 are shown. Here I plot color-magnitude and color-color diagrams for the north-eastern halo field (Field 1 in Rejkuba et al. 2001). Very similar diagrams, and conclusions, are reached for the second, southern halo field, as well. For more detailed analysis, the interested reader is referred to Rejkuba et al.(2003a, 2003b) and Rejkuba (2004).

2 RGB and AGB in Color-Magnitude Diagrams

Combination of the optical and near-IR colors in a color-magnitude diagram (CMD) allows the old and intermediate-age stellar populations to be probed.

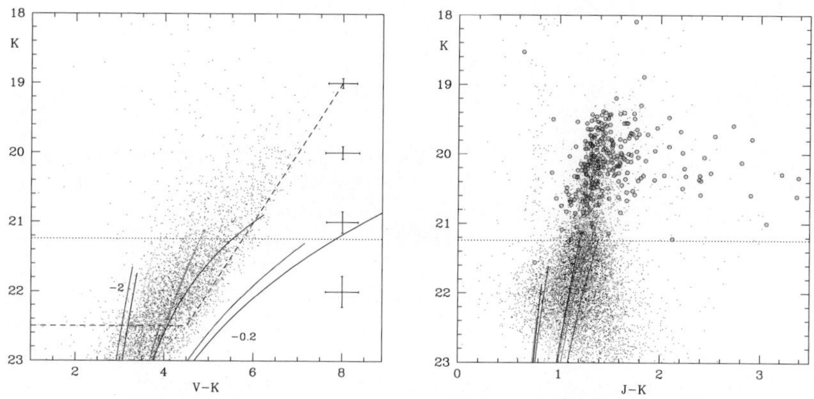

Fig. 1. Left: Optical-near IR CMD of halo stars in NGC 5128. **Right:** Near-IR CMD of the same field. The horizontal line indicates the position of the RGB tip in both CMDs. Overplotted are the fiducial RGBs for the following old Galactic globular clusters: NGC 6528, NGC 6553, M 69, 47 Tuc, M 107, M 4, M 55, M 30 and M 15, in order of decreasing metallicity from −0.2 to −1.9 dex. The dashed slanted line in the VK CMD indicates 50% completeness of the V-band photometry. The large number of sources brighter than the RGB tip are AGB variables. Those variables for which reliable periods can be obtained from our 3 yr-monitoring programme are marked with large circles.

Theoretically, more than two thirds of the light in K-band is dominated by cool RGB and AGB stars. The red dwarfs are too faint to be detected at the distance of NGC 5128, and thus the VK and JK CMDs are entirely dominated by RGB and AGB stars (Fig. 1).

The spread in color of the RGB is larger than the photometric uncertainties, indicating the presence of a spread in metallicity and/or age. The age spread is possible, and indeed most likely (see below), but it cannot entirely account for the range of colors of the RGB stars, which are much more sensitive to metallicity changes. From a comparison with the fiducial RGBs of the old Galactic globular clusters (GGC) with a range of metallicities, and assuming old ages for the stars in the NGC 5128, the most metal-poor stars have metallicities as low as −2 dex, while the metal-rich end on the VK CMD, set by the incompleteness of the V-band photometry (dashed slanted line), indicates approx. −0.5 dex. The most metal-rich giants, which are too faint in the V-band due to huge bolometric corrections, emit most of their energy in the near-IR and are thus easily observed in JK CMD (Fig. 1, right). The most metal-rich GGC fiducials overplotted are for NGC 6553 and 6528, and Bulge globulars with $-0.3 \lesssim [M/H] \lesssim -0.2$ dex. There is a small component of red giants in NGC 5128 with metallicities close to and slightly above solar. Their mean metallicity is -0.45 ± 0.05 dex. For comparison, Walsh et al. (1999) measured the mean oxygen-abundance of five planetary nebulae in NGC 5128 to be $[O/H] = -0.5 \pm 0.3$ dex, in agreement with the average metallicity inferred from the RGB color.

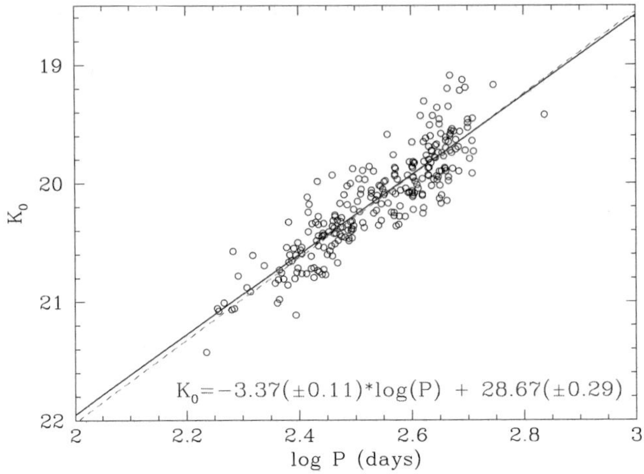

Fig. 2. Period-magnitude diagram for Mira LPVs in NGC 5128. The solid line is a least-squares fit to the data points (shown in the figure) and the dashed line is a fit with a fixed slope (−3.47) as measured in the LMC (Feast et al. 1989).

The dotted horizontal line in both CMDs in Fig. 1 is drawn at $K_s = 21.24$ mag, the position of the tip of the RGB as measured from the ISAAC data. The stars brighter than the RGB tip by up to ~ 2 mag are in the AGB evolutionary phase. There are additionally a few dozen relatively bright stars in the K-band images with no J or H-band counterparts. Their stellar profiles and magnitudes indicate that they might be highly dust obscured AGB stars.

3 AGB Variable Stars

Rejkuba et al. (2003a) measured periods for more than 1000 long period variables (LPV) in the NGC 5128 halo. The period-magnitude diagram for the LPVs with the most reliable periods is shown in Fig. 2. They form a sequence with the slope (indicated in the figure) which is very close to that of the LMC Mira variable stars (−3.47; Feast et al. 1989). Adopting the zero point of the Mira period-luminosity relation of 0.98 and a distance modulus of the LMC of 18.50, I have obtained a distance modulus to NGC 5128 of 27.96 ± 0.11, in excellent agreement with other literature values (see Rejkuba 2004 for more discussion).

Periods of Mira variables can be used to age date stellar populations assuming that their metallicities are known. Longer period Miras are expected to have higher mass progenitors, hence to be younger. However, the more metal-rich a star is, the older the star, assuming a constant mass. For example, according to models (e.g. Vassiliadis & Wood 1993), a 1 $M_\odot$ star will evolve to a Mira variable with a period of $\sim$ 400 days. In a Solar metallicity population, the turn-off age of a 1 $M_\odot$ star is 7.7 Gyr, while at $[M/H] = -0.7$ dex it is $\sim$ 4.5 Gyr.

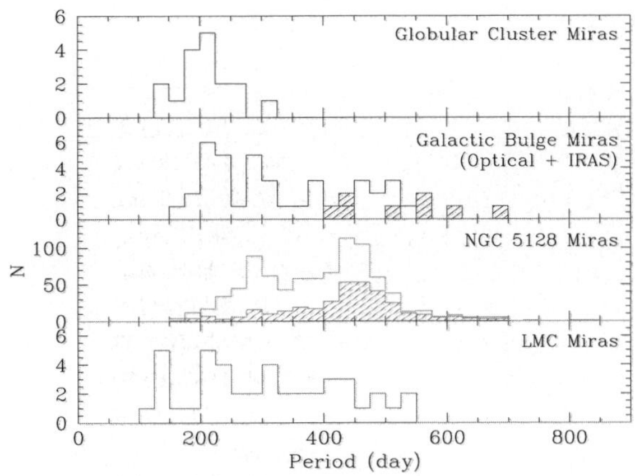

Fig. 3. Period distribution of LPVs in NGC 5128 (second panel from bottom) is compared with: Galactic globular clusters (age $\sim$ 12 Gyr; [M/H] > -1 dex) Miras; Milky Way Bulge (age $\sim$ 10 Gyr, [M/H] $\sim$ 0.0 dex); and the LMC (mostly age $<$ 3 Gyr, [M/H] ~ -0.7 dex) Miras. See Rejkuba (2003b) for further details.

A comparison of Mira period distributions in NGC 5128 and 3 other systems with different mean ages and metallicities is shown in Fig. 3. The brightest Miras in NGC 5128 reach $M_K = -8.65$ and have periods in excess of 800 days. The large majority has similar periods to those of Galactic bulge and old globular cluster Miras, but 10% of them have periods in excess of 500 days and are thus probably younger than $\sim$ 5 Gyr, unless they all have extremely high metallicities.

4 Are There Carbon Stars?

Carbon stars are typically found among intermediate-age metal-poor populations. Their presence is definite proof of an intermediate-age component. If present in NGC 5128 they may come from a recently accreted LMC-type or a small gas-rich spiral galaxy. Carbon stars in the LMC have $1.4 < (J - K_s) < 2$, but the stars redder than $J - K_s > 2$ can either be obscured oxygen or carbon-rich giants. The $H - K$ vs. $J - H$ color-color diagram has been a traditional tool for detecting these stars (see Rejkuba et al. 2003b for NGC 5128). In Fig. 4 I show a combined optical near-IR color-color diagram, $J - K$ vs. $V - K$ for all the stars brighter than the RGB tip from the CMD in Fig. 1. The recent models from Marigo (2002) fit the range of colors of late-type oxygen and carbon-rich giants in the Solar neighbourhood in this diagram well. In Fig. 4 NGC 5128 AGB stars are plotted with small filled symbols, Galactic M-stars are from Fluks et al. (1994; open large circles) and C-stars are from Bergeat et al. (2001; large open triangles). Most of the stars in NGC 5128 are located along the oxygen-rich sequence, and only a small number of them are found along the location of carbon-rich giants. This is expected in a relatively high metallicity environment

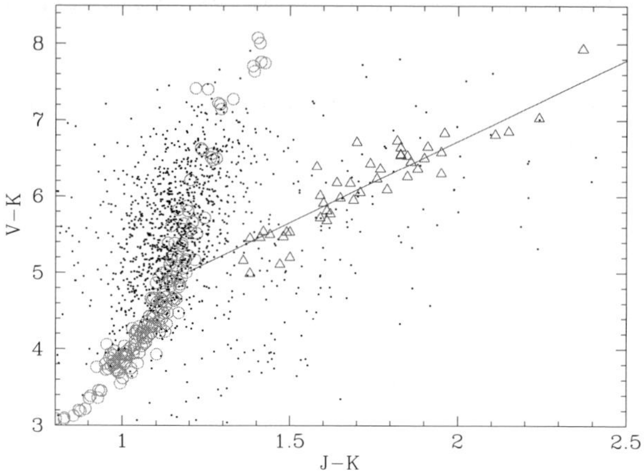

Fig. 4. Location of oxygen-rich (M-stars; large open circles; Fluks et al. (1994) data) and carbon-rich (C-stars; large open triangles; Bergeat et al. (2001) data) giants from the Solar neighbourhood is compared with AGB stars in NGC 5128 (small dots) in a optical near-IR color-color diagram. The large majority of the AGB stars ($K_s < 21.3$) in NGC 5128 are oxygen-rich and only few have colors consistent with C-stars. Due to limits of the V-band photometry, the reddest stars may be incomplete.

like that of the NGC 5128 halo. However, as shown above, there is also a metal-poor tail in this galaxy. Most of it appears to be older than ~ 5 Gyr, and only a very small component has intermediate-ages.

Much deeper photometry has been recently obtained with the ACS camera on board of HST. It reaches V-band magnitudes of ~ 30 and enables to detect additional age-sensitive features, like the AGB bump, red clump and the horizontal branch. The analysis of these new observations will be used to quantify the age distribution of the NGC 5128 halo stars (Rejkuba et al., in preparation).

References

1. Bergeat, J., Knapik, A, Rutily, B, 2001, A&A, **369**, 178
2. Feast, M. W., et al., 1989, MNRAS, **241**, 375
3. Fluks, M. A., et al., 1994, A&AS, **105**, 311
4. Marigo, P., 2002, A&A, **387**, 507
5. Rejkuba, M., 2004, A&A, **413**, 903
6. Rejkuba, M., Minniti, D., Silva, D. R., Bedding, T. R., 2001, A&A, **379**, 781
7. Rejkuba, M., Minniti, D., Silva, D. R., 2003a, A&A, **406**, 75
8. Rejkuba, M., Minniti, D., Silva, D. R., Bedding, T. R., 2003b, A&A, **411**, 351
9. Soria, R. et al., 1996, ApJ, **465**, 79
10. Vassiliadis, E., Wood, P. R., 1993, ApJ, **413**, 641
11. Walsh, J. R., Walton, N. A., Jacoby, G. H., Peletier, R. F., 1999, A&A, **346**, 754

Molecules in the AGB Stars of Nearby Galaxies

M. Matsuura[1], A.A. Zijlstra[1], J.Th. van Loon[2], and I. Yamamura[3]

[1] Department of Physics and Astronomy, The University of Manchester, P.O. Box 88, Sackville Street, Manchester M60 1QD, United Kingdom
[2] Astrophysics Group, School of Chemistry and Physics, Keele University, Staffordshire ST5 5BG, UK
[3] ISAS, JAXA Yoshino-dai 3-1-1, Sagamihara, Kanagawa 229-8510, Japan

Abstract. We obtained infrared spectra of Asymptotic Giant Branch (AGB) stars in nearby galaxies, and measured the equivalent widths of HCN and C_2H_2 bands for the carbon-rich stars, and SiO abundances for the oxygen-rich stars. We report how the abundances of the molecules depend on metallicity of the host galaxy.

1 Introduction

Asymptotic Giant Branch (AGB) stars are one of the important sources of the dust and the gas in galaxies. However, the dependence of the composition and amount of dust and gas on the metallicity of the host galaxy is not well studied.

The effective temperature of AGB stars is as cool as 2000–3000 K, so that various kinds of molecules are formed in their photosphere. In general, the abundance of molecules should decrease as the metallicity of the host galaxy decreases. Additionally, the third dredge-up during the AGB phase changes the elemental abundance of the AGB star, and especially for carbon. These elemental or molecular abundances of the extra-galactic AGB stars are not well studied yet. Here, we obtained the spectra of the AGB stars in the Large Magellanic Cloud (LMC; [Z/H]~ -0.3), the Small Magellanic Cloud (SMC; [Z/H]~ -0.6), and Sagittarius Dwarf Spheroidal Galaxy (SgrD; [Z/H]$\sim -0.1 - -0.55$). We report the metallicity dependence of C_2H_2 and HCN molecules in the carbon-rich stars, and SiO abundance in the oxygen-rich stars. C_2H_2 and SiO are particularly interesting because they are parent molecules for formation of the dust grains and should affect the AGB mass-loss rate and the gas to dust ratio.

2 Observations and Discussions

We select AGB stars in the LMC, the SMC and the SgrD from [10], [4], [7], [8]. We observed the L-band spectra with ISAAC/VLT at ESO Paranal, Chile, using two modes: low resolution (LR) and medium resolution (MR). With the LR mode, the entire L-band is covered, and this mode is used mainly for carbon-rich stars. The MR mode has a spectral coverage of 0.255 μm, and was used to resolve SiO bands in oxygen-rich stars. The wavelength resolutions (R= $\lambda/\Delta\lambda$) are 360 or 600 in the LR mode, and 2000 or 3300 in the MR mode, depending on the slit width.

2.1 Carbon-Rich Stars

Figure 1 is an example of the ISAAC spectra of the carbon-rich stars in the LMC, and the SMC, and the ISO/SWS spectrum of a carbon-rich star in the Solar neighbourhood. Several molecular bands are found in L-band: 3.1 μm HCN+C_2H_2 band, 3.3 μm and 3.6 μm CH bands, 3.5 μm HCN band, and 3.8 μm probably C_2H_2 band. We measured the equivalent widths (EWs) of the three molecular bands on carbon-rich stars (except CH). The EWs are plotted against an infrared colour (Fig. 2). For a comparison, we measure EWs of the carbon stars in the Solar neighbourhood using the ISO/SWS spectra, and obtained the infrared colour of these nearby stars from the publications.

The 3.5 μm EW is almost negligible in the extra-galactic AGB stars, except one star, IRAS 04496−6958 in the LMC, in contrast to the nearby AGB stars which show EWs up to 12 Å.

HCN abundance is lower at lower metallicity, because the abundance is restricted by the nitrogen abundance, which is smaller in these nearby galaxies.

The 3.8 μm EW, which is probably a measure of the C_2H_2 abundance, is high in red carbon stars ($H - K > 1.2$). This is because C_2H_2 is efficiently formed at lower temperature [5]. Among the red carbon stars, $EW_{3.8}$ is large in the LMC and SMC stars.

If all the elemental abundances are simply scaled to the metallicity, C_2H_2 abundance should be less at lower metallicity because fewer carbon atoms are available. However, high C_2H_2 abundance at lower metallicity could be explained if the C/O ratio in the carbon-rich stars is systematically higher at lower metallicity. Under the high C/O ratio, more carbon atoms remain after CO formation, and abundances of carbon-bearing molecules increase. The amount of carbon atoms which are dredged-up during the thermal pulse depends on the metallicity. The result is a higher C/O ratio [9].

The 3.1 μm band is a mixture of HCN and C_2H_2. In red stars $EW_{3.1}$ is larger in extra-galactic AGB stars, while in blue stars, there is no clear tendency.

2.2 Oxygen-Rich Stars

The 3.95–4.10 μm spectra of four oxygen-rich stars are shown in Fig. 3. As a reference, the spectrum of the galactic M giant star, g Her, which shows SiO bands, and model spectrum of the SiO bands are plotted. Besides a possible detection of Brα in IRAS 04407−7000, there is no feature found in the LMC stars. The aim of these observations is the detection of SiO bands from extra-galactic AGB stars, however, no SiO bands are found in any of these objects.

Figure 4 shows the EW of SiO bands, which is the sum of the equivalent widths of the three strong $\Delta v = 2$ ^{28}SiO features [2]. The LMC stars' EWs are almost zero, as expected from non-detection of the SiO bands. In this figure, we use LMC Si abundance of log(Si/H)+12$\approx$7.0 and Galactic Si abundance of 7.6 as found in the planetary nebulae [3].

Aringer et al. [1] calculated the EW of SiO bands using a hydrostatic model (Fig. 4). The EW is sensitive to both the effective temperature and the silicon

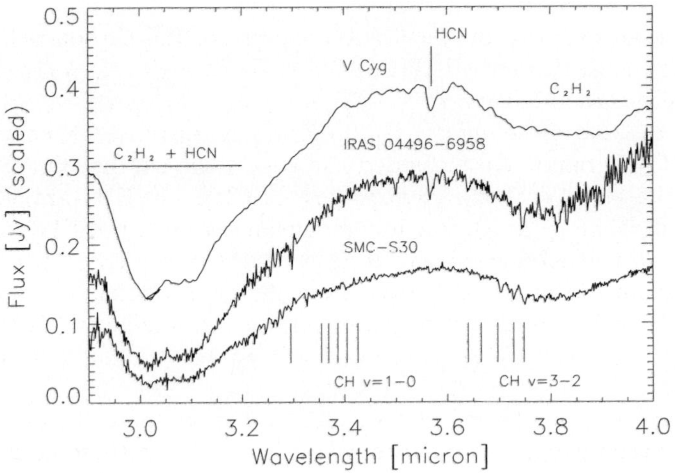

Fig. 1. Spectra of carbon-rich stars: the Galactic star V Cyg (R∼300), the LMC star IRAS 04496−6958, the SMC star, SMC-S30

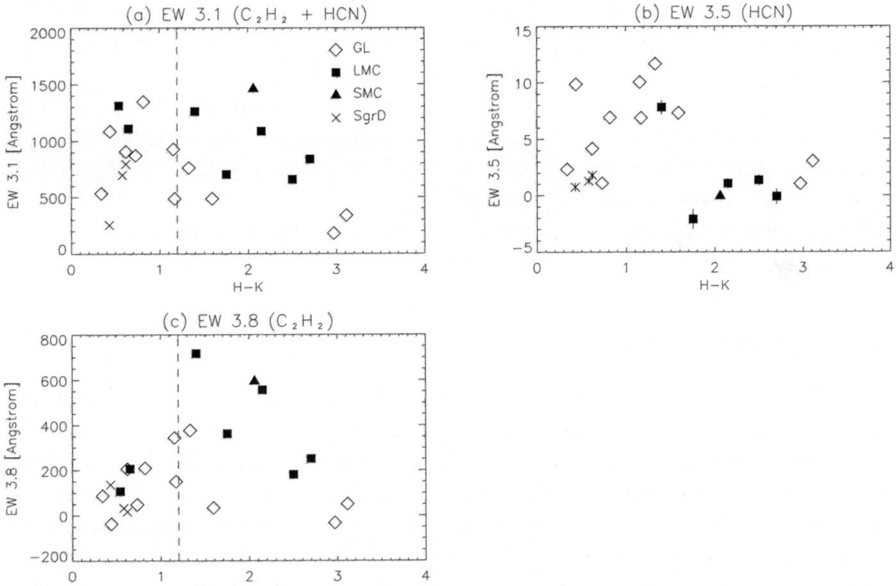

Fig. 2. The distribution of equivalent widths of the three molecular bands at 3.1, 3.5 and 3.8 μm with (H-K) in carbon-rich stars is shown in the three plots.

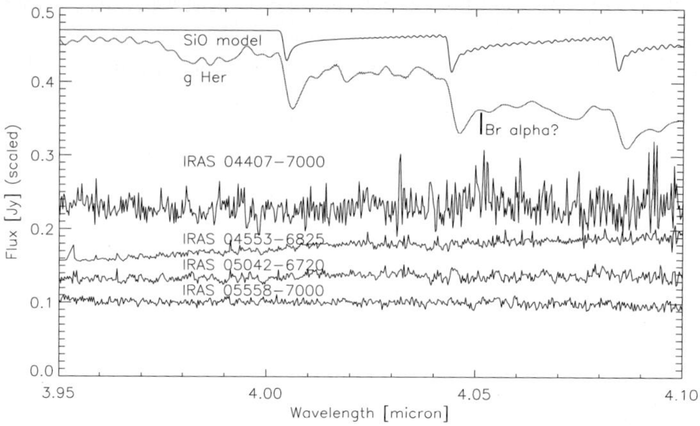

Fig. 3. The spectra of the oxygen-rich stars

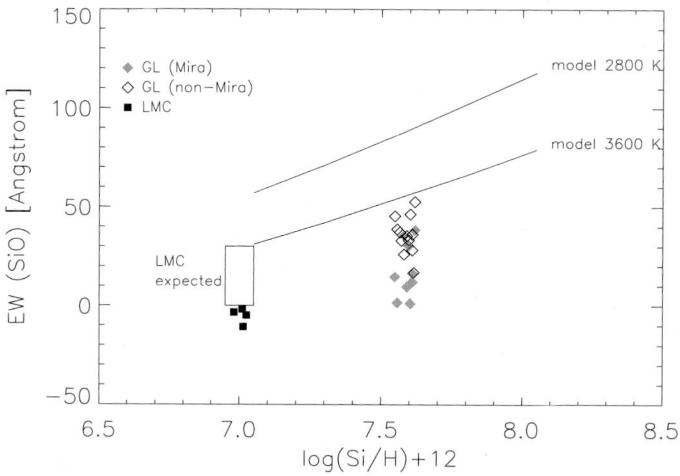

Fig. 4. The equivalent width of SiO bands is plotted as a function of metallicity of the host galaxies. The sample is randomly distributed around the metallicity for clarity, and does not show the Si abundance of individual stars. The model lines are from Aringer et al. [1].

abundance. The SiO abundance increases almost linearly as the Si abundance becomes higher. Aringer et al. [2] measured the SiO EW of the Galactic stars which ranges from 0–50 Å. If the EW of Galactic stars are linearly extrapolated towards lower metallicity, EW of 0–30 Å is expected for LMC stars (Fig. 4). Indeed, SiO EW is smaller at the LMC metallicity, as hydrostatic models predicted; however, our results shows that LMC EWs are even lower than the expectation.

Our low SiO abundance might indicate lower Si abundance in the AGB stars than log(Si/H)+12≈7.0; however, other possible reasons are more likely. Aringer et al. [2] reported that observed EWs of SiO bands are lower than theoretical EWs for Galactic stars. Mira variables, especially, show weaker bands than semi-regular variables. It has been suggested that the strong pulsations of Mira variables form the extra-atmosphere above the photosphere, and molecules in this extended atmosphere fill the absorption features of molecules in the photosphere. We may have observed the LMC stars when the extended atmosphere is the most prominent and the SiO bands are in their weakest phase.

3 Conclusions

Our observations show that at lower metallicity, C_2H_2 molecules are more abundant in carbon-rich stars, while SiO molecules are less abundant in oxygen-rich stars. These two kinds of molecules are parent molecules of grain formation, such as PAH [6] and silicate. (PAH may be cataloged as a large molecule, rather than a dust grain.) Our results imply that at lower metallicity, fewer silicate grains are formed around oxygen-rich stars. However, more carbon-rich grains, especially graphite, should be formed around carbon stars if the C/O ratio is higher. PAH may be formed in the same amount at Solar metallicity in carbon-rich stars and carbon-rich post-AGB stars. The mass-loss rate in the extra-galactic carbon-rich stars decreases at lower metals, which are essential for silicon carbide, but increases at the higher C/O ratio, which affects graphite abundance. The mass-loss from the entire AGB star population will be more carbon-rich at lower metallicity, because of a higher proportion of carbon-rich stars and more efficient dust formation in the carbon-rich stars.

References

1. Aringer, B., Jørgensen, U.G., Langhoff, S.R.: A&A 323, 202 (1997)
2. Aringer, B., Höfner, S., Wiedemann, G., Hron, J., et al.: A&A 342, 799 (1999)
3. Dopita, M. A., Vassiliadis, E., Wood, P.R., et al.: ApJ 474, 188 (1997)
4. Groenewegen, M.A.T., Blommaert, J.A.D.L.: A&A 332, 25 (1998)
5. Matsuura, M., Zijlstra, A.A., van Loon, J.Th., et al.; ApJ 580, L133 (2002)
6. Tielens, A.G.G.M., Waters, L.B.F.M., Molster, F.J., and Justtanont, K.: Ap&SS 255, 415 (1998)
7. Trams, N.R., van Loon, J.Th., Waters, L.B.F.M., et al.: A&A 346, 843 (1999)
8. van Loon, J.Th., Zijlstra, A.A., Groenewegen, M.A.T.: A&A 346, 805 (1999)
9. Vassiliadis, E., Wood, P.R.: A&A 413, 641 (1993)
10. Whitelock, P.A., Irwin, M., Catchpole, R.M.: New Astronomy 1, 57 (1996)

Clues on Post-Asymptotic Giant Branch Evolution and Planetary Nebulae Populations from the Magellanic Clouds

Letizia Stanghellini

National Optical Astronomy Observatory, 950 N. Cherry Ave., Tucson AZ 85719, USA

Abstract. The recent *HST* optical images, and the optical and ultraviolet spectra, of Magellanic Cloud planetary nebulae (PNe), together with the large database that has been collected in the past decade, allows unprecedented insight into the evolution of PNe and their central stars. In this paper we present a selection of recent results: The analysis of PN morphology, both in the optical and ultra-violet emission lines; the relation between nebular morphology and the chemistry produced by stellar evolution; the direct determination of the transition time from observations; and the study of the nature and evolutionary stage of the components of the planetary nebula luminosity function.

1 Introduction

Planetary nebulae (PNe) in the Magellanic Clouds (LMC, SMC) are the nearest extragalactic PNe known. They are close enough for detailed individual analysis, and they can be spatially resolved with current technology; yet, they do not suffer the distance biases and extreme differential interstellar absorption for which Galactic PNe are well known. Magellanic Cloud PNe (MCPne) are thus the ideal probes to study stellar evolution of low- and intermediate-mass stars. They also are ideal benchmarks to extend the knowledge of PNe to the extra-Galactic PN populations, with the additional bonus of covering a large metallicity range.

PNe in the Magellanic Clouds have been observed from the ground, as point sources, since the sixties. The advent of the Hubble Space Telescope (*HST*) afforded their spatial resolution, making it possible to study their size, morphology, and their central stars (CSs). The work presented in this paper is based on the most recent *HST* data-sets acquired by the MCPN team[1] in Cycles 8 through 10 with the Space Telescope Imaging Spectrograph (STIS, Programs 8271, 8663, 9077, and 9120) and the Wide Field Planetary Camera 2 (WFPC2, Program 8702). Older archived narrow-band images (Program 6407, 5185, 4821, 4075, and 1266) were also used to augment the morphological data-set.

[1] Team membership and other information are available in the MAST MCPN web page http://archive.stsci.edu/hst/mcpn/home.html

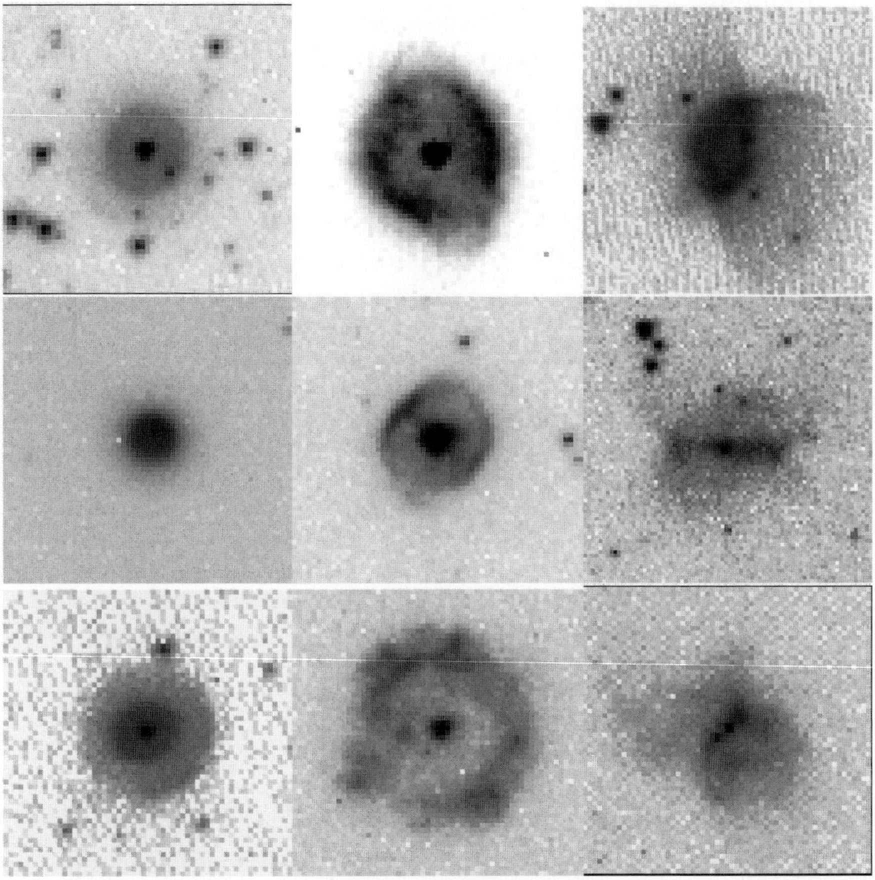

Fig. 1. Magellanic Cloud PNe. Left panel (each panel from top to bottom): R PNe LMC J 33, SMC SP 34, and LMC MG 40; Central panel: E PNe LMC SMP 101, SMC MG 8, and SMC MG 13; Right panel: B PNe LMC SMP 91, SMC MA 1682, and LMC MG 16. All thumbnails are 9 arcsec^2 sections of the STIS broad-band images

2 Planetary Nebula Morphology

The morphology of Galactic PNe has been studied thoroughly in the nineties, and it has been found that morphological type is related to the PN progenitor's evolutionary history and its mass. There is strong evidence that bipolar PNe are the progeny of the massive Asymptotic Giant Branch (AGB) stars (4-8 $M_\odot$). Bipolar PNe are nitrogen-rich and carbon-poor [1,2]. The morphological types, and their distribution in a given PN population, are then good indicators of age and evolutionary history.

Galactic PNe have been classified as round (R), elliptical (E), bipolar (B; includes quadrupolar and multipolar), bipolar core (BC; R or E PNe with a central *bilobate* concentration, or ring enhancement), and point-symmetric (P). A recent description of these classes is published by [1]. The majority of Galactic

PNe are elliptical, but the actual number of B PNe could be underestimated, given that they typically lie in the Galactic plane (i.e., they may suffer from high reddening). PNe in the Magellanic Clouds, when spatially resolved, show the same admixture of morphological types as the Galactic PNe, as illustrated in Fig. 1 [3, 4, 5]. While we do not attempt a statistical comparison of the Magellanic and Galactic PN morphological types, given the selection effects that hamper Galactic PNe, we can meaningfully compare the LMC and SMC samples. Both samples are characterized by low field extinction, and they have been preselected in more or less the same way.

Table 1. Morphological Distribution

Morphological type	% LMC	% SMC
Round (R)	29	35
Elliptical (E)	17	29
R & E	46	64
Bipolar (B)	34	6
Bipolar core (BC)	17	24
Point-symmetric (P)	3	6

The results of the morphological distribution of PNe in the Clouds is illustrated in Table 1. One striking difference between the LMC and SMC morphological type distributions is that the fraction of B PNe in the LMC is almost six times that of the SMC. Furthermore, half of the LMC PNe are either bipolar or have a morphology that suggests a central ring (BC), while only a third of the SMC PNe fall in these two aspherical morphological classes.

From this analysis it emerges that the different processes involved in the formation of the different PN shapes occur in all galaxies where morphology has been studied. It is also apparent that the SMC environment may disfavor the onset of bipolarity in PNe. Otherwise, the different morphological statistics may indicate different populations of stellar progenitors in the two Clouds. While it seems reasonable to conclude that a low metallicity environment is unfavorable to bipolar formation and evolution, the exact causes have not yet been studied. A detailed study of metallicity and mass-loss may clarify this point. On the other hand, the different morphology counts may simply imply that most of the progenitors of the SMC PNe have masses in the lower-mass range ($M<4M_{\odot}$), thus they do not produce B PNe. A comparison of the SMC and LMC CS masses does not disagree with such a possibility [6].

3 Results from UV Spectroscopy

The study of the chemical composition that characterizes PN ejecta is a way to determine the origin and evolution of the progenitor stars. PNe contain the

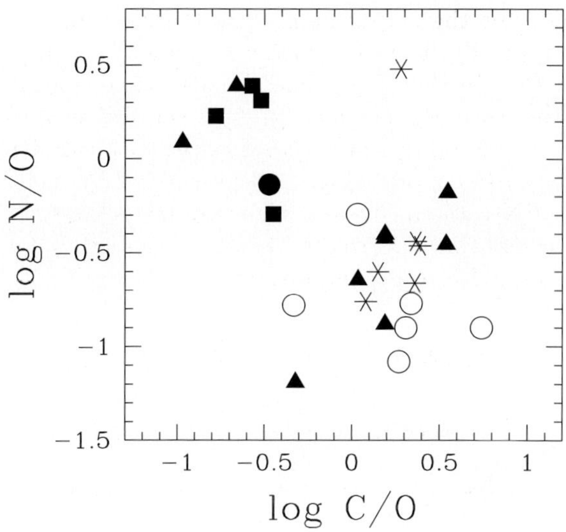

Fig. 2. The C/O vs. N/O relation for R (open circles), E (asterisks), BC (triangles), B (squares), and P (filled circles) PNe in the LMC (adapted from [11])

products of stellar nucleosynthesis, dredged-up at the stellar envelope during the late AGB evolution. The study of carbon and nitrogen abundance in PNe is especially interesting. The theory of stellar evolution suggests that Galactic and Magellanic AGB evolution produce different N and C concentrations depending on the AGB progenitor's mass. If $M_{MS} > 4\ M_{\odot}$, the so-called *hot-bottom burning* (HBB) process would convert much of the carbon content into nitrogen, resulting in a different chemical make-up of the remnant PN [7]. In order to disclose possible correlation between the morphology and the *evolutionary* chemistry of the LMC PNe we need to have at our disposal the carbon and nitrogen abundance of a sufficient number of PNe to populate the major morphological classes. We use the nitrogen and oxygen abundances in the literature [8, 10] and the carbon abundance in [9, 11] to plot, in Fig. 2, the relation between the (logarithmic) C/O and N/O abundances.

From this Figure we infer that the carbon stars progeny (log C/O>0) is limited to R, E, and BC PNe. B and P PNe originate from the evolution of stars that did not go though the carbon star phase, or, more likely, their progenitor suffered carbon depletion, possibly due to the occurrence of the HBB. It appears that R and E PNe are the progeny of lower-mass stars, those with $M_{MS} < 4\ M_{\odot}$, while the opposite holds for the most aspherical (B and P) PNe. From Fig. 2 we see that BC PNe might have evolved from carbon stars in some cases, while in other cases they had evolved from more massive stars. It remains to be seen whether some of the BC PNe are indeed bipolar with faint lobes (those with low carbon and high nitrogen content), and others are misclassified R and E PNe.

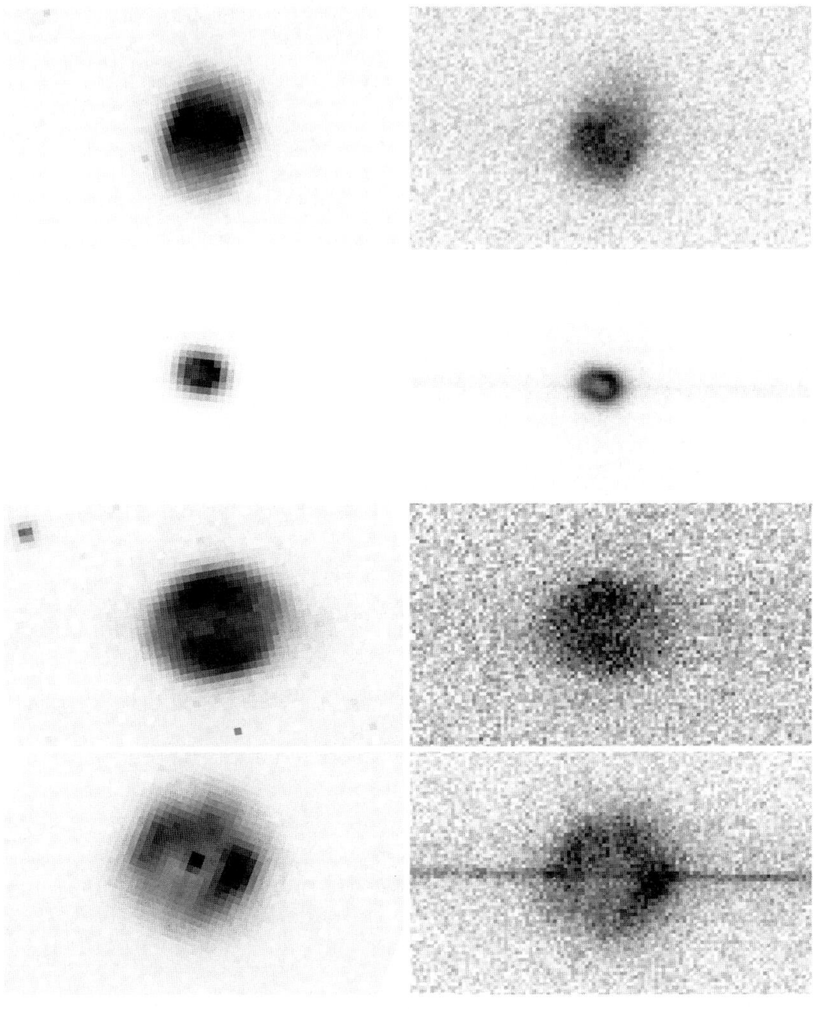

Fig. 3. Morphology of LMC PNe. Left column: clear STIS CCD images. Right column: C III] 1908 Å images from the G230L STIS spectra. The PNe are, from top to bottom, SMP 19, SMP 79, SMP 95, and SMP 102

In order to enlarge the sample of Magellanic Cloud PNe whose carbon abundance is known, we have used the *HST* data from program 9120, acquired with *STIS* grisms G140L and G230L [11]. The slitless spectra were obtained with a large aperture, showing the morphology of the PNe through the brightest UV emission line. It is then possible to compare the UV to optical morphology. In Fig. 3 we show the clear optical images through the *STIS* CCD (left column) and the parts of the G230L slitless spectra corresponding to the C III] emission line at λ1908 Å. The image sections are all to scale, thus the (relative) physical sizes are also to scale. The clear images have been rotated so their orientation coincides with the orientation of the slitless spectra. The UV emission deriving from the CIII] line originates in the same volume of the nebulae as the optical emission [4]. In Fig. 3 we can also clearly see the spectrum of the central star of SMP 102.

The fitting of the UV stellar continua have been used to derive stellar temperatures. In some cases, central stars of Magellanic Cloud PNe show P-Cygni profiles in the UV spectra [12].

4 Stellar Evolution Beyond the AGB and the Transition Time

The known distance of Magellanic Cloud PNe makes them the ideal probes to study stellar evolution. AGB and post-AGB evolution have been studied theoretically and observationally in many different fashions. Nonetheless, there are still several processes that need to be understood to make the comparison between theory and data meaningful. In this section we describe how Magellanic Cloud PNe are used to shed some light on the transition time problem. Other aspects of AGB evolution will be examined in future papers.

When comparing CS data to theoretical predictions, it is often assumed that the time lag between the envelope ejection quenching (EEQ) and the PN illumination, the so-called transition time (t_{tr}), has a duration either of a fixed value (generally 1000 yr), or it is set to zero. Both of these assumptions are inadequate to express a time-scale that is likely to assume different values in different PNe, and that represents a sizable fraction of the PN lifetime [13]. From a conceptual point of view, the transition time is mainly dependent on the residual envelope mass that survives the envelope ejection (M_e^R) and on the post-AGB mass-consumption process (or processes) that is in place in each specific CS. The dynamical time, $\tau_{dyn} = D_{PN}/v_{exp}$ (where D_{PN} is the diameter of the nebula and v_{exp} is the nebular expansion velocity), represents the dynamical nebular age in the assumption that the nebular shell does not suffer acceleration. If we measure the diameter of the main nebular shell, the dynamical time tracks the evolutionary time measured from the EEQ event. In order to compare the dynamical time to the stellar evolutionary time we need to use the equation $\tau_{dyn} = t_{tr} + t_{ev}$, where t_{ev} is the evolutionary time evaluated *after* the PN illumination. This equation holds as long as the evolutionary tracks mark the temporal zero-point at $T_{eff} \simeq 30,000$ K, as they usually do.

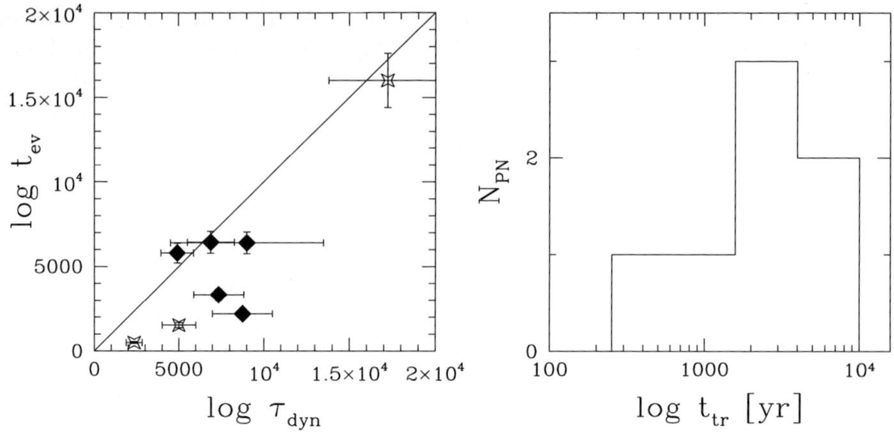

Fig. 4. (a): Evolutionary vs. dynamical time for LMC (diamonds) and SMC (stars) PNe. **(b)** : Transition time histogram

In order to estimate the transition time one needs to determine the residual envelope mass, the mass-loss during transition, and whether the CSs are burning hydrogen or helium. All these properties are rarely known, if ever. Another approach to determine, or at least constrain, the transition time is to use the Magellanic CSs. Since the distance to the Magellanic Cloud PNe is known, their CS mass can be estimated much more reliably than for Galactic PNe. In Fig. 4**(a)** we plot the evolutionary time of the Magellanic CSs, derived from their location in the HR diagram, against the dynamical time of their nebular shells. Stellar luminosities and temperatures used to determine the CSs location on the HR diagram with respect to the Magellanic post-AGB evolutionary tracks are from [6] and [14]; the nebular diameters are from [4] and [5], and the expansion velocities are taken from [15]. Note that the majority of LMC post-AGB evolutionary tracks are He-burning tracks [16], while there is an indication that these stars are H-burners (see below in the PNLF section). To compensate for this mismatch we have scaled the H- versus He-burning tracks and increased the evolutionary time-scales accordingly. Figure 4**(a)** shows that the dynamical and evolutionary times correlate with one another (R_{xy}=0.88), yet the dynamical time is usually higher, as predicted in most evolutionary cases [13]. There is one exception, LMC SMP 13, whose transition time is just below zero (t_{ev} and t_{dyn} are within their respective error bars). From the evolutionary and dynamical times we derive the transition time in all cases but for SMP 13, whose transition time is obviously close to zero, or, alternatively, whose evolution occurs in thermal time-scale [13]. In Fig. 4**(b)** we report the histogram of the transition times for the Magellanic Cloud PNe whose dynamical and evolutionary times are available. The values of the transition time spread from a few hundreds to several thousands years. In a few cases it goes to 10,000 yr, certainly a considerable fraction of the PN lifetime. The most important conclusion from this exercise is that transition time can not be ignored when dealing with the evolution of PNe and their central stars. This is the first time that transition time has been effectively estimated, and the Magellanic Cloud PNe are really the only PNe whose transition time may be estimated, given their known distances. A detailed analysis of the Magellanic Cloud PN estimated transition times and their models will be presented in a future paper.

5 The Planetary Nebula Luminosity Function

The planetary nebula luminosity function (PNLF) is a well-known secondary distance indicator (for a review and references see [17]). The high-luminosity cut-off of the PNLF seems to be invariant for galaxies with similar metallicities, while with a metallicity-correction it is also possible to unify the method for all galaxies. Despite its empirical success, it is still theoretically unclear why the PNLF is invariant for different PN population. Moreover, the evolutionary mechanisms behind the PNLF shape are still uncertain. In particular, the PNLF seems extremely hard to model. A realistic representation of the PNLF would involve several unknown and unconstrained parameters and processes related to the post-AGB and nebular evolution, such as the transition time, the mass-loss and its rates, the onset of PN morphology. The morphological database of Magellanic Cloud PNe may be used to gain insight into the PNLF. In Figs. 5 and 6 we plot the luminosity of the LMC and the SMC PNe against their photometric physical radii. The luminosity is given in [O III] λ5007 Å magnitudes, as typical for PNLFs, and the radii are in parsecs. All targets have been coded for morphological type, as described in the labels.

The snapshot survey targets plotted in Figs. 5 and 6 were selected from the brightest LMC and SMC PNe [4, 5], thus the completeness near the high luminosity cut-off should be adequate [17]. In both figures, but more clearly in Fig. 5, one can see two different *branches* of PNe, running from small to large radii. These branches decline in brightness for larger radii, and the B and P PNe populate the low brightness- large radii parts of the plots. The PNLF structure that derives from the figure has a dip at intermediate magnitudes, as predicted for H-burning post-AGB stars [13, 18].

More importantly, the high luminosity PNe in both figures are R, E, and BC. These are typically the low nitrogen and high carbon PNe, the progeny of the lower mass stars ($M<4\ M_\odot$). The usefulness of this information is multifold: when modeling the high luminosity parts of the PNLF one could set aside the more difficult modeling of bipolar PNe, since spherical and ellipsoidal approximation should suffice. Furthermore, the mass range of PN progenitors that involve the high luminosity cut-off of the PNLF may be limited to $M<4\ M_\odot$ in PNLF models.

The brightest PNe in the SMC are definitely R and E. The brightest B PNe are almost two magnitudes below the high luminosity cut-off. In the LMC, there is a considerable fraction of bright BC PNe. Are these the same BC PNe with high carbon and low nitrogen abundances as seen in Fig. 2? We have abundance information for three of the five bipolar core PNe in the brightest two magnitudes

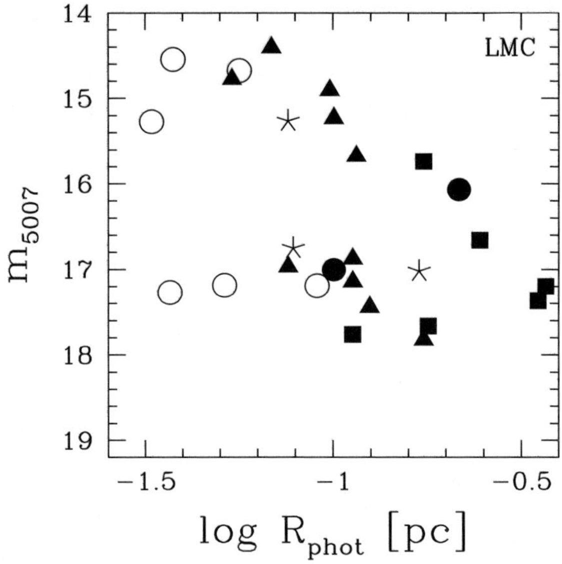

Fig. 5. Magnitude λ5007 Å versus photometric radius for R (open circles), E (asterisks), R and E with a bipolar core (triangles), B (squares), and P (filled circles) PNe in the LMC

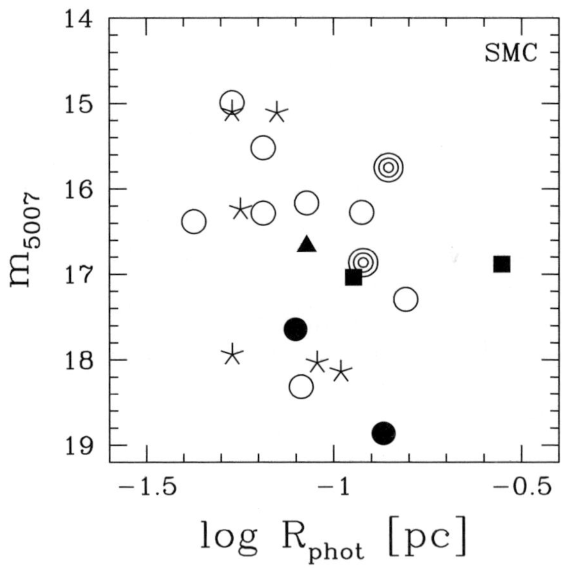

Fig. 6. Same as in Fig. 5, for SMC PNe. The location of the target-like symbols correspond to two previously misclassified ultra-compact H II regions

of the LMC PNLF, and these three objects have high carbon and low nitrogen content.

In Fig. 6 we plot all the known (spatially-resolved) SMC PNe and the two H II regions corresponding to young stellar clusters, that were previously misclassified as PNe [19], and were only identified as young stellar clusters after being spatially resolved with *HST*. The figure illustrates that misclassified PNe may occupy the same areas of the diagnostic m_{5007} - log R_{phot} plot. The brightest of these misclassified PNe is just one magnitude below the high luminosity cut-off of the SMC PNLF. In a galaxy particularly rich in these ultra-compact H II regions that are embedded in young stellar clusters, the distance determination based on the PNLF should be revisited after accounting for the contamination effects.

6 Finale

Magellanic Cloud PNe and their central stars have proven essential to probe post-AGB evolution and the ensemble properties of PNe, such as those related to the PNLF. The multi-wavelength database of Magellanic Cloud PNe is homogeneous, and now comprises spatially-resolved *HST* images of a representative fraction of PNe in these galaxies. This database allows unprecedented insight into stellar and nebula evolution. In this paper we showed that R, E, B and P nebular morphology exists in both the LMC and the SMC, and that their relative frequency is different, aspherical PNe being rare in the SMC. Morphology in the light of the ultraviolet C III] $\lambda5007$ Å line is similar to that of the major optical emission lines, such as Hα and [O III] $\lambda5007$ Å .

We have also shown that, as in the Milky Way, morphology correlates with nitrogen and carbon abundances, and that the aspherical PNe are typically carbon-poor and nitrogen-rich, as expected of the progeny of the high-mass AGB stars.

By comparing the rate of nebular and stellar evolution of Magellanic Cloud PNe it has been possible to measure the time-lag between the EEQ and the PN illumination. The resulting transition time is a sizable fraction of the nebular lifetime in some cases. We conclude that the transition time can not be ignored when comparing nebular and stellar models.

The PNe in the Magellanic Cloud PNLF can be studied in detail. The high luminosity PNe are generally R or E, and, in all the cases where abundances have been calculated, the high luminosity PNe are carbon-rich and nitrogen-poor. This result has bearing on the future modeling of the PNLF.

Some of the results outlined here will be more sound when the planned Magellanic Cloud PN data-set will reach completion. In particular, we plan to obtain UV spectra of several SMC PNe to extend the result of PN carbon abundances to the SMC. The transition time calculation will be extended to the whole sample of Magellanic Cloud PNe whose central stars have been observed, and more SMC central stars will be observed as planned. Finally, a thorough study of the Magellanic Cloud PNLF is about to be completed by the MCPN team with the inclusion of the optical images from recent *HST* observations. The characteristics of the central stars hosted by the brightest nebulae in each galaxy will be studied in detail for more astrophysical insight into the invariance of the PNLF high luminosity cut-off across galaxies.

It is a pleasure to thank the MCPN team for their contribution to this project. Arturo Manchado and the IAC are warmly thanked for their hospitality in July 2004, when this paper was written.

References

1. L. Stanghellini, E. Villaver, A. Manchado, & M. A. Guerrero: Ap. J. 576, 285 (2002)
2. M. Peimbert: IAU Symp. 76: Planetary Nebulae 76, ed. by Y. Terzian (Dordrecht: Reidel 1978) p. 215
3. L. Stanghellini, J. C. Blades, S. J. Osmer, M. J. Barlow, & X.-W. Liu: Ap. J. 510, 687 (1999)
4. R. A. Shaw, L. Stanghellini, M. Mutchler, B. Balick, & J. C. Blades: Ap. J. 548, 727 (2001)
5. L. Stanghellini, R. A. Shaw, B. Balick, M. Mutchler, J. C. Blades, & E. Villaver: Ap. J. 596, 997 (2003)
6. E. Villaver, L. Stanghellini, & R. A. Shaw: Ap. J. (in press)
7. L. B. van den Hoek & M. A. T. Groenewegen: Astron. Astrophys. Suppl. Ser. 123, 305 (1997)
8. G. Stasinska, M. G. Richer, and M. L. McCall: Astron. Astrophys. 336, 667 (1998)
9. P. Leisy & M. Dennefeld: Astron. Astrophys. Suppl. Ser. 116, 95 (1996)
10. P. Leisy & M. Dennefeld: Astron. Astrophys. Suppl. Ser. (submitted)
11. L. Stanghellini, R. A. Shaw, & D. Gilmore: Ap. J. (submitted)
12. A. Arrieta & L. Stanghellini: (this volume)
13. L. Stanghellini & A. Renzini: Ap. J. 542, 308 (2000)
14. E. Villaver, E., L. Stanghellini, & R. A. Shaw: Ap. J. 597, 298 (2003)
15. M. A. Dopita, S. J. Meatheringham, B. L. Webster, & H. C. Ford: Ap. J. 327, 639 (1988)
16. E. Vassiliadis & P. R. Wood: Ap. J. Suppl. Ser. 92, 125 (1994)
17. R. Ciardullo: (this volume)
18. R. H. Mendez & T. Soffner: Astron. Astrophys. 321, 898 (1997)
19. L. Stanghellini, E. Villaver, R. A. Shaw, & M. Mutchler: Ap. J. 598, 1000 (2003)

Central Stars of Planetary Nebulae in the Magellanic Clouds

Eva Villaver[1,2]

[1] Space Telescope Science Institute, 3700 San Martin Drive, Baltimore, MD 21218, USA; villaver@stsci.edu

[2] Affiliated with the Space Telescope Division of the European Space Agency

Abstract. In the last two decades a lot of effort has been dedicated to the study of Planetary Nebulae (PNe) in the two nearest and most easily observable Galactic companions: the Magellanic Clouds (MCs). However, there are only a few studies focused on the determination of the Central Star (CS) properties: luminosities, effective temperatures, and masses. A known distance to the Clouds offers a great advantage over the study of the CSs of Galactic PNe counterparts. However, the unresolved nature of the nebula from ground-based observations has so far prevented us from taking full advantage of a known distance in the determination of the CS parameters, since most ground-based studies have to rely heavily on photoionization modeling of the nebula in order to derive the CS parameters. Here I present a review of the studies of CSs of PNe taken in the MCs and the recent developments in the determination of the CS parameters using the high spatial resolution capabilities offered by the *Hubble Space Telescope* (HST).

1 Introduction

Stellar evolution predicts that stars in the mass range 1–8 $M_{\odot}$ will loose most of their stellar envelopes as they ascend the Asymptotic Giant Branch (AGB) phase. It is the mass-loss that establishes the mass boundary between those stars that will evolve off the AGB phase into Planetary Nebulae (PNe), and ultimately fade into white dwarfs, and those that will end their lives as type II Supernovae.

The final mass reached by low-and intermediate-mass stars, given a main-sequence mass (hereafter, the initial mass), depends mainly on the mass-loss taking place during the AGB phase which is barely understood. The mass-loss is thought to be driven mainly by a combination of dust formation, caused by the stellar pulsation and dust acceleration by radiation pressure [6,66]. The dust formation process depends on the chemical composition of the gas: the lower the metallicity, the smaller the amount of dust that is formed and the lower the expected mass-loss rate. Thus, low metallicity stars with dust-driven winds are expected to loose smaller amounts of matter [65]. The mass-loss during the AGB phase can occur even in the absence of dust, but its efficiency is also much lower [63,64]. As a consequence, a star born in a lower metallicity environment is expected to have reduced mass-loss rates and therefore a higher final mass than a star with the same initial mass born in a higher metallicity environment.

The Large and Small Magellanic Clouds (hereafter LMC and SMC respectively) metallicities are on average half and quarter that of the solar mix [44,45].

The accurate determination of CSs masses in both Clouds should, in principle, allow to directly probe stellar evolution and mass-loss under different metallicities environments.

In this review, I concentrate on the studies that determine the parameters that allow to locate the CSs of PNe in the Magellanic Clouds (MCs) in the Hertzsprung-Russell (H-R) diagram, and hence determine their masses. I discuss how photoionization models, required in the interpretation of ground-based spectroscopic studies have influenced previous determinations. Finally, I summarize the *HST* observing programs targeted at the CS, which have yielded the latest most accurate CS mass determinations.

2 Unresolved Observational Studies of CSs in the MCs

2.1 Ground-Based Optical Spectroscopy

The CS properties of PNe in the MCs are mostly derived from ground-based optical spectroscopic observations. A large fraction of PNe in the MCs have sizes smaller than 1".5 and thus are unresolved sources in ground-based observations. As a result, ground-based spectroscopy does not allow to separate the stellar continuum from the nebular one and so the CS luminosity and temperature have to be computed using modeling techniques that rely completely on nebular measurements.

Stellar temperatures of CSs in the MCs are normally measured using two different approaches: i.e. the energy-balance method [52], or the Zanstra method [67]. The energy-balance method only requires the measurement of nebular lines, whereas the Zanstra method requires stellar and nebular data.

The energy-balance method has been used to estimate CS temperatures of 44 CSs in the LMC and SMC from optical spectroscopic data [23,5]. The energy-balance method relies on the measurement of all collisionally excited lines, many of which fall on the ultraviolet (UV) and infrared (IR) spectral range. The contribution to the radiative cooling of unobserved UV and IR ions of C, O, S, and Ne can be significant. As a consequence, CS temperatures derived from the energy-balance method using optical spectra alone can be underestimated. The Zanstra method has been used with ground-based spectroscopic observations to obtain the CS temperature of a total of 85 PNe in the MCs [33,34]. The Zanstra method requires an estimate of the stellar continuum that in these observations could not be separated from the nebular continuum in the spectra. In order to estimate the stellar continuum they measured the total emission in the V-band from the spectra and then subtracted a nebular continuum emission computed from radiation transfer models.

Different approaches to derive the CS temperatures are also used. A modified energy-balance method that considers abundance variations has been used to determine the temperatures of 82 CSs in the MCs [12,13]. The stellar parameters of 9 CSs were determined using the crossover method [27], a modified Zanstra method [28,29].

In all the studies mentioned above the stellar luminosities are determined from the Hβ line flux of the nebula under the assumption that the nebula is optically thick. The nebular Hβ line flux is used as a measurement of the recombination and therefore the ionization flux from the star. The main problem of this approach to estimate the stellar luminosity is that nebulae are often optically thin.

To summarize, the determination of CS properties of PNe in the MCs from ground-based optical spectroscopic observations do not offer a significant advantage over the studies of CSs of PNe in the Galaxy. Although with MC PNe the uncertainty in the distance is removed, the problem that they cannot be resolved using ground-based observations still remains. As a consequence, most studies can only provide lower limits to the stellar luminosities. The situation for stellar temperature determinations is not better, as it is most likely that they have been underestimated when determined using the energy-balance method and overestimated when derived using the Zanstra method (or similar approaches).

2.2 UV-Based Determinations

UV spectroscopy allows to measure the emission of important nebular line coolants; i.e the carbon and silicon emission lines that are weak in the optical. In metal-poor systems, such us those in the MCs, the carbon emission lines might even be the main nebular coolants [50]. UV measurements allow to put additional constraints to the nebular photoionization models. This is a fundamental problem for the PNe in the MCs, because the determination of the CS parameters from optical data have to rely heavily on the photoionization modeling of the nebula. An additional advantage of using far-UV observations as opposed to optical observations, is that the stellar continuum dominates the nebular continuum.

PNe in the MCs are weak in the UV and therefore only a few bright objects have been detected with *International Ultraviolet Explorer* (IUE). The CS parameters for 12 PNe in the MCs have been determined [1], including the revision of the 3 objects published earlier [31,51]. Data for an additional SMC object is also available [32]. All these studies still need to use photoionization models to derive the CS parameters.

Recently, a more complete approach was under-taken in a study of 7 CSs in the LMC with the *Far Ultraviolet Spectroscopic Explorer* (FUSE) [24]. The FUSE spectra were modeled and the CS temperatures, gravities and luminosities were determined. However, in order to determine the CS parameters accurately, the nebular continuum (and the hydrogen absorption) needed to be estimated.

3 Resolved Studies: *HST* Observations

3.1 Targeting the Nebula: Narrow-Band Imaging

The first observations of PNe in the MCs taken with the *HST* were designed to measure the nebular emission and therefore were taken with narrow-band

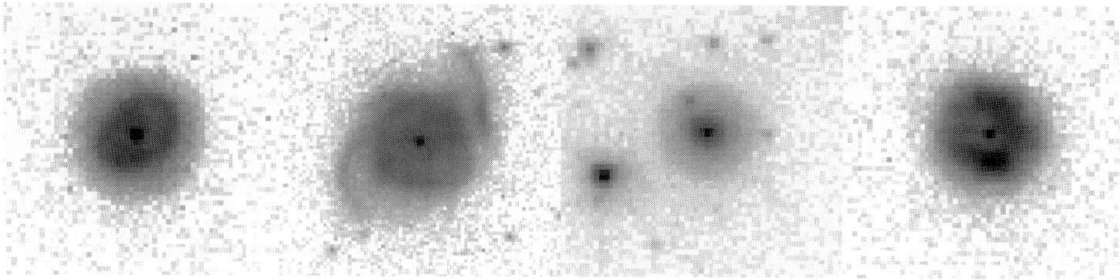

Fig. 1. A sample of *HST* images of LMC PNe [48]. From left to right: SMP 4, SMP 10, SMP 18, and SMP 102, Note that the CS is clearly detected. The images were obtained using the clear (50CCD) STIS bandpass. The field of view is 3 arcseconds and the intensity scale is logarithmic.

filters [14–16,57,17,58]. Although these observations allowed, for the first time, to resolve the PNe in the MCs, they did not yield a significant improvement in the determination of the CSs parameters compared to results based on ground-based measurements. This was because the determination of the CSs parameters still relies on the photoionization modeling of the ground-based spectra. The only improvement was that the resolved *HST* narrow-band images did help to constrain the photoionization models, mainly by imposing restrictions on the nebular size.

UV *HST* spectro-photometry was obtained for 14 objects using the *Faint Object Spectrograph* [57]. For these objects, the stellar temperatures from color indexes could not be determined because of the large uncertainties of the method. For one of the objects, the stellar wind parameters were measured from the P-Cygni emission line profile.

3.2 Targeting the Star: Broad-Band Photometry

The first sample of *HST* data of PNe in the MCs, aimed at determining the stellar continuum proved very successful [61,62]. Broad-band images with the *Space Telescope Imaging Spectrograph* (STIS), and medium-band F547M (Strömgren y) filter with the *Wide Field Planetary Camera 2* (WFPC2) were obtained for a sample of 27 and 35 PNe targets in the LMC and SMC respectively [48,50]. The high spatial resolution offered by *HST* allowed the nebular and stellar contributions to the emission to be separated. The stellar flux was then determined using standard aperture photometry techniques. Briefly, the flux within a circular aperture centered on the star was summed, and corrected for the line-of-sight nebular emission by subtracting the nebular flux obtained from an annulus surrounding the aperture.

The CS temperatures were determined using the Zanstra method. The Zanstra method determines the total ionizing flux of the star by comparing the flux of a nebular recombination line of hydrogen or helium with the stellar continuum flux in the V-band. The advantage over previous measurements is that the flux in the V-band can be determined accurately and independent of nebular modeling [61,62] . Figure 1 shows a sample of the data.

Table 1 shows the stellar parameters determined for a sample of CSs in the LMC [61] where in column (1) the PN name is given; in column (2) the morphological classification of the nebula (Round [R], elliptical [E], bipolar [B], quadrupolar [Q], bipolar core [bc] and point-symmetric [P]); in (3) the CS effective temperatures (in units of 10^3 K) derived from the Zanstra method for the He II line; in (4) the stellar luminosity, and in column (5) the CS mass. The average CS mass of the sample in Table 1 is $0.65\,M_{\odot}$, which is close to (although slightly higher than) the average mass of white dwarfs in the Galaxy [18,2]. Although, there does not seem to be any significant correlation between morphology and CS mass in the sample [61], it should be noted that there are very few asymmetric PNe with detected CSs.

Table 1. Stellar parameters for the LMC sample [61]

Name	Morphology	$T_{\rm eff}$ [10^3K]	$\log L_*/L_{\odot}$	M_c [$M_{\odot}$]
J 41	E(bc)	60.1±3.5	3.37±0.07	0.59
SMP 4	E	89.9±7.2	3.32±0.10	0.58
SMP 10	P	74.6±4.3	3.29±0.07	0.58
SMP 13	R(bc)	129.1±11.9	3.51±0.13	0.63
SMP 19	E(bc)	143.5±17.5	3.45±0.23	0.63
SMP 30	B	149.3±37.8	2.65±0.32	0.67
SMP 31	R	28.6± 2.4	3.62±0.10	0.59
SMP 34	E	67.8± 3.6	4.29±0.07	0.84
SMP 42	Q	66.4± 0.8	3.82±0.02	0.67
SMP 50	E(bc)	80.4± 0.9	4.11±0.03	0.75
SMP 52	R(bc)	100.9± 2.6	4.00±0.07	0.70
SMP 56	R	45.9± 2.0	3.89±0.05	0.68
SMP 59	Q	98.2± 6.8	3.86±0.09	0.65
SMP 63	R	38.8± 0.4	3.77±0.02	0.64
SMP 100	Q	127.1±11.7	3.47±0.13	0.63
SMP 102	E(Bc)	131.8±12.4	3.39±0.14	0.60

The advantage of using this data in the determination of the stellar parameters is that it allows one to eliminate the dependency on photoionization models in the determination of the stellar flux. Although it should be noted that significant uncertainty still remains in the determination of effective temperatures through the Zanstra method. A comparison between the CS parameters derived

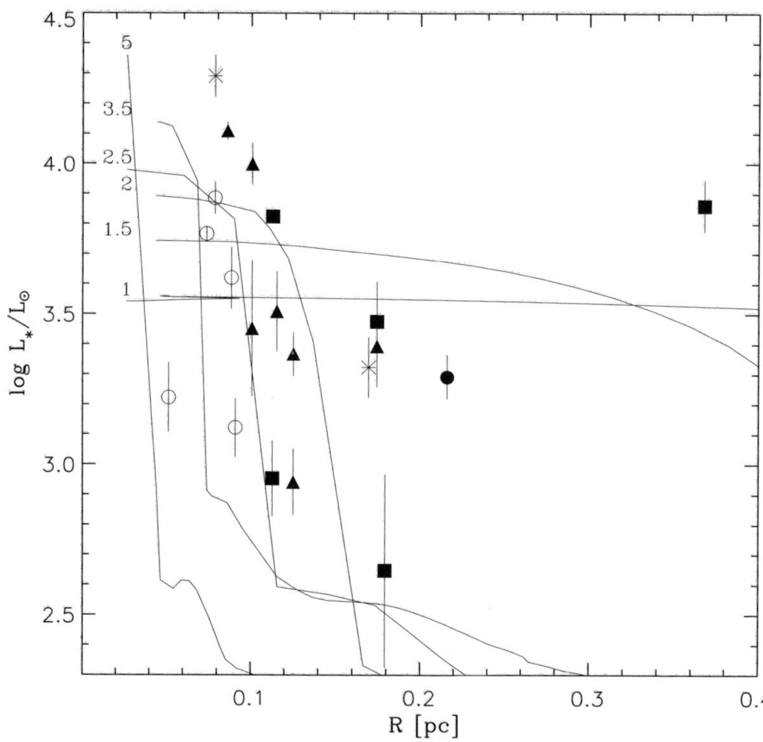

Fig. 2. The points represent the logarithm of the CS observed luminosity versus the physical radius of the nebulae [61] for LMC objects. The symbols represent the morphological type of the nebulae: round (open circles), elliptical (asterisks), bipolar and quadrupolar (squares), bipolar core (triangles) and point-symmetric (filled circles). The solid lines represent the evolution of the nebular radius with the stellar luminosity from PN models [59,60]. Each line has been marked with the initial mass of the progenitor used in the hydrodynamical simulations.

in [61,62] with those determined in the literature from optical ground-based spectroscopy shows that, as expected, lower stellar luminosities [23] and higher stellar temperatures [33,34] were determined using ground-based spectroscopy.

The Star-Nebula Relation: As an example of the kind of study these data allow, Figure 2 shows the relation between CSs luminosity and the PNe photometric radius for the LMC sample studied in [61]. The radius of the nebula and the CS luminosity are good indicators of the evolutionary status of the nebula and the CS respectively.

On the one hand, the understanding of the evolution of the nebular radius requires the solution of a gas dynamic problem, which depends, amongst other things, on the energy that the stellar wind injects into the gas, which is a function of the evolution of the stellar luminosity and the core mass. On the other hand, the stellar luminosity, after a constant luminosity phase for hydrogen-burners, decreases during the evolution at a rate that depends mainly on the core mass. Therefore, it is expected that both quantities plotted in Figure 2 decrease with time, and that they do not evolve independently.

As an indication of how the nebular evolution relates to the evolution of the CS, in Figure 2 the results from the numerical simulations of [59,60] have been superimposed. A qualitative comparison of the models and data shows that high luminosity CSs with large nebulae may have a low-mass progenitor. The CS luminosity decreases very fast for a high-mass progenitor and the radius remains small. Although a high-mass progenitor has a powerful wind that injects large amounts of energy into the gas, it does so for only a short length of time, because of the fast evolution of the CS in the HR diagram. As a consequence, the nebular radius remains small. Identical patterns of evolution as shown in Figure 2 are found for the CSs analyzed in the SMC [62].

A Comparison Between the LMC and SMC CS Samples Figure 3 shows the normalized histogram of the mass distribution obtained for the SMC and the LMC samples [62]. The samples are of comparable size. The number of measured CS masses is 14 and 16 for the SMC and LMC respectively. From the mass distributions it seems that the SMC hosts a larger fraction of low-mass CSs compared to the LMC and the SMC sample lacks a population of stars between 0.65 $M_{\odot}$ and 0.75 $M_{\odot}$ [62]. The Kolmogorov-Smirnov test (KS-test) shows that LMC and SMC CSs mass distributions are significantly different. If the initial mass distribution of CSs in the galaxies in the 1-5 $M_{\odot}$ range were the same, one would expect to find higher final masses in the SMC compared to in the LMC, a consequence of the reduced mass-loss rate expected in a lower metallicity environment, since there is no evidence for a dependency of the initial mass function with metallicity [46,47]. However, this is not the case, and it seems that the SMC hosts more CSs with lower masses.

The history of star formation in the LMC and SMC are different [38]. The LMC experienced an episode of star formation ∼3-5 Gyr ago [3], while the SMC seems to have been forming stars at a constant rate during the last 2-12 Gyr [9]. It appears that stars with ages between 7.8×10^{8} yr and 2.7×10^{8} yr are abundant in the LMC, but missing in the SMC [56,62]. This latter population may correspond to a burst of star formation in the LMC. Thus, if the initial mass-function is the same, then the star formation history differences may account for the differences in the observed mass distribution.

The kinematics of a given stellar population provides an additional age indicator. The kinematics of the SMC PNe is that of a spheroidal population without rotation [10] an indication that the PNe population in the SMC is kinematically old. A study of Carbon stars, another intermediate-age stellar population, affords a similar scenario [21,22], in agreement with the small CS masses found for SMC PNe.

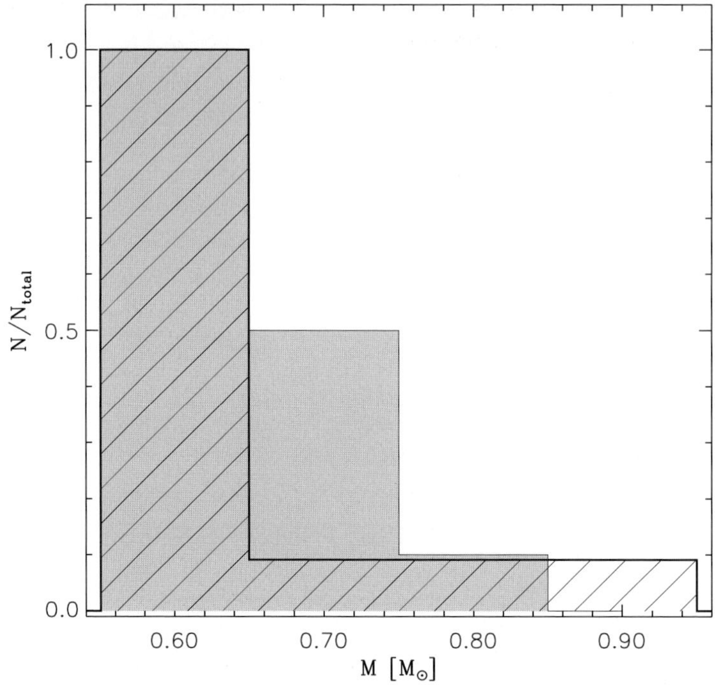

Fig. 3. Normalized mass distribution for the CSs in the SMC (dashed) and LMC (in gray) [62].

4 CSs with Wolf-Rayet Nuclei

Wolf-Rayet (WR) type spectra have been reported only in a handful of the CSs of PNe in the MCs. These are: SMC SMP 6, and MG 8 [36]; LMC SMP 38, SMP 58 [36]; SMP 61 [4]; LM1-64 [41]; MA 17 [19]; and SMP 83 [54]. Extensive studies of the CSs with IUE and optical spectroscopy [42] show that most of them have medium to low excitation classes (WC-4 to WC-8 spectral-types). Recently, two possible evolutionary scenarios have been proposed to account for the WR type spectra in CSs of PNe; either the star experienced a late thermal pulse during the post-AGB evolution or a thermal pulse at the end of the AGB evolution [26,25].

It is worth mentioning that the distribution of CS WR spectral types is different between the Galaxy and the MCs. The intermediate [WC] CS spectral class, common in the MCs, are not found among the WR CSs in the Galaxy [35,55]. Differences have also been found between the massive WR stars in the Galaxy and the MCs [7]. In particular, the massive WR stars have been found to be underabundant in the SMC. The under-abundance of massive WR stars in the SMC has been interpreted as an effect of a reduction of the mass-loss rates due to the lower metallicity in the SMC environment [30]. It would be interesting to see if the observed difference between the Galaxy and the MCs still holds, when a larger number of WR-CS in the MCs are analyzed.

4.1 LMC SMC 83

SMP 83 (N66) in the LMC has a very puzzling nature and probably is the only extra-Galactic CS of a PN that has had devoted studies over the last years. Early optical ground-based spectroscopy [11] revealed its extreme nitrogen abundances, classifying it as a Peimbert's type I object [39] and suggesting that it might have evolved from a massive star. The mass of the CS was found to be close to the Chandrasekhar mass limit, one of the largest masses found in a PNe CS, $\sim 1.2 M_{\odot}$. The effective temperature was derived from the ionization degree of the nebula ($T_{\rm eff}$ = 180,000 K) and luminosity (logL/$L_{\odot}$ = 4.7) [11], although a smaller value for the effective temperature was found using the Zanstra method (80,000 K) [11].

A high excitation type for the nebula and CS parameters in clear disagreement with the previous measurements have been reported [40]. The mass was found to be 0.64 $M_{\odot}$ (which is in the usual range for CSs of PNe and much smaller than the previous determination), $T_{\rm eff}$ = 90,000 K and $\log L/L_{\odot}$ =3.97. The effective temperature in this case was determined using the Stoy energy-balance method [43]. *HST* imaging and optical and UV spectroscopy were combined to generate a photoionization model [14]. The *HST* images constrained the model dependent parameters (i.e the nebular radius) slightly better and very similar parameters for the CS were obtained as in a previous study [14].

The extremely unusual nature of this object was first revealed in reported changes in the optical spectrum of the CS [53]. Between 1993 and 1994 the brightness of the CS increase dramatically, returning to its original level a few years later. SMP 83 seems to be a extremely massive object which has been classified as a Wolf-Rayet nuclei with nitrogen-rich composition (WN-type) [54], which is another of the peculiarities of this object, as most CSs of PNe with Wolf-Rayet spectra show carbon-rich composition (WC-type).

Serendipitous observations of this PN with *HST* confirmed the constant nature of the nebular flux and the exponential increase of the stellar flux with time [57]. More recently, from the analysis of the time-dependent line spectra obtained with STIS, the CS has been found to be most likely a binary system, in which a white dwarf accretes matter from a non-degenerate companion at a high rate [20]. Studies of the CSs of objects such us SMP 83 are fundamental to shed light on the possible evolutionary paths of common stars.

5 Summary

Studies of CSs in the Galaxy are hampered by the poor knowledge of their distances, a problem that can be overcome by observing PNe in the Magellanic Clouds. However, one problem that still remained is the determination of the

CS parameters, due to the unresolved nature of MC PNe from the ground-based observations. Ground-based optical spectroscopy needs to be combined with photoionization modeling in order to yield the stellar parameters. As a consequence, most of the determination of the CSs of PNe masses are highly model-dependent, i.e., they are based on CSs luminosities derived solely from Hβ nebular fluxes [23,36,29,16,17,58].

In all, only four CS masses were previously determined from direct measurement of the stellar flux, and then only from the UV spectrum of the CS [14,24] where nebular modeling is also needed in order to estimate the stellar continua.

Recent *HST* observations, designed to measure the stellar continuum directly, allow the dependence on photoionization modeling in the determination of the stellar flux to be eliminated. Two samples of CS masses in the SMC and LMC, free from photoionization modeling uncertainty, are available [61,62]. The number of objects is small, and therefore the distribution of the objects in the HR diagram has limited statistical value. However, these masses are extremely valuable and can be used in statistical studies, because the original samples were selected in a similar fashion and were observed with the same *HST* instrument and observing configuration. The average mass of the CSs in the SMC is 0.63 $M_{\odot}$, which is similar to the mass obtained for a sample of CSs in the LMC (0.65 $M_{\odot}$) and similar to the CSs in the Galaxy. However, the SMC and LMC CS mass distributions differ, the SMC sample lacks an intermediate-mass stellar population (0.65 to 0.75 $M_{\odot}$), which cannot be explained by a mass-loss dependency with metallicity.

It is a pleasure to thank Letizia Stanghellini and R.A. Shaw for their invaluable contribution to this work.

References

1. Aller, L. H., Keyes, C. D., Maran, S. P., Gull, T. R., Michalitsianos, A. G., & Stecher, T. P. 1987, ApJ, 320, 159
2. Bergeron, P., Saffer, R. A., & Liebert, J. 1992, ApJ, 394, 228
3. Bertelli, G., Mateo, M., Chiosi, C., & Bressan, A. 1992, ApJ,
4. Bianchi, L., Vassiliadis, E., & Dopita, M. 1997, ApJ, 480, 290
5. Boroson, T. A., & Liebert, J. 1989, ApJ, 339, 844
6. Bowen, G. H. 1988, ApJ, 329, 299
7. Conti, P. S. & Massey, P. 1989, ApJ, 337, 251
8. Crowl, H. H., Sarajedini, A., Piatti, A. E., Geisler, D., Bica, E., Clariá, J. J., & Santos, J. F. C. 2001, AJ, 122, 220
9. Dolphin, A. E., Walker, A. R., Hodge, P. W., Mateo, M., Olszewski, E. W., Schommer, R. A., & Suntzeff, N. B. 2001, ApJ, 562, 303
10. Dopita, M. A., Lawrence, C. J., Ford, H. C., & Webster, B. L. 1985, ApJ, 296, 390
11. Dopita, M. A.,Ford, H. C., & Webster, B. L. 1985, ApJ, 297, 593
12. Dopita, M. A. & Meatheringham, S. J. 1991a, ApJ, 367, 115
13. Dopita, M. A. & Meatheringham, S. J. 1991b, ApJ, 377, 480
14. Dopita, M. A., Ford, H. C., Bohlin, R., Evans, I. N., & Meatheringham, S. J. 1993, ApJ, 418, 804

15. Dopita, M. A., et al. 1994, ApJ, 426, 150
16. Dopita, M. A., et al. 1996, ApJ, 460, 320
17. Dopita, M. A., et al. 1997, ApJ, 474, 188
18. Finley, D. S., Koester, D., & Basri, G. 1997, ApJ, 488, 375
19. Hamann, W.-R. 1996, ASP Conf. Ser. 96: Hydrogen Deficient Stars, 127
20. Hamann, W.-R., Peña, M., Gräfener, G., & Ruiz, M. T. 2003, A&A, 409, 969
21. Hardy, E., Suntzeff, N. B., & Azzopardi, M. 1989, ApJ, 344, 210
22. Hatzidimitriou, D., Croke, B. F., Morgan, D. H., & Cannon, R. D. 1997, A&AS, 122, 507
23. Henry, R. B. C., Liebert, J., & Boroson, T. A. 1989, ApJ, 339, 872
24. Herald, J. E. & Bianchi, L. 2004, ApJ, 611, 294
25. Herwig, F. 2001, Ap&SS, 275, 15
26. Herwig, F., Blöcker, T., Langer, N., & Driebe, T. 1999, A&A, 349, L5
27. Jacoby, G. H., & Kaler, J. B. 1993, ApJ, 417, 209
28. Kaler, J. B. & Jacoby, G. H. 1989, ApJ, 345, 871
29. Kaler, J. B. & Jacoby, G. H. 1989, ApJ, 362, 491
30. Maeder, A. & Conti, P. S. 1994, ARA&A, 32, 227
31. Maran, S. P., Aller, L. H., Gull, T. R., & Stecher, T. P. 1982, ApJ, 253, L43
32. Meatheringham, S. J., Maran, S. P., Stecher, T. P., Michalitsianos, A. G., Gull, T. R., Aller, L. H., & Keyes, C. D. 1990, ApJ, 361, 101
33. Meatheringham, S. J. & Dopita, M. A. 1991, ApJS, 76, 1085
34. Meatheringham, S. J. & Dopita, M. A. 1991, ApJS, 75, 407
35. Mendez, R. H. & Niemela, V. S. 1982, IAU Symp. 99: Wolf-Rayet Stars: Observations, Physics, Evolution, 99, 457
36. Monk, D. J., Barlow, M. J., & Clegg, R. E. S. 1988, MNRAS, 234, 583
37. Morgan, D. H., Vassiliadis, E., & Dopita, M. A. 1991, MNRAS, 251, 51P
38. Olszewski, E. W., Suntzeff, N. B., & Mateo, M. 1996, ARA&A, 34, 511
39. Peimbert, M. 1978, IAU Symp. 76: Planetary Nebulae, 76, 215
40. Peña, M. & Ruiz, M. T. 1988, Revista Mexicana de Astronomia y Astrofisica, 16, 55
41. Peña, M., Olguin, L., Ruiz, M. T., & Torres-Peimbert, S. 1994, Revista Mexicana de Astronomia y Astrofisica, 28, 27
42. Peña, M., Ruiz, M. T., & Torres-Peimbert, S. 1997, A&A, 324, 674
43. Preite-Martinez, A, & Pottasch, S. R. 1983, A&A 126, 31
44. Russell, S. C. & Bessell, M. S. 1989, ApJS, 70, 865
45. Russell, S. C. & Dopita, M. A. 1990, ApJS, 74, 93
46. Salpeter, E. E. 1955, ApJ, 121, 161
47. Scalo, J. 1998, ASP Conf. Ser. 142: The Stellar Initial Mass Function (38th Herstmonceux Conference), 201
48. Shaw, R. A., Stanghellini, L., Mutchler, M., Balick, B., & Blades, J. C. 2001, ApJ, 548, 727
49. Stanghellini, L., Shaw, R. A., Mutchler, M., Palen, S., Balick, B., & Blades, J. C. 2002, ApJ, 575, 178
50. Stanghellini, L., Shaw, R. A., Balick, B., Mutchler, M., Balick, B., & Blades, J. C., Villaver, E. 2003, ApJ, 596, 997
51. Stecher, T. P., Savedoff, M. P., Maran, S. P., Gull, T. R., & Aller, L. H. 1982, ApJ, 262, L41
52. Stoy, R. H. 1933, MNRAS, 93, 588
53. Torres-Peimbert, S., Peimbert, M., Ruiz, M. T., & Pena, M. 1993, IAU Symp. 155: Planetary Nebulae, 155, 584

54. Torres-Peimbert, S., Ruiz, M. T., & Maza, J. 1995, ApJ, 441, 343
55. Tylenda, R., Acker, A., & Stenholm, B. 1993, A&AS, 102, 595
56. Vassiliadis, E.,& Wood, P. R. 1994, ApJS, 92, 125
57. Vassiliadis, E., et al. 1996, ApJS, 105, 375
58. Vassiliadis, E. et al. 1998, ApJ, 503, 253
59. Villaver, E., García-Segura, G., & Manchado, A. 2002a, ApJ, 571, 880
60. Villaver, E., Manchado, A., & García-Segura, G. 2002b, ApJ, 581, 1204
61. Villaver, E., Stanghellini, L., & Shaw, R. A. 2003, ApJ, 597, 298
62. Villaver, E., Stanghellini, L., & Shaw, R. A. 2004, ApJ, in press
63. Willson, L. A., Bowen, G. H., & Struck, C. 1996, ASP Conf. Ser. 98: From Stars to Galaxies: the Impact of Stellar Physics on Galaxy Evolution, 197
64. Willson, L. A. 2000, ARA&A, 38, 573
65. Winters, J. M., Le Bertre, T., Jeong, K. S., Helling, C., & Sedlmayr, E. 2000, A&A, 361, 641
66. Wood, P. R. 1979, ApJ, 227, 220
67. Zanstra, H. 1931, Publ. Dom. Astrophys. Obs. Victoria, 4, 209

Stellar Winds of Central Stars of the LMC PNe

Anabel Arrieta[1] and Letizia Stanghellini[2]

[1] Universidad Iberoamericana, Departamento Física y Matemáticas, Prol. Paseo de la Reforma 880, México, D.F.

[2] National Optical Astronomy Observatory, 950 North Cherry Avenue, Tucson, AZ 85719

1 Introduction

The radiation driven wind theory states that the modified momentum $\Pi = \dot{M}V_{\infty}R^{0.5}$ is proportional to the stellar luminosity. If this relation is confirmed, it can be used as a distance indicator for the central stars (CSs) of the Galactic planetary nebulae. This relation for the massive O stars was established in [3] and [4]. An attempt to extend the relation to the Galactic central stars in planetary nebulae (CSPNe) did not show any convincing relation [6].

In this work we investigate the relation between the modified wind momentum Π versus the luminosity of the CSPNe in the Large Magellanic Cloud (LMC) taking advantage of the less uncertain distances to these objects. We show that these winds obey a similar relation as the Galactic O stars, thus are subject to the same physics, and that their momentum can be used as a distance indicator.

2 Observations and Analysis

The present work is based on observations designed to study the nebular physics of planetary nebulae in the LMC. The observations were made with Space Telescope Imaging Spectrograph (STIS) in ultraviolet slitless spectroscopy mode, with gratings G140L, G230. The data are described in [5]. Here we have selected 12 of the target nebulae, those that show a stellar continuum in the UV spectra.

The 1D extraction from the 2D spectral images were performed using a small aperture around the location of the central star spectra. The spectra were then corrected for the Galactic and LMC extinction. The temperature and radius were derived by the approximation of the stellar continuum with a black body using a method described in [1]. The luminosities were calculated base on these parameters and they are in agreement with those derived by Zanstra analysis for the same objects [7]. The nebular contribution to the continuum was estimated via CLOUDY modeling.

Some of the objects show P Cygni profiles corresponding to the C IV 1548/1551 Å, Si IV 1394/1403 Å and N V 1239/1243 Å lines. We fit these lines using a line transfer code together with a genetic minimization algorithm [2]. As a result we obtained the mass loss rates and the terminal velocities for these stars.

3 Results

At this point of our study is encouraging to observe that the modified wind momenta and luminosities of the LMC CSs broadly agree with the same relation of the Galactic massive O stars. Refinement of our analysis and a larger sample of LMC CSs will very likely nail down this relation in CSPNs. If one assume that the winds of the CSPNe in LMC should be similar to the winds in our galaxy, then the relation between the wind momentum and the stellar luminosity shown in Fig. 1 can be used as a distance indicator. We are inclined to infer from our results that the absence of such a correlation in the work by [6] may be due to the large errors that characterize Galactic PN distances, thus the derived luminosity of their central stars.

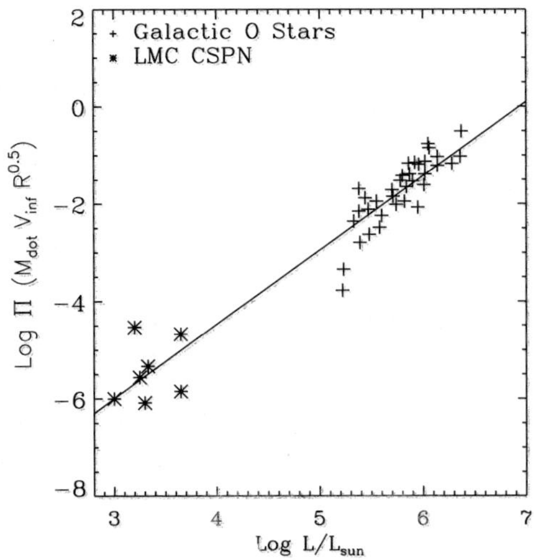

Fig. 1. The modified wind momentum versus the stellar luminosity for Galactic O stars and central stars of the LMC PNe

References

1. C. de Jager: *The brightest stars* (Dordrecht, Boston U.S.A. 1980)
2. L. Georgiev, X. Hernández: Rev.Mex.A.A. submitted (2004)
3. R.P. Kudritzki et al.: 'Winds in the atmospheres of central stars of Planetary Nebulae'. In: *Planetary nebulae,180th Symp. of the IAU, The Netherlands, August 26-30,1996*. ed. by H.J. Habing, H.J.G.L.M. Lamers (Dordrecht,Boston 1997) pp.64–73
4. H.J.G.L.M. Lamers, C. Leitherer: ApJ **412**, 771 (1993)
5. L. Stanghellini, R.A. Shaw, D. Karakla: in preparation (2004)
6. C.M. Tinkler, H.J.G.L.M. Lamers: A&A **384**, 987 (2002)
7. E. Villaver, L. Stanghellini, R.A. Shaw: ApJ **593**, 298 (2003)

Parameters of PNe: Constant Density versus Density Distribution

Michaela Kraus

Astronomical Institute, Utrecht University, Princetonplein 5,
3584 CC Utrecht, The Netherlands

Abstract. We derive the stellar and circumstellar parameters of the Galactic compact planetary nebula Hen 2-90 with two models: the classical constant density nebula model, where the emission (especially of the forbidden lines) is formed in a sphere or shell of constant density; and a model where the emission is formed in a shell with a r^{-2} density distribution. Depending on the density $N_{\mathrm{e},0}$ in the second model, the resulting range of valid values can deviate from the classical model results. This seems to be especially true for elemental abundances as is shown for example for S^+. We find that the abundance of S^+ is larger for the models with density distribution than for a constant density. A careful analysis of data with an appropriate input model is advisable when determining stellar and circumstellar parameters of planetary nebulae, not only in our galaxy.

1 Introduction

The parameters of a star and its circumstellar material are difficult to determine, especially if the star itself is hidden by its circumstellar material. In this case, the parameters must be derived indirectly, for instance from the Balmer and other nebular emission lines. Especially for planetary nebulae (PNe) there exist some simple approximations to derive the effective temperature, distance or elemental abundances. But these approximations are normally based on the assumption of a constant density spherically symmetric nebula. Although the theory of interacting stellar winds that shape the nebulae and lead to a density distribution rather than to a constant density, which has first been proposed by Kwok et al. (1978), is nowadays accepted and well established, people still use the assumption of a constant density nebula to derive stellar and circumstellar parameters (see e.g. Costa et al. 1993; Ali 1999; Phillips 2003; Gruenwald & Viegas 2000; Villaver et al. 2003, Stanghellini et al. 2004). We investigate differences in the stellar and circumstellar parameters when applying a constant density model and a model with a density distribution.

2 Observations and Models

To determine the stellar and nebular parameters for the Galactic compact planetary nebula Hen 2-90 we make use of a set of optical observations specified in Table 1. These data have been corrected for an interstellar extinction of

Table 1. Line integrated fluxes of some optical emission lines from the Galactic compact planetary nebula Hen 2-90 and continuum flux at the wavelength of the Hβ line. The data have been extinction corrected with $A_{\rm V} = 4.2^{\rm mag}$ as derived from the Balmer line ratio for $T_{\rm e} = 10^4$ K. The distance has been set to 2 kpc.

Line	λ [Å]	line integrated flux
Hα	6563	5.01×10^{-10} erg s^{-1} cm^{-2}
Hβ	4861	1.73×10^{-10} erg s^{-1} cm^{-2}
[SII]	6731	1.47×10^{-11} erg s^{-1} cm^{-2}
[SII]	6716	6.54×10^{-12} erg s^{-1} cm^{-2}
continuum	4861	1.89×10^{-24} erg s^{-1} cm^{-2} Hz^{-1}

$A_{\rm V} = 4.2^{\rm mag}$ derived from the Balmer line ratio for $T_{\rm e} = 10^4$ K. We concentrate on two specific models: the constant density nebula model (CDNM) and a density distribution model (DDM). Both models have the following assumptions in common:

- spherically symmetric fully ionized shell or nebula,
- constant electron temperature fixed at 10^4 K,
- all lines in Table 1 are optically thin, and
- hydrogen follows case B recombination.

In addition, for CDNM we use a constant electron density throughout the nebula, while for DDM we assume a density distribution of the form

$$N_{\rm e}(r) = N_{\rm e,0}\frac{R_0^2}{r^2} \tag{1}$$

$N_{\rm e,0}$ is the electron density at the inner edge, R_0, of the emission shell and is kept as a free parameter. Such a density distribution has been found from hydrodynamic models to be valid over large regions within the wind interaction zone (see e.g. Steffen & Schönberner, 2003).

3 Results

3.1 The Effective Temperature

The effective temperature of the central star of a PN is usually derived using the Zanstra method for hydrogen (see e.g. Osterbrock 1989; Pottasch 1984) which results in

$$\frac{L_\nu}{\int\limits_{\nu_0}^{\infty}\frac{L_\nu}{h\nu}\,d\nu} = \frac{L_\nu}{\frac{L_{\rm H\beta}}{h\nu_{\rm H\beta}}}\,\frac{\frac{L_{\rm H\beta}}{h\nu_{\rm H\beta}}}{\int\limits_{\nu_0}^{\infty}\frac{L_\nu}{h\nu}\,d\nu} = h\nu_{\rm H\beta}\frac{F_\nu^{\rm obs}}{F_{\rm H\beta}^{\rm obs}}\,f(N_{\rm e}) \tag{2}$$

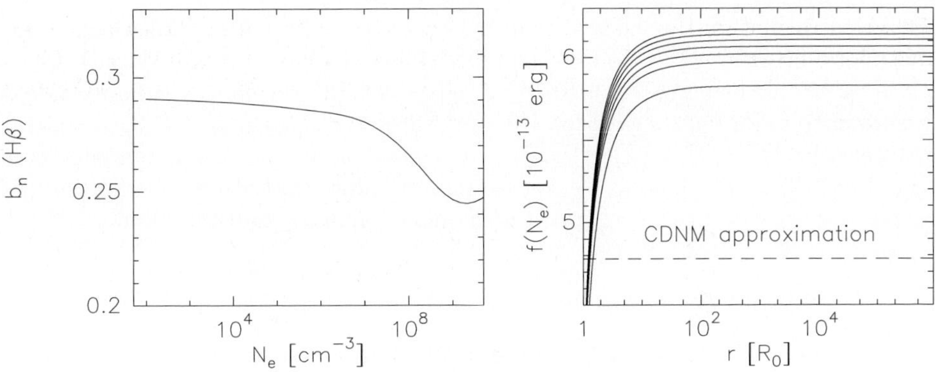

Fig. 1. Left: b_n factor for Hβ as a function of electron density. For low values of N_e, it is fairly constant. Right: The function $f(N_e)$ of Eq. (2) for values of $N_{e,0}$ ranging from $10^2\,cm^{-3}$ (top) to $10^8\,cm^{-3}$ (bottom). Also shown is the CDNM value (dashed line)

The left hand side is the continuum luminosity (e.g. at the wavelength of Hβ) over the total number of ionizing photons. This is a pure function of the stellar continuum flux, i.e. a function of effective temperature and surface gravity. The right hand side contains the (reddening independent) fraction of observed fluxes times a function, $f(N_e)$, which is a constant in case of CDNM, but a function of electron density distribution in case of DDM because the b_n values that describe the deviation from LTE of the Balmer line level populations are no longer constant for high electron densities (left panel of Fig. 1). The right panel of Fig. 1 shows $f(N_e)$ for different values of $N_e(R_0)$ ranging from $10^2\,cm^{-3}$ (top) to $10^8\,cm^{-3}$ (bottom) as a function of r. The range of possible effective temperature and surface gravity combinations of the central star derived with

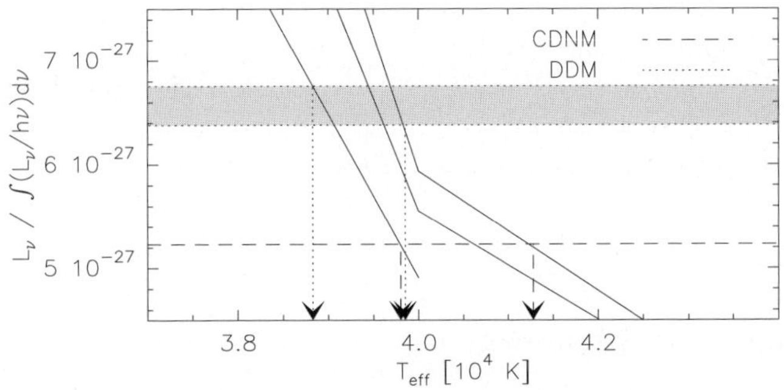

Fig. 2. Range of effective temperatures derived with the Zanstra method for H in case of CDNM (dashed arrows) and DDM (dotted arrows) for the $N_{e,0}$ values of Fig. 1. The top boundary of the shaded region is for $N_{e,0} = 10^2\,cm^{-3}$, the bottom boundary for $N_{e,0} = 10^8\,cm^{-3}$. The solid lines are Kurucz models for the stellar $\log g = 5.0; 4.5; 4.0$ (from right to left)

the Zanstra method for H is shown in Fig. 2 where we used Kurucz model atmospheres (Kurucz 1979) to calculate the left hand side of Eq. (2). While CDNM gives values in the range of $39\,800 \leq T_{\rm eff}/{\rm K} \leq 41\,300$, the range found for DDM is shifted to lower values covering the range $38\,800 \leq T_{\rm eff}/{\rm K} \leq 39\,850$; the lowest temperature is for the lowest (chosen) $N_{\rm e,0}$ value of $10^2\,{\rm cm}^{-3}$. The temperatures found with the two models are not extremely different; both methods result in a spectral type O for the central star of this compact planetary nebula.

3.2 Elemental Abundances

We take for example the [S II] lines and calculate for both models the intensity ratio of a [S II] line over Hβ (for details see e.g. Pottasch et al. 2003). For CDNM we need to know the constant electron density. This can be found from the sulfur line ratio (left panel of Fig. 3) and turns out to be $4 \times 10^4\,{\rm cm}^{-3}$. The resulting ratio of the two [S II] lines over Hβ is shown in the left panel of Fig. 4. From this plot, an S^+ abundance of 8.75×10^{-8} is found.

In case of DDM the observed sulfur line ratio defines the lower limit of $N_{\rm e,0}$ (right panel of Fig. 3) which is about $10^5\,{\rm cm}^{-3}$. Models with $N_{\rm e,0} < 10^5\,{\rm cm}^{-3}$ cannot reproduce the observed line ratio of 2.25. In addition, the crossing point of the theoretical line ratio curve with the observed ratio defines the outer edge (in terms of R_0) of the emission shell for each density distribution model. This outer edge has to be taken into account when calculating the intensity ratio of [S II] over Hβ. The results are shown in the right panel of Fig. 4. We find that the S^+ abundance values calculated with DDM are all larger than the value found with CDNM, with the lowest value of 1.64×10^{-7} for $N_{\rm e,0} = 10^5\,{\rm cm}^{-3}$ which is about twice the CDNM value.

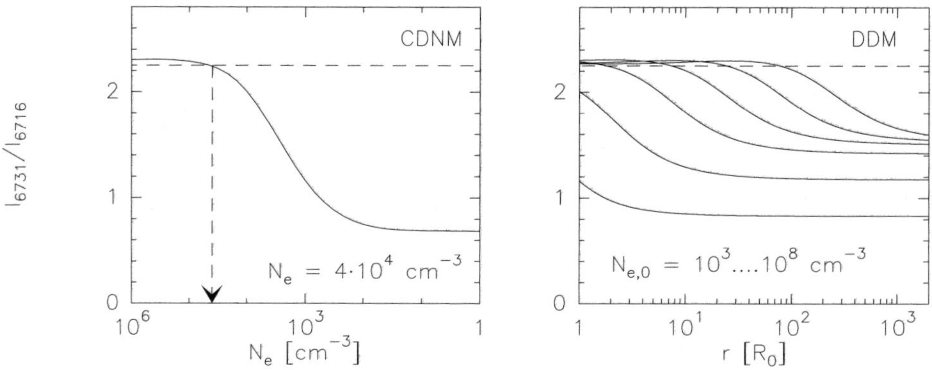

Fig. 3. Left: Constant electron density in CDNM derived from the observed (dashed line) sulfur line ratio. Right: The sulfur line ratios in case of DDM for $N_{\rm e,0}$ ranging from $10^3\,{\rm cm}^{-3}$ (lower left) to $10^8\,{\rm cm}^{-3}$ (upper right). The observed ratio (dashed line) limits $N_{\rm e,0}$ to values $\gtrsim 10^5\,{\rm cm}^{-3}$. The crossing point between model and observation defines the outer edge of the emitting shell for each model

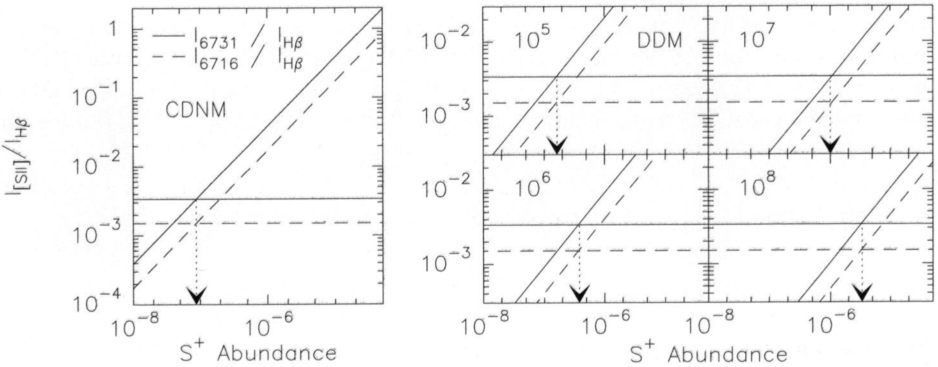

Fig. 4. [S II] over Hβ intensity ratio as a function of S^+ abundance for CDNM (left) and DDM (right) for the different possible $N_{e,0}$ values as indicated. Horizontal lines indicate observed ratios, arrows indicate the derived S^+ abundance. All abundances found with DDM are clearly larger than that for CDNM.

4 Conclusions

We derived the effective temperature and the S^+ abundance for the Galactic compact PN Hen 2-90 using the constant density nebula approach and a shell model with an r^{-2} density distribution. We find that the effective temperature of the central star derived with DDM is slightly (but not significantly) lower than with CDNM, classifying the star as spectral type O. The most interesting result is, however, that the S^+ abundance found with DDM is definitely larger than the value found with CDNM. Even though we neglected additional ionization stages, this result might hold for all abundances derived from forbidden lines and might solve the problem of the underabundances sometimes found in PNe.

References

1. A. Ali: New Astronomy **4**, 95 (1999)
2. R.D.D. Costa, J.A. de Freitas Pacheco, W.J. Maciel: A&A **276**, 184 (1993)
3. R. Gruenwald, S.M. Viegas: ApJ **543**, 889 (2000)
4. R.L. Kurucz: ApJS **40**, 1 (1979)
5. S. Kwok, S., C.R. Purton, P.M. FitzGerald: ApJ **219**, L 125 (1978)
6. D.E. Osterbrock: *Physics of gaseous nebulae and active galactic nuclei* (University science books, Mill Valley 1989)
7. J.P. Phillips: MNRAS **344**, 501 (2003)
8. S.R. Pottasch: *Planetary nebulae* (D.Reidel Publishing Company, Dordrecht 1984)
9. S.R. Pottasch at al: A&A **409**, 599 (2003)
10. L. Stanghellini et al: ApJ *in press*, astro-ph/0411631 (2004)
11. M. Steffen, D. Schönberner: *Planetary Nebulae: Their Evolution and Role in the Universe*, IAU Symposium 209, ed: S. Kwok, M. Dopita, R. Sutherland, (ASP, San Francisco, 2003), 439
12. E. Villaver, L. Stanghellini, R.A. Shaw: ApJ **597**, 298 (2003)

Part V

Physical Conditions of PNe

Plasma Diagnostics and Elemental Abundance Determinations for PNe – Current Status

Xiao-wei Liu

Department of Astronomy, Peking University, Beijing 100871, P. R. China

1 Overview

Much of our knowledge of elemental abundances in our own and other galaxies rests on photoionized gaseous nebulae, such as H II regions and planetary nebulae (PNe). The current paper deals with the two related long-standing problems in nebular astrophysics, whereby electron temperatures derived from the hydrogen recombination spectrum Balmer discontinuity are systematically lower than values deduced from the collisionally excited [O III] nebular to auroral line ratio, and whereby heavy element abundances relative to hydrogen deduced from optical recombination lines (ORLs) are systematically higher than those determined from collision excited lines (CELs). We concentrate on results from bright Galactic PNe. Observations of extragalactic PNe are scarce due to the difficulty of detecting faint ORLs from heavy element ions. The situation is however changing rapidly with the wide deployment of 10m class telescopes and sophisticated spectrographs.

Deep optical spectrophotometry, allowing plasma diagnostics and abundance determinations using weak nebular continuum emission and ORLs emitted by heavy element ions, has now been carried out for several dozen Galactic PNe and a few Magellanic Cloud PNe. The results show that the ORL to CEL temperature and abundance determination disparities are ubiquitous. Many PNe have also been observed in the IR using the Infrared Space Observatory (ISO), giving access to fine-structure lines (FSLs), crucial to discriminate various scenarios for the causes of the disparities, including temperature fluctuations, density inhomogeneities and abundance variations. It is shown that temperature fluctuations and density inhomogeneities fail to provide a self-consistent explanation of observations and that there is strong evidence pointing towards the presence of a new ultra-cold plasma, highly enriched in helium and heavy elements and probably in the form of H-deficient knots, embedded in the nebula. Section 2 reviews some basic properties of CELs and ORLs. Observations of ORLs are summarized in Section 3 and elemental abundances derived from ORLs are compared to those derived from CELs in Section 4. Temperature, density and abundance inhomogeneities as the possible causes of the ORL to CEL discrepancies are discussed in Section 5. The nature and origins of H-deficient knots are discussed in the last Section.

Recent reviews on this topic are given by Liu (2002, 2003), Esteban (2002), Peimbert (2002), Torres-Peimbert & Peimbert (2003). A general review on the physics of emission line nebulae is given by Ferland (2003).

2 Recombination and Collisional Excitation

Ionized and heated by UV radiation fields, photoionized nebulae glow by emitting strong emission lines. Ionized H and He emit by recombinations with free electrons, producing familiar lines such as Hα, Hβ, He I λ5876 and He II λ4686 in the optical. Strong emission lines from heavy element ions are almost invariably CELs. For example, excited by impacts with thermal electrons, doubly ionized O^{2+} emits [O III] $\lambda\lambda$4959,5007 nebular lines in the optical, [O III] 52- and 88-µm fine-structure lines in the far-IR and O III] $\lambda\lambda$1661,1666 intercombination lines in the UV. Recombinations of heavy element ions followed by cascades to ground levels also produce emission lines, such as the O II M 1 at 4649 Å emitted by recombinations of O^{2+}. Due to the abundance effects, except for a few lines excited by di-electronic recombination, recombination lines of heavy element ions are generally quite faint and detectable only in the optical. Consequently these lines are often loosely referred to as ORLs.

Most emission lines of interest for plasma diagnostics and abundance determinations are optically thin, thus their observed intensities are proportional to emissivity, defined as the energy emitted per ion per electron per unit time, $I(\lambda) \propto \epsilon(\lambda) N(\mathrm{X}) N_\mathrm{e}$, where $N(\mathrm{X})$ is the number density of the parent ionic species responsible for the emission of line λ. For example, both the [O III] λ5007 CEL and O II λ4649 ORL originate from doubly-ionized oxygen, thus $N(\mathrm{X}) = N(\mathrm{O}^{2+})$ in both cases. By measuring the intensity ratio of an ionic line relative to Hβ, $I(\lambda)/I(\mathrm{H}\beta)$, the ionic abundance ratio, $N(\mathrm{X})/N(\mathrm{H}^+)$, can be determined. Thus $N(\mathrm{O}^{2+})/N(\mathrm{H}^+)$ can be determined by measuring either the [O III] λ5007 CEL or the O II λ4649 ORL.

The behaviour of emissivity of a CEL as a function of T_e and N_e depends on two fundamental quantities – excitation energy E_ex and critical density N_c of the upper level of the transition (c.f. Osterbrock 1989): $\epsilon(\lambda) \propto T_\mathrm{e}^{-1/2} \exp(-E_\mathrm{ex}/kT_\mathrm{e})$, and $\epsilon(\lambda)$ is independent of N_e for $N_\mathrm{e} << N_\mathrm{c}$ and $\epsilon(\lambda) \propto N_\mathrm{e}^{-1}$ for $N_\mathrm{e} >> N_\mathrm{c}$. For [O III] λ5007, $E_\mathrm{ex} = 29,170$ K and $N_\mathrm{c} = 6.9 \times 10^5$ cm^{-3}, thus for a typical PN of $T_\mathrm{e} \sim 10,000$ K and $N_\mathrm{e} \leq 10,000$ cm^{-3}, $\epsilon(\lambda5007)$ is a strong function of T_e but essentially independent of N_e. By contrast, [O III] 52µm has $E_\mathrm{ex} = 440$ K and $N_\mathrm{c} = 3,500$ cm^{-3}, thus intensity of this line is almost independent of T_e but inversely proportional to N_e. Emissivities of all recombination lines have only a similar, weak power-law dependence on T_e and are essentially independent of N_e under typical nebular conditions. Emissivity of free to bound recombinations of ionized H and He with electrons, responsible for the weak nebular continuum observable in the optical and UV, also has a weak power-law dependence on T_e.

Figure 1 plots the emissivities of several ORLs and CELs as a function of T_e. Under typical nebular conditions, i.e. $T_\mathrm{e} \sim 10,000$ K, collisional excitation is several orders of magnitude more efficient in emitting a photon which cools the nebula. As T_e declines, the efficiency drops exponentially. Owing to their very low E_ex's, far-IR FSLs such as [O III] 52µm remain efficient coolants over a wide range of T_e, and thus can cool the nebula down to a few hundred Kelvin. At such low temperatures, emission of recombination lines becomes significant, whereas optical CELs, such as [O III] λ5007, disappear completely.

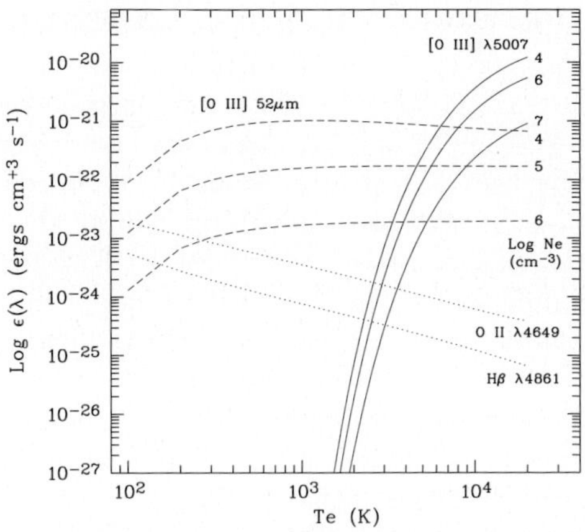

Fig. 1. Emissivities of Hβ and O II λ4649 ORLs and [O III] λ5007 and 52μm CELs as functions of T_e. Emissivities of ORLs are essentially independent of N_e. For CELs, curves for several values of N_e's are shown.

3 Observations of ORLs

The high quality measurement and comprehensive analysis by Liu et al. (1995a) of the extraordinary rich and prominent O II permitted line spectrum of the Saturn nebula NGC 7009, which has been known for decades since the pioneering work of Wyse (1942), showed that $N(O^{2+})/N(H^+)$ derived from O II ORLs is a factor of 5 higher than derived from the [O III] λλ4959,5007 CELs. Discrepancies of similar magnitude are also found for C and N and later for Ne (Luo et al. 2001). The problem is not new and has been known since the 1980s, as in the case of $N(C^{2+})/N(H^+)$ when comparing the values derived from the C II λ4267 ORL and from the C III] λ1908 CEL. The study of Liu et al. (1995a) firmly established the ORL origin of those weak permitted lines and their legitimacy as abundance indicators, and ruled out measurement errors, blending by unknown lines and/or contamination of the relatively weak ORLs by excitation mechanisms other than recombination (e.g. fluorescence) as the cause of the discrepancies.

Since then, a deep spectroscopic survey for a large sample of Galactic PNe, aimed at determining nebular CNONe abundances using ORLs and contrasting the results with values deduced from the traditional method based on CELs, has been carried out by the UCL group. Over a hundred PNe have been observed. For a 2 meter class telescope equipped with a long-slit spectrograph coupled with a high quantum efficiency CCD, ORLs from abundant second row elements can be measured in a few hours with a resolution of 3000 or better for bright Galactic PNe. The survey leads to the discovery of a rare group of PNe that exhibit extraordinary rich and prominent ORLs (Liu et al. 2000, 2001). Analyses of these PNe yield large discrepancies between the ORL and CEL abundances. In the most extreme case, the ORL abundances are nearly two orders of magnitude higher than the CEL values (c.f. Liu 2003 and references therein). Several dozen

PNe in the UCL sample have now been analyzed (Tsamis et al. 2003b, 2004; Liu et al. 2004ab; Wesson et al. 2004). Analysis for yet another subsample, comprising of over 30 Galactic bugle PNe, is currently in progress.

Meanwhile, improved effective recombination coefficients have been calculated and published by the UCL group led by P. J. Storey for most abundant ions of second-row elements (O II, Storey 1994, Liu et al. (1995a); C II, Davey et al. 2000; N II, Kisielius & Storey 2002; Ne II, Kisielius et al. 1998, P. J. Storey, private communications). In these calculations, the effects of resonance recombinations are treated explicitly, therefore the published coefficients include contributions from both radiative and di-electronic recombinations. Radiative recombination and di-electronic recombination coefficients calculated with less sophisticated techniques for CNO ions are available from Péquignot et al. (1991) and Nussbaumer & Storey (1984), respectively.

Owing to the observational difficulty caused by low abundances as well as the lack of reliable atomic data, systematic abundance analyses using ORLs for third-row elements have yet to start. The only exception is magnesium. Due to its unusual large span in ionization potential, Mg^{2+} is often the most abundant ionic species of magnesium in a typical nebula. This combined with the fact that the recombined ion Mg II has only one active electron and thus a relatively simple spectrum, make the Mg II 3d–4f λ4481 line relatively easy to detect. To a good approximation, one can assume that it has an effective recombination coefficient equal to that of the C II 3d–4f λ4267 line. ORL abundances of magnesium have now been determined from the λ4481 line for a number of PNe.

Dinerstein and Garnett (2001) surveyed two dozen PNe at a relatively low spectral resolution ($R \sim 2000$) and detected O II ORLs in ten of them. High quality echelle spectra aimed at ORL abundance analysis are published by the UNAM and IAC group for a number of bright Galactic H II regions and PNe (M 42, Esteban et al. 1998; M 17, Esteban et al. 1999a; M 8, Esteban et al. 1999b; NGC 5307, Ruiz et al. 2003; NGC 5315, Peimbert et al. 2004). Tsamis et al. (2003a) measured ORLs from CNO ions for three Galactic H II regions (M 42; M 17 and NGC 3576) and the giant H II regions 30 Doradus and N 11B in LMC and N 66 in SMC. Pushing to the limit of a four-metre class telescope, Esteban et al. (2002) measured O II ORLs from two giant H II regions in M 101 (NGC 5461 and 5471), one in M 33 (NGC 604) and another in NGC 2363. Detecting heavy element ORLs from H II regions is generally more difficult than from PNe, given the generally lower surface brightness of H II regions. The measurements are often complicated by contaminations of scattered starlight from early type stars in the region. ORLs used for nebular abundance determination are also present in spectra of early type stars, but in absorption rather than in emission.

4 Abundances Deduced from ORLs and CELs

Figure 2a plots the histogram of O^{2+}/H^{+} logarithmic abundance discrepancy factor (adf), defined as the ratio of the ionic abundances derived from ORLs and from CELs, adf(O^{2+})$\equiv$ $[O^{2+}/H^{+}]_{ORL}/[O^{2+}/H^{+}]_{CEL}$, for 60 PNe. The sample

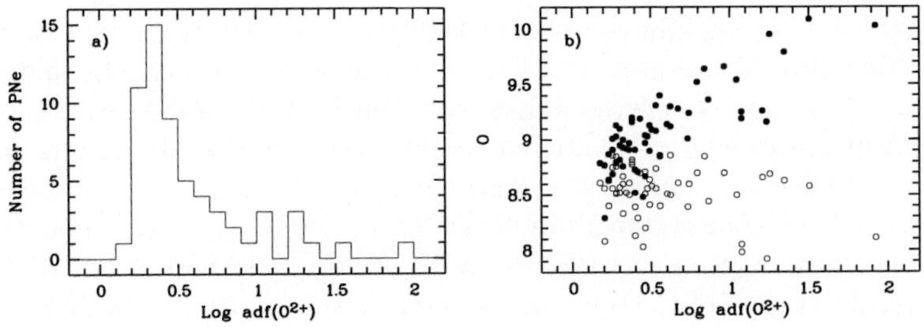

Fig. 2. a) Histogram of logarithmic abundance discrepancy factor (adf) for O^{2+}/H^{+} ionic abundance ratio; b) Oxygen elemental abundance (in units where H = 12) derived from ORLs (filled circles) and from CELs (open circles) plotted against Log adf(O^{2+}).

includes 12 Galactic, 2 LMC and 1 SMC PNe analyzed by Tsamis et al. (2004), 10 Galactic PNe by Liu et al. (2004b), 22 by Wesson et al. (2004), M 2-24 and NGC 7027 by Zhang & Liu (2003) and Zhang, Liu & Luo (2004), NGC 6543 by Wesson & Liu (2004), Me 1-1 by Shen, Liu & Danziger (2004), NGC 5307 by Ruiz et al. (2003), NGC 5315 by Peimbert et al. (2004), NGC 1501 by Ercolano et al. (2004) and several Galactic PNe by Liu et al. (NGC 7009, M 1-42, M 2-36, NGC 6153, Hf 2-2; c.f. Liu 2003 and references therein).

Figure 2a shows that for all of the sample PNe, Log adf(O^{2+}) > 0, i.e. O^{2+}/H^{+} derived from ORLs is always higher than the value given by CELs. For the current sample, adf(O^{2+}) ranges from a factor of 1.5 to 84, peaking strongly at 0.35 dex (or factor of 2). About 10% of the sample of PNe have an adf(O^{2+}) in excess of 10.

Two out of the three Magellanic Cloud PNe (LMC N 141 and SMC N 87) have adf(O^{2+}) = 2.6 and 2.8, respectively, whereas the extraordinary LMC N 66, whose central star recently underwent a violent mass loss event, shows a large adf(O^{2+}) of 11. Although the sample is too small to make a general statement, it appears that Magellanic Cloud PNe also show a wide range of adf's. With the wide deployment of 10-m class telescopes, measuring adf's for a large sample of Magellanic Cloud PNe should become feasible.

H II regions have also been found to have adf's always larger than unity, they seem to exhibit smaller adf's compared to PNe. For the limited number of Galactic and extragalactic H II regions analyzed so far (c.f. references given in the previous section), the measured adf(O^{2+})'s fall in a relatively narrow range from 1.3 to 2.4 except for the low excitation LMC N 11B, which shows a large adf of 6 (Tsamis et al. 2003a) but with a large uncertainty.

In Fig. 2b, total oxygen elemental abundances derived from ORLs and from CELs are plotted against adf. Albeit with significant scatter, oxygen abundances deduced from CELs fall in a relatively narrow range, from 7.92 to 8.86, with an average value of 8.54 and a standard deviation of 0.23 in logarithmic scale where H = 12. Excluding the three Magellanic Cloud PNe, the halo PN DdDm 1 (O = 8.05; Wesson et al. 2004) and the peculiar PN M 2-24 (O = 7.92, the lowest amongst the sampled PNe; Zhang & Liu 2003), the mean CEL oxygen abundance for the remaining 57 PNe is 8.57, slightly lower than the solar photospheric value of 8.69. This is consistent with the traditional understanding that PNe are descendants of low- and intermediate-mass Pop II stars and that oxygen abundance is largely unaffected by nucleosynthesis and various dredge-up processes that have occurred in progenitor stars.

In contrast, ORL oxygen abundances show a much wider range, reaching 10.09 in the most extreme case (NGC 1501). Amongst the 60 sampled PNe, four have oxygen ORL abundances in excess of ten times solar (M 1-42, NGC 40, Hf 2-2 and NGC 1501), yet except for Hf 2-2 which has a relatively low oxygen CEL abundance of 8.11, the oxygen CEL abundances of the other three PNe fall in a narrow range between 8.58 and 8.69. All the four PNe have an adf(O^{2+}) larger than 10. It is clear that the very high adf's of these extreme PNe are entirely caused by their very high ORL abundances. Since ionic abundances derived from ORLs are essentially independent of the nebular thermal and density structures (Fig. 1), physically they should be more reliable than the CEL results, which depend strongly on $T_{\rm e}$ and $N_{\rm e}$ and are sensitive to possible existence of temperature and density inhomogeneities. On the other hand, the very high oxygen ORL abundances derived for some nebulae cast doubt on their legitimacy if the high abundances are indeed representative of the whole ejected PN envelopes.

It has been realized since the earlier study of NGC 7009 (Liu et al. 1995a; Luo et al. 2001) that adf's larger than unity are observed in all abundant second-row elements, C, N, O and Ne, with comparable magnitudes for a given nebula. Interestingly, similar enhancement in ORL abundances has not been detected in magnesium, the only third-row element studied so far using an ORL (Barlow 2003). Of the 60 sample PNe, Mg abundances have been determined from the Mg II λ4481 ORL for 26 of them. Excluding the peculiar bulge PN M 2-24 which has an abnormally high Mg abundance of 8.38, the Mg abundances of the remaining 25 PNe range from 7.27 to 7.88, with an average vale of 7.51 and a standard deviation of 0.17. The average abundance is thus almost identical to the solar value of 7.54. Barlow et al. (2003) argue that depletion of magnesium onto dust grains is unlikely to be important for ionized regions of PNe. The stark contrast between Mg and second row elements should hold important clues to the nature and origin of the postulated ultra-cold ionized H-deficient clumps suspected to be embedded in PNe (c.f. next section for more details).

Garnett & Dinerstein (2001) find that larger, lower-surface brightness PNe tend to have adf's greater than compact, dense ones. The result is supported by our high quality measurements for a large sample of PNe. It is found that adf is anticorrelated with Hβ surface brightness, electron density and excitation class and positively correlated with nebular diameter. The scatter is large however in each case. Clearly while the problem of ORL to CEL abundance discrepancy is ubiquitous amongst PNe, it seems to become more profound or visible during the later evolutionary stages of PNe.

Only three PNe have been mapped in ORLs using long-slit spectroscopy. In all cases, adf is found to be strongly peaked at nebular centre, c.f. Liu (2003) for a recent review. No new data have been published since then. It is crucial to extend observations to more PNe using higher angular resolutions. The recent deployment of 3D spectrographs on 10-m class telescopes should have a major impact on this area.

5 Temperature, Density and Abundance Inhomogeneities

Our extensive deep spectroscopic survey for a large sample of Galactic PNe has led to the discovery of a number of nebulae exhibiting unusual rich and prominent ORLs from ionized CNONe ions. Detailed abundance analyses show that the heavy elemental abundances of these nebulae deduced from ORLs are often over an order of magnitude higher than those derived from CELs (c.f. Fig. 2a). All these nebulae have very low T_e(BJ)'s, the Balmer jump temperatures derived from the nebular continuum Balmer discontinuity of hydrogen recombination spectrum, much lower than T_e([O III])'s, the temperatures derived from the [O II] nebular to auroral line ratio (Liu et al. 2000, 2001b). The ORL to CEL abundance ratio is found to be tightly correlated with the difference between the two temperatures, T_e([O III]) − T_e(BJ)) (Liu et al. 2001).

Detailed UV to far-IR multi-waveband abundance studies contrasting ionic abundances derived from ORLs and from CELs, first carried out by Liu et al. for NGC 6153 and then for two bulge PN M 1-42 and M 2-36, have now been carried out for a large number of PNe (Tsamis et al. 2004; Liu et al. 2004b; Wesson et al. 2004). The analyses confirm the earlier findings that FSLs, which have $E_{ex} \leq 1000$ K, generally yield ionic abundances in good agreement with values given by UV and/or optical CELs and therefore rule out temperature fluctuations of the type envisaged by Peimbert (1967) as the major cause of the ORL to CEL abundance discrepancy (Liu et al. 2000).

The [O III] 52- and 88-μm lines have relatively low critical densities (3470 and 500 $\mathrm{cm^{-3}}$, respectively), therefore their emission can be suppressed if the nebula has large density inhomogeneities and contains a substantial amount of ionized gas in dense clumps with N_e significantly higher than a few thousand $\mathrm{cm^{-3}}$. Density inhomogeneities at modest levels are known amongst PNe (Liu et al. 2001a). On the other hand, a number of the sample PNe have N_e well below the critical density of the [O III] 52-μm line, yet showing large values of adf. Examples of such PNe include M 1-42 [$N_e = 1200\,\mathrm{cm^{-3}}$, adf($O^{2+}$) = 22; Liu et al. 2001b], NGC 40 [$N_e = 1200\,\mathrm{cm^{-3}}$, adf($O^{2+}$) = 18; Liu et al. 2004b], NGC 2022 [$N_e = 1500\,\mathrm{cm^{-3}}$, adf($O^{2+}$) = 16; Tsamis et al. 2004], NGC 3132 [$N_e = 550\,\mathrm{cm^{-3}}$, adf($O^{2+}$) = 2.4; Tsamis et al. 2004], NGC 6720 [$N_e = 500\,\mathrm{cm^{-3}}$, adf($O^{2+}$) = 2.4; Liu et al. 2004b]. The large adf's observed in these low-density PNe are difficult to be explained by the combined effects of density inhomogeneities and temperature fluctuations, as proposed by Ruiz et al. (2003) and Peimbert et al. (2004) for the high-density nebulae NGC 5307 and 5315. In addition, there are other FSLs of high N_c's. For example, the [Ne III] 15.5-μm line has a N_c of $2.0 \times 10^5\,\mathrm{cm^{-3}}$, yet in almost all PNe studied, it yields a Ne^{2+}/H^+ abundance ratio consistent with that derived from the [Ne III] λ3868 CEL, rather than with the higher value derived from Ne II ORLs.

Zhang et al. (2004) determine T_e and N_e simultaneously for a large sample of PNe by fitting the H I recombination spectrum (including line and continuum emission such as the Balmer discontinuity and Balmer decrement) and compare the results with those derived from various optical/IR CEL diagnostic ratios. While the results yield strong evidence suggesting that both temperature fluctuations and density inhomogeneities at modest levels are generally present in PNe, it is clear that the differences between T_e([O III]) and T_e(BJ) observed in PNe of extreme adf values are so large that they cannot be explained in terms of temperature fluctuations of the type proposed by Peimbert (1967).

Several attempts to detect temperature fluctuations by mapping the spatial variations of T_e([O III]) via either direct monochromatic imaging or long-slit spectroscopy have been published. Using HST WFPC2 narrow-band images in the light of the [O III] λ5007 and λ4363 lines, after careful subtraction of contamination by nebular continuum and nearby Hγ, Rubin et al. (2002) and Wesson & Liu (2004) mapped the $\lambda 5007/\lambda 4363$ ratio and the resultant T_e([O III]) across the surface of NGC 7009 and NGC 6543, respectively. The resultant T_e varies in a narrow range, yielding a surface temperature fluctuations parameter t_A^2 of less than 0.01. For NGC 7009, the results are further corroborated by STIS long-slit spectroscopy. Note that although the measurements yield only mean electron temperatures integrated along the line of sight, under the reasonable assumption of random temperature fluctuations, as implied by the concept, one expects t_A^2 to be of similar magnitude of the true temperature fluctuation parameter t^2. Thus for both nebulae, the observed t_A^2 values are too small to reconcile the large ORL to CEL abundance discrepancies, by a factor of 4.7 and 4.2 in NGC 7009 and 6543, respectively (Liu et al. 1995a; Wesson & Liu 2004).

For the Orion Nebula, O'Dell, Peimbert & Peimbert (2003) determine t_A^2 using HST WFPC2 images, and find that it varies between 0.005 and 0.016. From HST STIS long-slit spectra, Rubin et al. (2003) measure T_e([O III]) along several slit positions of the Orion Nebula and find a similar range of t_A^2. Much earlier, from ground-based slit spectroscopy, Liu et al. (1995b) find that T_e([O III]) $\approx$ T_e(BJ) and that both temperatures are nearly constant across the Orion Nebula, implying small temperature fluctuations. Note that the Orion Nebula has a small adf(O^{2+}) of only 1.3 (Esteban et al. 1998; Tsamis et al. 2003a), so it is not the best object in which to study the abundance discrepancy problem.

Liu et al. (2000) show that a two-abundance model incorporating a small amount of cold ionized H-deficient material can account for many of the observed patterns of NGC 6153 which has a large adf(O^{2+}) of 9.2. This is corroborated by the detailed photoionization models of Péquignot et al. (2002; 2003) and Tylenda (2003). In this scenario, the H-deficient gas is cooled to a few hundred Kelvin by recombinations and far-IR FSLs, compared to $\sim 10^4$ K in the main nebular component of "normal" chemical composition (i.e. about Solar). At such low temperatures, optical and UV CELs disappear almost entirely due to their very high E_{ex}'s, whereas emissivities of ORLs are enhanced. Only a small amount of H-deficient gas is required to account for the observed strengths of ORLs – in the particular model of NGC 6153 constructed by Péquignot et al., the H-deficient

Table 1. Comparison of ORL and CEL temperatures (K)

Nebula	T_e([O III])	T_e(H I BJ)	T_e(He I ORLs)	T_e(O II ORLs)	adf(O^{2+})
Hf 2-2	8820	900	775	630	84
M 1-42	9220	3560	2310	≤ 300	22
NGC 6153	9120	6080	3370	360	9.2
M 2-36	8380	5900	4160	520	6.9
NGC 7009	9980	8150	5380	420	4.7

component has a mass of only 0.0031 $M_\odot$, about 1% of the total ionized mass of 0.38 $M_\odot$ of the nebula. The H-deficient gas is enhanced in He and CNONe heavy elements by a factor of 6 and 100, respectively. In this model, UV and optical CELs with high E_{ex}'s arise almost entirely from the hot nebular component, with a small contribution, at the level of a few per cent, *by recombinations*, from the cold H-deficient component. H I recombination lines are also dominated by emission from the normal component, with about 5% from the cold component leading to a lower value of T_e(BJ) compared to T_e([O III]). For He I ORLs, about 30% of the emission arises from the cold H-deficient component, whereas emission of heavy element ORLs arises almost entirely from the cold H-deficient component. An interesting aspect of this model is that because of the tiny amount of gas in the H-deficient component, the average metallicity for the whole nebula is very close to that of the "normal" component as yielded by the traditional method of CEL abundance analysis. In the picture of two-abundance model, the small values of t^2_A yielded by HST imaging or spectroscopy have a natural explanation since as the H-deficient gas which has a very low T_e is "invisible" in the [O III] λ5007 and λ4363 lines.

The two-abundance (or H-deficient knot) model receives substantial support with the discovery of PNe exhibiting extremely low T_e(BJ)'s, such as M 1-42 (3560 K; Liu et al. 2001) and Hf 2-2 (900 K; Liu 2003). In both PNe, T_e([O III]) $\sim$ 9000 K. This clearly shows that the nebula contains two phases of ionized gas, each with distinct physical conditions. Further evidence in support of the two-abundance model was presented by Liu (2003) who determined the *average* emission temperatures of He I and heavy element ORLs by comparing the observed intensity ratios with those predicted by recombination theory. Given the weak dependence of intensity ratios of ORLs on T_e, measurements of very high accuracy are required. Analyses show that T_e(He I ORLs) $<$ T_e(BJ), suggesting that He I lines have a lower average emission temperature than H I lines. Even lower T_e's are yielded by O II and C II ORL ratios, reaching just a few hundred Kevin in extreme cases such as Hf 2-2 and M 1-42. Similar analyses have now been extended to a number of PNe (Tsamis et al. 2004; Liu et al. 2004b and Wesson et al. 2004), again yielding similar results. Table 1 presents example measurements of ORL and CEL temperatures for several PNe. Values of T_e(O II

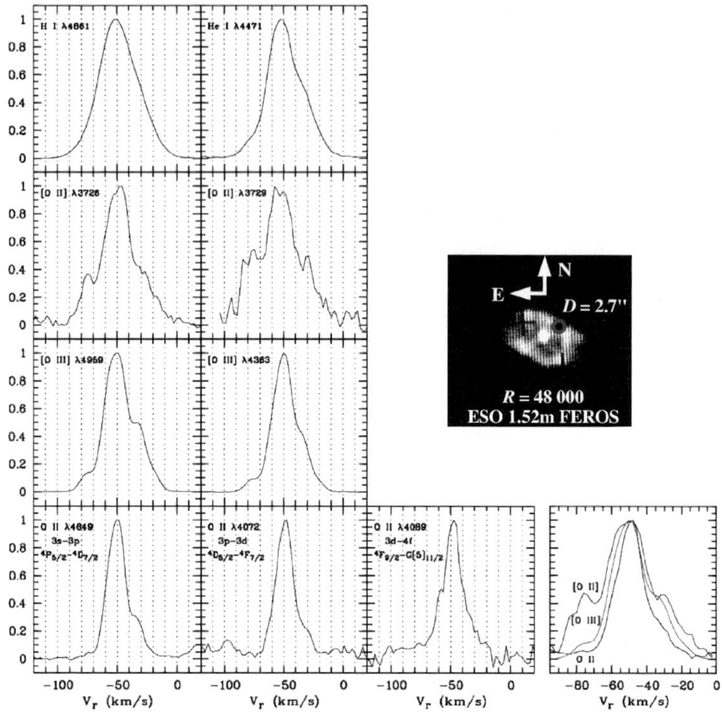

Fig. 3. CEL and ORL profiles in NGC 7009. The [O III] CELs and O II ORLs originate from the same O^{2+} ion, yet the O II ORLs are noticeably narrower.

ORLs) have been rederived from the observed O II $\lambda4089/\lambda4649$ ratio using the most recent calculations of the effective recombination coefficients for O II, down to temperatures as low as 300 K, by P. J. Storey (c.f. Fig. 7 in Tsamis et al. 2004). The observed temperature sequence is in complete agreement with the expectation of the two-abundance model, which predicts that T_e([O III]) $>$ T_e(H I BJ) $>$ T_e(He I ORLs) $>$ T_e(CNONe ORLs) (Liu 2003; Péquignot et al. 2003). Note that in the two-abundance model, which proposes a two phase nebula, each phase of ionized gas has a distinct chemical composition and temperature, is physically different to the picture of temperature fluctuations originally proposed by Peimbert (1967). As pointed out by Zhang, Liu & Liu (2004), in the scenario of temperature fluctuations one expects T_e(H I BJ) $<$ T_e(He I ORLs), exactly opposite to what observations yield.

Evidence in favour of a new component of ultra cold plasma emitting heavy element ORLs is also provided by high spectral resolution observations of weak ORLs, now starting to become available. In Fig. 3 we compare the observed line profiles of [O II] and [O III] CELs and H I, He I and O II ORLs, measured at a resolution of $R = 48,000$ with the FEROS spectrometer on the ESO 1.52m telescope. The very broad widths of H I and He I lines are apparently caused by thermal broadening which is significant for light element ions. The [O II] CELs are broader than the [O III] CELs, in line with the expectation that the former arise from the outer, lower ionization regions. The most interesting feature revealed by Fig. 3 is that although both the [O III] and O II lines originate from the same O^{2+} ion, the O II ORLs are noticeably narrower, suggesting that the O II ORLs arise either from much cooler plasmas (thus have a smaller thermal broadening) or from spatially distinct regions than the [O III] lines. Ultra deep echelle spectrum of the low ionization PN IC 418 at a resolution of 9 km s^{-1} has recently been obtained by Sharpee et al. (2004). They find that while O II lines are narrower than [O II] lines, they have widths comparable to those of [O III] lines. Given the very small adf's found for this high surface brightness PN [adf(O^{2+}) = 1.3], and the fact that the spectrum was an integration over a 6.5 arcsec slit length of 1 arcsec wide, it is probably of no surprise that [O III] CELs and O II ORLs have comparable line widths. In the likely case that the large adf's observed in other PNe are caused by the accumulated effects of a large number of small H-deficient knots, high spectral resolution *and* high spatial resolution spectroscopic observations are needed to detect possible differences in line width between [O III] CELs and O II ORLs.

6 Origins of the Ultra-Cold H-Deficient Plasmas

H-deficient knots are known to exist in a rare class of old PNe, including Abell 30, 58 and 78, IRAS 15154-5258 and 18333-2357 (e.g. Harrington 1996). In Abell 30, the H-deficient knots include two point-symmetric polar knots in PA = 331° at angular distances of 6.66 and 7.44 arcsec from the central star, plus a number of knots loosely outlining an inner equatorial ring. The knots are embedded in a round limb-brightened faint nebula of angular diameter of 127 arcsec. Deep spectroscopy of the two polar knots J 1 and J 3 by Wesson, Liu & Barlow (2003) reveals prominent ORLs from CNONe ions almost identical to those observed in PNe such as NGC 7009 and 6153. Analyses yield adf(O^{2+}) = 766, T_e([O III]) = 17,960 K, T_e(He I) = 4600 K, T_e([O II]) = 450 K for J 1 and adf(O^{2+}) = 598, T_e([O III]) = 16,680 K, T_e(He I) = 8840 K, T_e([O II]) = 2450 K for J 3. In both knots oxygen has an abundance comparable to hydrogen by number, whereas helium abundance is about an order of magnitude higher. Thus the knots in Abell 30 exhibit many characteristics similar to those H-deficient knots hypothesized to exist in nebulae such as NGC 6153 and M 1-42, except that hydrogen in the knots of Abell 30 appears to be almost completely depleted.

A photoionization model for the H-deficient knot J 3 in Abell 30 has been constructed by Ercolano et al. (2003) using the newly developed 3D Monte Carlo photoionization code MOCASSIN. In spite of the much simplified nature of the model, such as neglecting the diffuse radiation field of the main nebula in which the knot is embedded, the model reproduces many features of the observations. In particular it shows that the metal-rich inner core of the knot has a T_e of only 500 K, in good agreement with the result of diagnostic analysis using O II ORLs. Interestingly, both the empirical analysis and the photoionization model show that the knot is O-rich, rather than C-rich as suggested by earlier studies using CELs, in contradiction with the expectation of the "born-again" scenario proposed for the origin of PNe with H-deficient knots.

High resolution HST imaging has revealed that micro structures such as knots, filaments and clumps are ubiquitous in PNe (e.g. O'Dell et al. 2002). The nearby Helix nebula contains thousands of cometary globules, currently being ionized and evaporated by the strong radiation fields from the central star. Each of the knots contains a mass of gas of the order of the Earth (i.e. about $1 \times 10^{-6}\,M_{\odot}$), and has a dusty molecular core of H_2 and CO. The nature and origin of these knots remain a matter of debate. In particular the question of whether they are enhanced in heavy elements, and thus represent nearby examples of the H-deficient knots hypothesized to be located in other PNe such as NGC 6153, M 1-42 and Hf 2-2 that cause the large adf values observed in those nebulae, remains to be investigated by deep spectroscopic observations. At the moment, direct observational evidence pointing towards the existence of such H-deficient knots in these extreme PNe is still lacking. Torres-Peimbert, Peimbert & Peña (1990) suggested that the large adf(C^{2+}) observed in the halo PN NGC 4361 could be caused by a C-rich inner shell where the C/H abundance is 100 times than the outer region, resulting a lower T_e of 11,400 K in the inner shell compared to 20,200 K in the outer region. Using deep long-slit spectroscopy, Liu (1998) mapped carbon ORLs across the nebula at a spatial resolution of $4.9 \times 2\,\mathrm{arcsec}^2$ and failed to detect a C-rich inner zone. While the observation ruled out the two-zone model proposed by Torres-Peimbert et al., it did not rule out the possibilities of small metal-rich clumps well embedded in the nebula.

While there seems good evidence suggesting that there is another component of ultra-cold ionized H-deficient gas in PNe, probably in the form of numerous H-deficient knots, the nature and origin of such knots remain a mystery. It is not clear whether they are nucleo-processed material, as proposed for the H-deficient knots in Abell 30. It seems that the knots are enhanced in He and CNONe, but not in Mg. An alternative to the scenario of nucleo-processed material ejected by the progenitor star is that the knots are planets, planetesimals and comets of the planetary system of the progenitor star being ionized, stripped off and evaporated by the radiation field and stellar wind from the host star that is now evolving to become a white dwarf (Liu 2003). Analyses of PNe such as NGC 6153 and M 1-42 show that only a few Jupiter masses of metal-rich material is required to explain the strengths of ORLs observed in these nebulae, so the idea may not be so eccentric as it first looks. Evaporating proto-planetary disks around newly formed stars in H II regions (such as the proplyds in M 42) may likewise provide a natural solution to a similar ORL/CEL abundance discrepancy problem observed to be widely present in H II regions (Liu 2003).

The possible existence of H-deficient knots in ionized nebulae could potentially have a profound consequence on He abundances of PNe and H II regions, and on the primordial He abundance determined by extrapolating He abundances of metal-poor H II galaxies to zero metallicity (Zhang et al. 2004). Péquignot et al. (2002; 2003) and Liu (2003) noticed that in the two-abundance scenario, He abundance is least constrained by observations, since all strong He lines measurable in the optical and UV are dominated by recombination excitation, and thus can be strongly contaminated by emission from the small amount of H-deficient ultra-cold plasma located in the nebula. For NGC 6153, two-abundance photoionization modelling by Péquignot et al. (2002; 2003) shows that, after averaging over the entire volume of ionized gas, the nebula has a helium abundance of 0.101, essentially identical to the Solar value but much lower than the value of 0.136 derived from empirical analysis assuming a chemically homogeneous nebula (Liu et al. 2000). The lower helium abundance in NGC 6153 is supported by our measurement of the He I λ10830 line, for which collision by electron impacts from the 2s ^{3}S meta-stable level has a major contribution to its observed flux.

References

1. M. J. Barlow, X.-W. Liu, D. Péquignot, P. J. Storey, Y. G. Tsamis, C. Morisset: In: *Planetary Nebulae: Their Evolution and Role in the Universe, IAU Symp. 209, Canberra, Australia, 19-23 Nov. 2001*, ed. by S. Kwok, M. Dopita, R. Sutherland (Pub. Astr. Soc. Pacific, 2003), p.373
2. A. R. Davey, P. J. Storey, R. Kisielius: A&AS, **142**, 85 (2000)
3. B. Ercolano, M. J. Barlow, P. J. Storey, et al.: MNRAS, **344**, 1145 (2003)
4. B. Ercolano, R. Wesson, Y. Zhang, et al.: MNRAS, in press (astro-ph/0407230; 2004)
5. C. Esteban: RMxAA (Conf. Ser.), **12**, 56 (2002)
6. C. Esteban, M. Peimbert, S. Torres-Peimbert, V. Escalante: MNRAS, **295**, 401 (1998)
7. C. Esteban, M. Peimbert, S. Torres-Peimbert, J. García-Rojas: RMxAA, **35**, 65 (1999a)
8. C. Esteban, M. Peimbert, S. Torres-Peimbert, et al.: ApJS, **120**, 113 (1999b)
9. C. Esteban, M. Peimbert, S. Torres-Peimbert, M. Rodríguez: ApJ, **581**, 241 (2002)
10. G. Ferland: ARAA, **41**, 517 (2003)
11. D. R. Garnett, H. L. Dinerstein: RMxAA (Conf. Ser.), **10**, 13 (2001)
12. J. P. Harrington: In: *Hydrogen-Deficient Stars, ASP Conf. Ser. Vol. 96*, ed. by C. S. Jeffery, U. Hebera (Pub. Astron. Soc. Pac., San Francisco, 1996), p.193
13. R. Kisielius, P. J. Storey: A&A, **387**, 1135 (2002)
14. R. Kisielius, P. J. Storey, A. R. Davey, L. T. Neale: A&AS, **133**, 257 (1998)
15. X.-W. Liu: MNRAS, **295**, 699 (1998)
16. X.-W. Liu: RMxAA (Conf. Ser.), **12**, 70 (2002)
17. X.-W. Liu: In: *Planetary Nebulae: Their Evolution and Role in the Universe, IAU Symp. 209*, ed. by S. Kwok, M. Dopita, R. Sutherland (Pub. Astr. Soc. Pacific, 2003), p.339
18. X.-W. Liu, M. J. Barlow, M. Cohen, I. J. Danziger, S.-G. Luo, J. P. Baluteau, et al.: MNRAS, **323**, 343-361 (2001a)
19. X.-W. Liu, M. J. Barlow, I. J. Danziger, P. J. Storey: ApJ, **450**, L59 (1995b)
20. X.-W. Liu, S.-G. Luo, M. J. Barlow, et al.: MNRAS, **327**, 141 (2001b)
21. X.-W. Liu, P. J. Storey, M. J. Barlow, R. E. S. Clegg: MNRAS, **272**, 369 (1995a)
22. X.-W. Liu, P. J. Storey, M. J. Barlow, et al.: MNRAS, **312**, 585 (2000)
23. Y. Liu, X.-W. Liu, S.-G. Luo, Barlow M. J.: MNRAS, in press (2004a)

24. Y. Liu, X.-W. Liu, M. J. Barlow, S.-G. Luo: MNRAS, in press (2004b)
25. S.-G. Luo, X.-W. Liu, M. J. Barlow: MNRAS, **326**, 1049 (2001)
26. H. Nussbaumer, P. J. Storey: A&AS, **56**, 293 (1984)
27. C. R. O'Dell, B. Balick, A. R. Hajian, et al.: AJ, **123**, 3329 (2002)
28. C. R. O'Dell, M. Peimbert, A. Peimbert: AJ, **125**, 2590 (2003)
29. D. Osterbrock: *Astrophysics of Gaseous Nebulae and Active Galactic Nuclei* (University Science Books, Mill Valey, 1989)
30. M. Peimbert: ApJ, **150**, 825 (1967)
31. M. Peimbert: RMxAA (Conf. Ser.), **12**, 275 (2002)
32. M. Peimbert, A. Peimbert, M. T. Ruiz, C. Esteban: ApJS, **150**, 431 (2004)
33. D. Péquignot, M. Amara, X.-W. Liu, et al.: RMxAA (Conf. Ser.), **12**, 142 (2002)
34. D. Péquignot, X.-W. Liu, M. J. Barlow, et al.: In: *Planetary Nebulae: Their Evolution and Role in the Universe, IAU Symp. 209*, ed. by S. Kwok, M. Dopita, R. Sutherland (Pub. Astr. Soc. Pacific, 2003), p.347
35. D. Péquignot, P. Petitjean, C. Boisson: A&A, **251**, 680 (1991)
36. R. Rubin, N. J. Bhatt, R. J. Dufour, et al.: MNRAS, **334**, 777 (2002)
37. R. Rubin, P. G. Martin, R. J. Dufour, et al.: MNRAS, **340**, 362 (2003)
38. M. T. Ruiz, A. Peimbert, M. Peimbert, C. Esteban: ApJ, **595**, 247 (2003)
39. B. Sharpee, J. A. Baldwin, R. Williams: ApJ, in press (astro-ph/0407186; 2004)
40. Z.-X. Shen, X.-W. Liu, I. J. Danziger: A&A, in press (2004)
41. P. J. Storey: A&A, **282**, 999
42. S. Torres-Peimbert, M. Peimbert, M. Peńa: A&A, **233**, 540
43. Y. G. Tsamis, M. J. Barlow, X.-W. Liu, et al.: MNRAS, **338**, 186 (2003a)
44. Y. G. Tsamis, M. J. Barlow, X.-W. Liu, I. J. Danziger, P. J. Storey: MNRAS, **353**, 953 (2004)
45. S. Torres-Peimbert, M. Peimbert:, In: *Planetary Nebulae: Their Evolution and Role in the Universe, IAU Symp. 209*, ed. by S. Kwok, M. Dopita, R. Sutherland (Pub. Astr. Soc. Pacific, 2003), p.363
46. R. Tylenda: In: *Planetary Nebulae: Their Evolution and Role in the Universe, IAU Symp. 209*, ed. by S. Kwok, M. Dopita, R. Sutherland (Pub. Astr. Soc. Pacific, 2003), p.159
47. R. Wesson, X.-W. Liu: MNRAS, **351**, 1026 (2004)
48. R. Wesson, X.-W. Liu, M. J. Barlow: MNRAS, **340**, 253 (2003)
49. R. Wesson, X.-W. Liu, M. J. Barlow: MNRAS, submitted (2004)
50. A. B. Wyse: ApJ, **95**, 356 (1942)
51. Y. Zhang, X.-W. Liu: A&A, **404**, 545 (2003)
52. Y. Zhang, X.-W. Liu, Y. Liu: in preparation (2004)
53. Y. Zhang, X.-W. Liu, S.-G. Luo: in preparation (2004)
54. Y. Zhang, X.-W. Liu, R. Wesson, et al.: MNRAS, **351**, 935 (2004)

Probing Nebular Physical Conditions Using the Hydrogen Recombination Spectrum

Y. Zhang[1], X.-W. Liu[1], R. Wesson[2], P.J. Storey[2], Y. Liu[1], and I.J. Danziger[3]

[1] Department of Astronomy, Peking University, Beijing 100871, China
[2] Department of Physics and Astronomy, University College London, Gower Street, London WC1E 6BT, UK
[3] Osservatorio Astronomico di Trieste, Via G.B. Tiepolo 11, 1-34131 Trieste, Italy

The hydrogen recombination spectrum is important in the study of nebular physical conditions. We present a method to simultaneously determine electron temperatures and densities of planetary nebulae (PNe), using the observed hydrogen recombination spectrum near the Balmer jump at 3646 Å. The method has been applied to spectra of 48 Galactic PNe.

Comparison between electron temperatures and densities derived from the hydrogen recombination spectrum and those derived from collisionally excited line (CEL) ratios provides an excellent probe to investigate temperature and density variations within nebulae. Electron temperatures derived from the hydrogen recombination spectrum [hereafter $T_{\rm e}$(Bal)] are found to be systematically lower than those derived from the [O III] $\lambda4959/\lambda4363$ and the [O III] $(88\mu{\rm m}+52\mu{\rm m})/\lambda4959$ ratios [hereafter $T_{\rm e}([{\rm O\ III}]_{\rm na})$ and $T_{\rm e}([{\rm O\ III}]_{\rm fn})$]. And electron densities determined from the hydrogen Balmer decrement are found to be systematically higher than those derived from the [O II] $\lambda3729/\lambda3726$, [S II] $\lambda6731/\lambda6716$, [Cl III] $\lambda5537/\lambda5517$, [Ar IV] $\lambda4740/\lambda4711$ and [O III] 88μm/52μm ratios. These results suggest that temperature and density variations are generally present within nebulae. However, not all the observed temperature disparities can be attributed to the presence of temperature and density variations, especially in some extreme nebulae where the temperature disparities are particularly large. Detailed multi-waveband studies of these extreme PNe have shown that temperature and density variations fail to explain the very large temperature disparities and the known discrepancy between the ionic abundances of heavy elements derived from CELs and those derived from recombination lines. Rather, they point to the presence of new gas component, cold and H-deficient, embedded in diffuse nebula,

We find that He/H abundances of PNe are positively correlated with the difference between $T_{\rm e}([{\rm O\ III}]_{\rm na})$ and $T_{\rm e}$(Bal), indicating that He/H abundances might have been overestimated in general because of the possible existence of cold H-deficient material. The discrepancies between $T_{\rm e}([{\rm O\ III}]_{\rm na})$ and $T_{\rm e}$(Bal) are found to be anti-correlated with electron densities derived from various density indicators, suggesting that temperature discrepancy is a phenomenon related to nebular evolution, in the sense that temperature discrepancy increases as the nebula expands and ages.

Details of these results are given in Zhang et al. 2004, MNRAS, 351, 935.

A Reexamination of Electron Density Diagnostics for Ionized Gaseous Nebulae

W. Wang[1], X.-W. Liu[1], Y. Zhang[1], and M.J. Barlow[2]

[1] Department of Astronomy, Peking University, Beijing 100871, P.R. China
[2] Department of Physics and Astronomy, University College London, Gower Street, London WC1E 6BT, U.K.

1 Introduction

Electron density, N_e, is one of the key physical parameters characterizing an ionized gaseous nebula. Accurate measurement of N_e is a prerequisite for the determination of nebular chemical abundances and for the calculation of the mass of ionized gas. Information on N_e can also be used to estimate the distance to a nebula. In the optical, the most commonly used density-diagnostic ratios are [O II] $\lambda3729/\lambda3726$, [S II] $\lambda6716/\lambda6731$, [Cl III] $\lambda5517/\lambda5537$ and [Ar IV] $\lambda4711/\lambda4740$. Previous study are somehow controversial: Stanghellini & Kaler [14] found that N_e([O II]) $<$ N_e([S II]) $<$ N_e([Cl III])$\approx$ N_e([Ar IV]); Meatheringham & Dopita [9] found that for 44 PNe in the Magellanic Clouds N_e([O II]) $>$ N_e([S II]). For these four $2p^3$ ions, the intensity ratio of emission from the two upper levels $^2D_{5/2}$, $^2D_{3/2}$ to the ground $^4S_{3/2}$ level is fixed by the ratio of corresponding collisional strengths at low densities [$N_e << N_c(\lambda; T_e)$, where $N_c(\lambda; T_e)$ is the critical density of the diagnostic line; c.f. Osterbrock [18]], and is determined entirely by the ratio of Einstein transition probabilities of the two upper levels at high densities. Therefore, the accuracy of calculations of those atomic parameters can be validated by observations. In this paper, we present measurements of intensity ratios of the four optical N_e-diagnostics and N_e deduced from them for a large sample of over a hundred PNe.

2 Observations and Results

Most of the observations were obtained with the ESO 1.52 m telescope using the Boller & Chivens (B&C) long-slit ($2''$ width) spectrograph in three observing runs in 1995, 1996 and 2001. A number of northern hemisphere PNe were later observed with the 2.5 m Isaac Newton Telescope (INT) using the Intermediate Dispersion Spectrograph (IDS) and the 235 mm camera equipped with an EEV CCD.

Intensity ratios for the four optical N_e-diagnostic doublets, derived from our new observations, are presented in Table 2 in Wang et al. [20], along with their uncertainties. Electron densities deduced from the ratios are also given in the Table assuming an electron temperature of 10 000 K, and calculated using EQUIB, a Fortran code to solve level populations for a multi-level atom. A five level atomic model was assumed for the four ions of interest here.

3 Discussion

3.1 [O II] and [S II]

O^0 and S^0 have comparable ionization potentials. In addition, the [O II] and [S II] doublet lines have very similar critical densities and are therefore sensitive to N_e variations over a similar density regime. Figure 1 shows that for our default atomic data set the measured [O II] and [S II] ratios are consistent with each other over a wide N_e range, from less than $100\,cm^{-3}$ to over $10^4\,cm^{-3}$. At densities higher than $\sim 10^4\,cm^{-3}$, there is however some evidence that the [O II] doublet ratio indicates densities systematically lower than implied by the [S II] ratio. Our result thus is well consistent with earlier [O II] and [S II] density survey by Kingsburgh & English ([7], hereafter KE92), if both densities are calculated using our default atomic parameters (cf. Wang et al [20]).

Low Density Limit Copetti & Writzl ([2], hereafter CW02) compared the observed [O II] ratios and the [S II] ratios for a large sample of PNe and found that the locus calculated using the [O II] collision strengths of McLaughlin & Bell ([8], hereafter MB98) disagrees with observations near the low-N_e limit. Based on this, they concluded that the results of MB98 are probably spurious. A similar plot is shown in Fig. 1 using our own data. As noted by CW02, the locus calculated using the [O II] collision strengths of McLaughlin & Bell deviates from observations at low densities. [S II] has a larger atomic mass number, and has a larger fine-structure splitting than [O II], thus one would expect [S II] to be affected more by relativistic effects. Consequently the [S II] doublet ratio ought to show an even larger deviation from the ratio of statistical weights should the effects invoked by MB98 apply. So the dash-dotted line in Fig. 1 can in principle either be brought into agreement with observations or curve downwards as the low-N_e limit is approached (top-right corner of the graph). Thus the apparent disagreement noticed by CW02 and shown here in Fig. 1 is not sufficient to reject the results of MB98.

The key evidence against the results of MB98 remains the fact that none of the nebulae in our sample, including PNe and H II regions, shows a line ratio in excess of 1.5 (Fig. 1). The largest $\lambda 3729/\lambda 3726$ ratio determined for a PN in our sample is 1.23 ± 0.01 (M 1-64, c.f. Table 2 in Wang et al. [20]), and that for an H II region is 1.43 (SMC N66, Tsamis et al. 2003). Amongst the 64 Galactic PNe observed by KE92 , IC 5148-50 has the largest $\lambda 3729/\lambda 3726$ ratio, 1.45 ± 0.04. In addition, none of the nebulae in the extensive literature survey by CW02 shows a ratio approaching 1.92.

Similarly for [S II], none of the recent high quality measurements for a PN or an H II region of the $\lambda 6716/\lambda 6731$ doublet ratio yield a value in excess of the canonical value of 1.5 under *LS*-coupling. For the sample of PNe, the largest ratio recorded is 1.21 ± 0.03, whereas for those observed by KE92, the maximum value is 1.35 ± 0.04.

In Table 1 we list all the recent measurements that we are aware of for PN halos, SNRs, H II regions and diffuse ionized gas (DIG) that yield an [O II] and [S II]

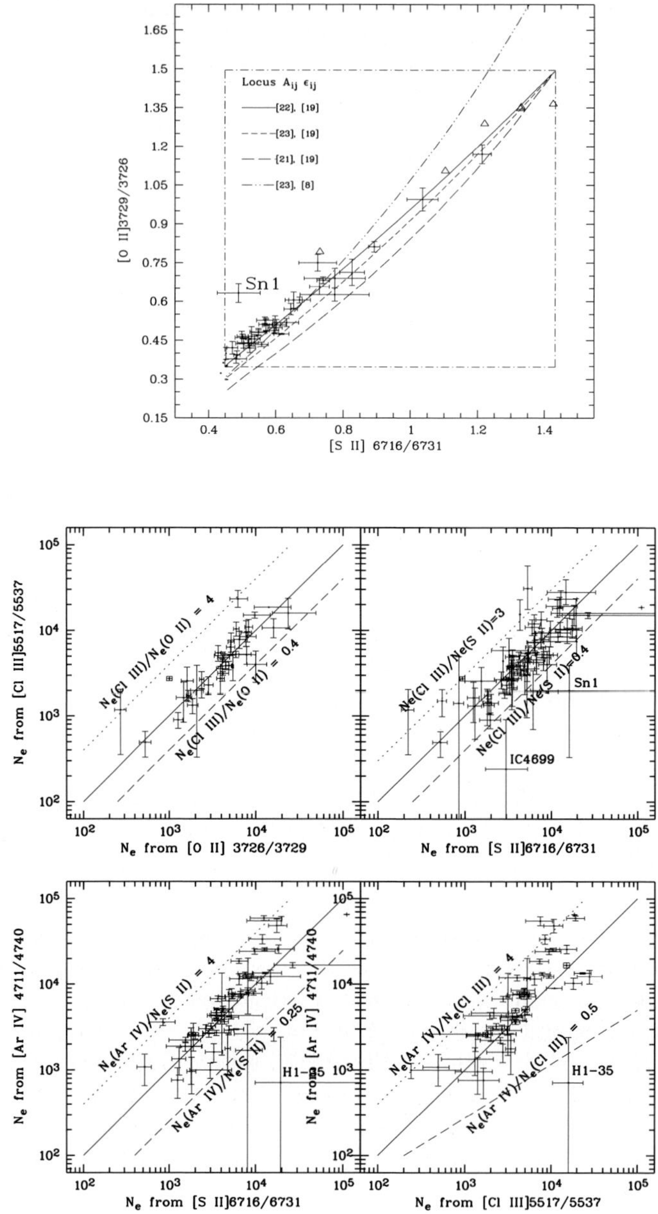

Fig.1 The observed [O II] ratio is plotted against the [S II] ratio (upper) and comparisons between densities deduced from each indicator (lower). The loci in the upper panel were derived using different atomic parameters of [O II] listed in the figure assuming $T_e = 10\ 000\,\mathrm{K}$, where A_{ij} and ϵ_{ij} represent transition probabilities and collision strengths respectively. References [22] and [19] are our default atomic data set for [O II].

Table 1. [O II] and [S II] doublet ratios in excess of 1.4 reported in the literature. A "::" indicates an uncertainty larger than 20%

Source	$I(\lambda3729)/I(\lambda3726)$	Ref.	Source	$I(\lambda6716)/I(\lambda6731)$	Source	$I(\lambda6716)/I(\lambda6731)$
IC418	$1.79\,^{2.83}_{1.30}$	[13]	NGC 3195 [a]	1.98::	Mz 2[a]	$1.44\,^{1.55}_{1.33}$
PX Pup	$1.43\,^{1.54}_{1.33}$	[13]	Fg 1 [a]	$1.45\,^{1.56}_{1.34}$	IC 4634 [a]	2.14::
He 2-146	$1.43\,^{1.45}_{1.41}$	[7]	Pb 8 [a]	$1.40\,^{1.48}_{1.32}$	IC 4637 [a]	1.75::
NGC 7293	$1.41\,^{1.43}_{1.39}$	[7]	He 2-104 [a]	$1.71\,^{1.92}_{1.50}$	K772 [b]	$2.3\,^{2.4}_{2.2}$
IC 5148-50	$1.45\,^{1.49}_{1.41}$	[7]	He 2-141 [a]	$1.62\,^{1.84}_{1.40}$	K70 [b]	$1.7\,^{1.8}_{1.6}$
NGC6543	$1.44\,^{1.56}_{1.32}$	[12]	He 2-146 [a]	$1.52\,^{1.67}_{1.37}$		
NGC 2371-2	$1.47\,^{1.73}_{1.21}$	[6]	NGC 6072 [a]	$1.61\,^{1.80}_{1.42}$		

[a] Extended diffuse emission regions around PNe from Kingsburgh & English [7].
[b] K 772 and K70 are two H II regions in M 31 from Galarza et al. [3]

ratio in excess of 1.4. Clearly the collision strengths of Pradhan [19] are widely supported by observations; that is to say, *LS*-coupling is a good approximation for the five lowest levels of [O II]. On the other hand, the fact that the [S II] ratios of two H II regions (K 772 and K 70) in M 31 exceed 1.5 by more than 2σ potentially pose a challenge to the validity of *LS*-coupling for the [S II] doublet. Given the very low surface brightness of such emission regions, measurements are always difficult. Any new independent measurements to corroborate these results will be invaluable. Observations for the [O II] $\lambda\lambda3729$, 3726 doublet ratio should also be attempted, although in this case, given the smaller wavelength difference between the two [O II] lines, higher spectral resolution is required to resolve the doublet.

High Density Limit As described earlier, at high electron densities, the [O II] ratio is determined entirely by the ratio of the Einstein transition probabilities. Thus measurements of the doublet ratio at the high-N_e limit can be used to observationally constrain the $A(^2D_{5/2})/A(^2D_{3/2})$ ratio. The derivation in Fig. 1 of theoretical loci calculated using radiative transition probabilities of [O II] show that the values of [22], [23] and [21] indicate about a 6, 20 and 25-50 per cent rise in the A ratio, respectively, based on the assumption that the transition probabilities for [S II] are accurate. Evidence in favour of this interpretation will be discussed later in the paper.

3.2 [Cl III]

On the whole, densities deduced from the [Cl III] doublet ratios are similar to those deduced from the [O II] and [S II] doublet ratios. The agreement between the densities derived from the [S II] ratio and from the [Cl III] ratio is particularly good (c.f. Fig. f1). At densities higher than $\gtrsim 6\times10^3$ cm^{-3}, there is evidence that the [O II] doublet ratio yields densities slightly lower than those given by the [S II] ratio and by the [Cl III] ratio, when using our default atomic data set (c.f. Fig. 1). The O^+ ionization zone is expected to have a larger overlap with that of Cl^{2+} than S^+ with Cl^{2+}, as judged from their ionization potentials. This fact, combined with the realization that the [O II] and [S II] doublets have comparable critical densities which are smaller than those for the [Cl III] doublet, suggest that the systematically lower densities found from the [O II] doublet compared to those found from the [S II] and [Cl III] doublets at densities higher than $\gtrsim 6 \times 10^3$ cm^{-3} are unlikely to be caused by density inhomogeneities or ionization stratification. The discrepancy is more likely caused by uncertainties in the transition probabilities of the [O II] forbidden lines. As estimated in the previous Section, the [O II] $A(^2D_{5/2})/A(^2D_{3/2})$ ratio given by Zeippen ([22]) may need to be revised upwards by about 6 per cent.

3.3 [Ar IV]

Our current optical study shows some evidence that the [Ar IV] doublet ratio tends to yield densities systematically higher than those deduced from the [Cl III] doublet ratio and [O II] or [S II] ratios . The deviation is most profound at densities $\gtrsim 10^4$ cm^{-3} and increases as N_e increases (c.f. Fig. 1). Since Ar^{2+} has an ionization potential of 40.7 eV, much higher than 23.8 eV of Cl^+, it is possible that part of the discrepancy between the [Cl III] and [Ar IV] densities shown in Fig. 1 is caused by ionization stratification, whereby the inner, higher ionization Ar^{3+} zone has a higher N_e than the outer, lower ionization Cl^{2+} zone. Alternatively, nebulae with densities exceeding $\sim 10^4$ cm^{-3} may contain ionized condensations with N_e exceeding the critical densities for [Cl III], [O II] and [S II], where the [Ar IV] emission is less suppressed by collisional de-excitation.

References

1. K. Butler, C. J. Zeippen: Astron. Astrophys. **208**, 337 (1989)
2. M. V. F. Copetti, B. C. Writzl: Astron. Astrophys. **382**, 282 (2002)
3. V. C. Galarza, R. A. M. Walterbos et al.: Astrophys. J. **118**, 2775 (1999)
4. F. P. Keenan, A. Hibbert et al.: Phys. Scr. **48**, 129 (1993)
5. F. P. Keenan, L. H. Aller et al.: Mon. Not. R. Astron. Soc. **281**, 1073 (1996)
6. R. L. Kingsburgh, M. J. Barlow: Mon. Not. R. Astron. Soc. **257**, 317 (1992)
7. R. L. Kingsburgh, J. English: Mon. Not. R. Astron. Soc. **259**, 63 (1992)
8. B. M. McLaughlin, K. L. Bell: J. Phys. B **31**, 4317 (1998)
9. S. J. Meatheringham, M. A. Dopita: Astrophys. J. Suppl. **76**, 1085 (1991)
10. C. Mendoza , C. J. Zeippen: Mon. Not. R. Astron. Soc. **198**, 127 (1982)
11. C. Mendoza: ' Transition probabilities for forbidden lines in the 3p4 configuration. III' In: *Planetary Nebulae.* ed. by Flower D. R. (Kluwer, Dordrecht 1982) pp. 143
12. D. Middlemass, R. E. S. Clegg et al.: Mon. Not. R. Astron. Soc. **239**, 1 (1989)
13. D. J. Monk, M. J. Barlow: Mon. Not. R. Astron. Soc., **242**, 457 (1990)
14. L. Stanghellini, J. B. Kaler: Astrophys. J. **343**, 81 (1989)
15. C. A. Ramsbottom, K. L. Bell et al.: Mon. Not. R. Astron. Soc. **284**, 754 (1997)

16. Y. G. Tsamis, M. J. Barlow et al.: Mon. Not. R. Astron. Soc. **338**, 687 (2003)
17. C. Esteban, M. Peimbert et al.: Astrophys. J. **581**, 24 (2002)
18. D. E. Osterbrock: *Astrophysics of Gaseous Nebulae and Active Galactic Nuclei* (University Science Books, Mill Valley, CA 1989)
19. A. K. Pradhan: Mon. Not. R. Astron. Soc. **177**, 31 (1976)
20. W. Wang, X.-W. Liu et al.: Astron. Astrophys. submitted (2004)
21. W. L. Wiese, J. R. Fuhr et al.: Am. Chem. Soc. for the N.I.S.T. **199** (1996)
22. C. J. Zeippen: Mon. Not. R. Astron. Soc. **198**, 111 (1982)
23. C. J. Zeippen: Astron. Astrophys. **173**, 410 (1987)
24. C. J. Zeippen, K. Butler et al.: Astron. Astrophys. **188**, 251 (1987)

Probing the Nebular Thermal Structure Using the He I Recombination Spectrum and Evidence Against Temperature Fluctuations and Density Inhomogeneities

Y. Zhang, X.-W. Liu, and Y. Liu

Department of Astronomy, Peking University, Beijing 100871, China

1 Introduction

A long-standing problem in nebular astrophysics has been that heavy element abundances derived from collisionally excited lines (CELs) are systematically lower than those derived from optical recombination lines (ORLs). A chemically homogeneous nebula with temperature and density variations fails to account for the discrepancy (c.f. [3]). Instead detailed multi-waveband analyses of a large sample point to the presence of H-deficient condensations embedded in diffuse nebulae [2,8,9]. The 2-abundance model, first proposed by Liu et al. [4], predicts that

$$\mathrm{T_e(CNONe\ ORLs)} \leq \mathrm{T_e(He\ I)} \leq \mathrm{T_e(H\ I)} \leq \mathrm{T_e(CEL)}.$$

In this scenario, measurements of T_e(He I) are of great interest as they provide a test of the two-abundance model and can be used to constrain the mass of H-deficient material embedded in the nebula.

2 Method

Using the analytical fits of He I line emissivities given by Benjamin et al. [1], which include the effects of collisional excitation from the $2s^3S$ and $2s^1S$ levels, the intensity ratio of two He I lines can be written as

$$\frac{I_1}{I_2} = \frac{a_1}{a_2} \mathrm{T_e}^{b_1 - b_2} \exp \frac{c_1 - c_2}{\mathrm{T_e}}.$$

The fitting parameters a, b and c are derived using the recombination data of Smits [7] and the collision strengths of Sawey & Berrington [6]. Therefore, the above equation can be used to determine T_e from the observed He I line ratios. However, it is a difficult task as the dependence of He I line ratios on T_e is relatively weak. Furthermore, the metastable $2s^3S$ level may be significantly populated, leading to additional uncertainties caused by self-absorption. Very accurate measurements of line ratios and reliable atomic data are required to determine T_e(He I) from the observations.

For temperature determinations, we find that the He I $\lambda7281/\lambda6678$ ratio has certain advantages over the other He I line ratios: 1) the He I $\lambda\lambda7281, 6678$ lines are strong and not blended with any other known features in a typical nebular spectrum; 2) compared to other He I lines, uncertainties in the atomic data associated with the He I $\lambda\lambda7281, 6678$ lines are also believed to be small; 3) the He I $\lambda7281/\lambda6678$ ratio is more sensitive to T_e and less sensitive to N_e compared to other He I line ratios; 4) both the He I $\lambda6678$ and $\lambda7281$ lines are from singlet states and thus are independent of the optical depth effects of the $2s^3S$ level; 5) and given the small wavelength span between the He I $\lambda\lambda7281, 6678$ lines, T_e derived from their intensity ratio is insensitive to uncertainties in reddening correction and flux calibration.

3 Results and Discussion

42 PNe are analysed, with data taken from Wesson et al. [9], Liu et al. [2] and Tsamis et al. [8]. The results show that T_e(He I) is significantly lower than T_e(H I) with an average difference of $\langle T_e(\text{H I})-T_e(\text{He I})\rangle = 4000\,\text{K}$.

In contrast, according to the scenario of temperature fluctuations originally proposed by Peimbert [5], we deduce

$$T_e(\text{He I}) \geq T_e(\text{H I}).$$

Similarly, the scenario of density inhomogeneities predicts

$$T_e(\text{He I}) \simeq T_e(\text{H I}),$$

since both He I and H I recombination lines have very high critical densities. Our results thus provide strong evidence against the paradigms of temperature fluctuations and density inhomogeneities.

We show that the presence of a tiny amount of H-deficient knots ($\sim 0.01\%$) can successfully explain the discrepancy. Emission from the H-deficient knots may dramatically enhance intensities of He I recombination lines. If this is the case, traditional dredge-up models may need to be revised. One implication is that the possible presence of H-deficient knots within H II regions could lead to an overestimate of the primordial helium abundance.

References

1. R. A. Benjamin, E. D. Skiliman, D. P. Smits: ApJ, **514**, 307 (1999)
2. Y. Liu, X.-W. Liu, S.-G. Luo: MNRAS, submitted (2004)
3. X.-W. Liu: IAU Symp. 209, Planetary Nebulae, eds. S. Kwok, M. Dopita, R. Sutherland (San Francisco: ASP), p.339 (2003)
4. X.-W. Liu, P. J. Storey et al.: MNRAS, **312**, 585 (2000)
5. M. Peimbert: ApJ, **150**, 825 (1967)
6. P. M. J. Sawey, K. A. Berrington: At. Data Nucl. Data Tables, **55**, 81 (1993)
7. D. P. Smits: MNRAS, **278**, 683 (1996)
8. Y. G. Tsamis, M. J. Barlow et al.: MNRAS, **345**, 196 (2003)
9. R. Wesson, X.-W. Liu, M. J. Barlow: MNRAS, submitted (2004)

Chemically Inhomogeneous Photoionization Modelling of the Planetary Nebula SMC N87

Yiannis G. Tsamis and Daniel Péquignot

LUTH, Laboratoire l'Univers et ses Théories, associé au CNRS (FRE 2462) et à l'Université Paris 7, Observatoire de Paris-Meudon, F-92195 Meudon Cédex, France

We constructed dual abundance photoionization models of SMC N87 along the principles laid out by Péquignot et al. (2002) in their modelling of the Galactic PN NGC 6153. Recent spectroscopic studies have necessitated this approach if one wants to interpret correctly the faint, heavy element optical recombination lines (ORLs) seen in the spectra of PNe and H II regions. The semi-empirical analysis of Tsamis et al. (2003, 2004) has shown that in N87 the abundances of C and O, relative to H, derived from ORLs are a factor of 2–3 larger than those derived from collisionally excited lines (CELs). This discrepancy can be explained by introducing chemical inhomogeneities in the nebula in the form of hydrogen-deficient, cold plasma regions.

SMC N87 is an almost round nebula with no hints of significant internal structure and a very faint nucleus (cf. *HST* imaging by Stanghellini et al. 2003). We modelled the PN by combining 4 spherically symmetric computations with the code NEBU, using *IUE* and optical line fluxes from the NTT 3.5-m (Tsamis et al. 2003), AAT 3.9-m (Barlow 1987), and *HST* (Stanghellini et al. 2003).

We consider a scenario in which the central star at some point in its evolution ejected into the nebula H-deficient material, which originated as nucleosynthetic products brought up to the post-AGB stellar surface by the 3rd dredge up. This H-poor (metal-rich) gas was immersed in the more 'normal' abundance inner regions of the PN, condensing into relatively dense optically thin knots. These, owing to their high metallicity, rapidly cooled down to temperatures much lower than the ambient gas, largely via the emission of infrared fine structure transitions. The model comprises: (1) an H-deficient sector (C1), with covering factor $\omega/4\pi$=0.70, optical depth τ=0.31 at 1 Ryd, outer radius 7.14×10^{16} cm, and a small filling factor, f; (2) a normal composition shell (C2), with $\omega/4\pi$=0.87, τ=5.5, outer radius 3.07×10^{17} cm, and f=1 from which a region coinciding with C1 is removed; (3) the gas pressure is the same in C1 and C2 corresponding to conditions of isobaric cooling for C1; (4) the nebula is radiation bounded overall. Emission from low ionization species like [O I] and [O II] led us to include peripheral clumps in a sector with $\omega/4\pi = 0.13$, and 'infinite' optical depth; (5) the primary spectrum is a black body (L_{BB}=2.05×10^{37} erg s^{-1}, T_{BB}=63.8 kK) with a flux cut-off for frequencies above the He^+ limit (4 Ryd) since no He II lines are observed.

The mean electron densities and temperatures in C1 and C2 are: (T_e, N_e) = (6300 K, 33 710 cm^{-3}) and (12 950 K, 2900 cm^{-3}), respectively. The mass of the H-poor gas is 0.0027 M$_\odot$ vs. 0.365 M$_\odot$ in the normal component. Therefore, even though C1 has enhanced abundances of oxygen and carbon relative to C2

(6 and 21 times Solar respectively, by number, relative to H), the mean elemental abundances in the PN approximate those returned by the semi-empirical method for $t^2 = 0$ (cf. Table 1). The model reproduces exactly the C and O ORLs, which are severely underestimated in chemically homogeneous models.

Table 1. Elemental abundances in SMC N87 (relative to H)

	He	C	N	O	Ne	S	Ar
avg. Model	0.089	8.56	7.14	8.02	7.24	6.00	5.39
$t^2 = 0$	0.097	8.58	7.04	8.03	7.03	*	5.32

N87 is poor in carbon, nitrogen and argon with respect to non-type I SMC PNe and has a rather large ionized mass. Only 65% of the original C was converted to N during the 1st dredge-up, instead of the 100% expected SMC conversion efficiency. The surface enhancement of C during the 3rd dredge-up was only ΔC/H $= 3.56\times10^{-4}$, which is atypical for SMC PNe, resembling rather that of a Galactic PN. For ΔHe/ΔC = 10 (Boothroyd & Sackmann 1988) however, the predicted He/H ratio is 0.087, in agreement with our model value of 0.089.

A potential difficulty with the scenario of a late stellar ejection as the origin of the knots is that the C/O ratio in the H-poor gas (=1.8) is smaller than that in the normal gas (=3.4), when the opposite might be expected. We also find that while the He/H number ratio in the knots cannot be larger than ~0.14 (otherwise the He I fluxes are overestimated), it could take values as low as 0.083, indicating that He, along with H, might also be depleted in the knots. This opens the way for alternative explanations as to the origins of the metal-rich gas, which could include a scenario involving a family of photoevaporating planetesimals in the inner regions of SMC N87.

References

1. Barlow M. J.: MNRAS **227**, 161 (1987)
2. Boothroyd A. I., Sackmann I.-J.: ApJ **328**, 653 (1988)
3. Péquignot D., Amara M., Liu X.-W., et al.: RMxAA (Ser. de Conf.) **12**, 142 (2002)
4. Stanghellini L., Shaw R. A., Balick B., et al.: ApJ **596**, 997 (2003)
5. Tsamis Y. G., Barlow M. J., Liu X.-W., et al.: MNRAS **345**, 186 (2003)
6. Tsamis Y. G., Barlow M. J., Liu X.-W., et al: MNRAS **353**, 953 (2004)

An Optical Recombination Line Abundance Survey for Galactic Bulge Planetary Nebulae

W. Wang[1], X.-W. Liu[1], Y. Zhang[1], and Y. Liu[2]

[1] Department of Astronomy, Peking University, Beijing 100871, P.R. China
[2] Center for Astrophysics, Guangzhou University, Guangzhou 510400, P.R. China

We have obtained deep optical spectra for a sample of 32 Galactic Bulge Planetary Nebulae (GBPNe) at medium spectral resolution. Detailed abundance analyses should allow:

- To determine the chemical composition and evolution of the GBPNe, based on both the traditional method using Collisionally Excited Lines (CELs) and the new method using Optical Recombination Lines (ORLs);
- To have a better understanding of Abundance Discrepancy Factors ADF's (defined as Z(ORLs)/Z(CELs)) and their possible causes;
- To obtain reliable abundances of two key elements, i.e. carbon and magnesium, using ORLs

So far two objects, PN M 3-32, and NGC 6567 have been analyzed. The results are compared to those of the bulge PNe M 1-42 and M 2-36 (Liu et al. 2003) and M 2-24 (Zhang & Liu 2003) and to results of disk PNe and the Sun.

Values of ADF's for C^{2+}, N^{2+}, O^{2+}, and Ne^{2+} for the five GBPNe analyzed so far are given in Table 1. They span a range from 2 to > 20 and are positively correlated withΔT_e, defined as T_e([O III])− T_e(BJ), where BJ is the Balmer Jump, consistent with the trend found by Liu et al. (2000) for a sample of Galactic PNe. Table 1 also shows that ADF's increase with increasing ΔN_e [defined as log N_e(BD) - log N_e([S II]), where N_e(BD) is electron density derived from the Balmer decrement]. These trends suggest that both T_e and N_e variations may play a role in the abundance discrepancies, although detailed multi-waveband analyses for extreme PNe such as NGC 6153 (Liu et al. 2000) and M 1-42 (Liu et al. 2001) and for a large sample of disk PNe (Tsamis et al. 2004; Liu et al. 2004; Wesson et al. 2004) show that both T_e and N_e variations fail to explain the large ADF's. The peculiar PN M 2-24 seems to harbour a dense emission core and no T_e([O III]) is available.

Total elemental abundances derived from ORLs and from CELs are obtained using ICF formulas given by Kingsburgh & Barlow (1994) and are listed in Table 2. We find that almost in all cases abundances of C, N, O and Ne derived from ORLs are above solar, and C, N, Ne from CELs are normally over-abundant, while O from CELs are sub-solar. The two types of line have been found to yield similar C/O and N/O ratios for disk PNe and H II regions (Liu et al. 2000, 2001; Tsamis et al. 2003a,b; Liu et al. 2004; Wesson et al. 2004), which is confirmed by the present study.

Table 1. Δ T_e, Δ (log N_e) and ADFs for five bulge PNe

Object	Δ T_e (K)	Δ (log N_e) (cm^{-3})	ADF O^{2+}	N^{2+}	C^{2+}	Ne^{2+}
NGC 6567	-3400	0.2	2.17	0.96	1.98	5.84
M 2-36	2480	0.4	6.86	5.61	4.80	7.66
M 3-32	4460	0.6	17.5	5.70	15.6	18.5
M 1-42	5660	0.8	22.4	8.22	22.8	17.4
M 2-24		3	17.4	59.9		45.5

Table 2. Total elemental abundances (log N(H)=12.0) and abundance ratios

object	ORLs method				CELs method				ORL/CEL Ratio	
	C	N	O	Ne	C	N	O	Ne	C/O	N/O
NGC 6567	9.25	8.56	8.71	8.44	8.95	8.58	8.47	7.67	1.14	0.55
M 2-36	9.41	9.17	9.64	9.16	8.73	8.42	8.85	8.57	0.78	0.91
M 3-32	9.53	9.41	9.89	9.35	8.34	8.65	8.64	8.09	0.89	0.33
M 1-42	9.35	9.59	9.79	9.40	7.80	8.68	8.63	8.12	2.45	0.56
M 2-24	8.77	8.55	9.16	9.03		8.09	7.92	8.37		0.17
Solar[a]					8.52	7.92	8.83	8.08		
Type I[b]					8.48	8.72	8.65	8.09		
non-Type I[b]					8.81	8.14	8.69	8.10		

[a] Grevesse & Sauval (1998)
[b] The average abundances of Galactic PNe from Kingsburgh & Barlow (1994)

References

1. N. Grevesse & A. J. Sauval: Astro. Jou. **85**, 161 (1998)
2. R. L. Kingsburgh, & M. J. Barlow: Mon. Not. R. Astron. Soc. **271**, 257 (1994)
3. X.-W. Liu, S.-G. Luo et al.: Mon. Not. R. Astron. Soc. **327**, 141 (2001)
4. X.-W. Liu, P. J. Storey et al.: Mon. Not. R. Astron. Soc. **312**, 585 (2000)
5. Y. Liu, X.-W. Liu et al.: Mon. Not. R. Astron. Soc., in press (2004)
6. Y. G. Tsamis, M. J. Barlow et al.: Mon. Not. R. Astron. Soc., in press (2004)
7. Y. G. Tsamis, M. J. Barlow et al.: Mon. Not. R. Astron. Soc. **338**, 687 (2003a)
8. Y. G. Tsamis, M. J. Barlow et al.: Mon. Not. R. Astron. Soc. **345**, 186 (2003b)
9. R. Wesson, X.-W. Liu et al.: Mon. Not. R. Astron. Soc., in press (2004)
10. Y. Zhang, X.-W. Liu: Astron. & Astrop. **404**, 545 (2003)

MOCASSIN: 3D Photoionisation and Dust Radiative Transfer Modelling of PNe

Barbara Ercolano[1], M.J. Barlow[1], P.J. Storey[1], and X.-W. Liu[2]

[1] Department of Physics and Astronomy, University College London, UK
[2] Department of Astronomy, Peking University, Beijing, China

Abstract. We present the first 3D Monte Carlo photoionisation code to include a fully self-consistent treatment of dust radiative transfer (RT) within the photoionised region. This is the latest development of the recently published pure photoionisation code MOCASSIN (Ercolano et al., 2003a) and it is currently undergoing several benchmarking tests. The preliminary results of these tests are presented. The new code provides the ideal tool for the analysis of dusty ionised regions showing asymmetries and/or density and chemical inhomogeneities

1 Introduction

Recent advances in instrumentation allow the geometry of many PNe (including those in the SMC and LMC) to be resolved. It is now clear that the assumption of spherical symmetry is not justified in most cases. This imposes serious limitations on the application of 1D codes, with the risk of obtaining incorrect abundance determinations, resulting in misleading conclusions. The 3D photoionisation code MOCASSIN (MOnte CArlo SimulationS of Ionised Nebulae, Ercolano et al. 2003a) was designed to remedy these serious shortcomings and has been successfully applied to the study of several PNe (Ercolano et al. 2003b,c; Ercolano et al. 2004).

The presence of dust grains in ionised environments can have significant effects on the radiative transfer, as the grains compete with the gas for the absorption of UV photons, as well as being heated by nebular resonance line photons. These processes can only be treated properly by incorporating the scattering, absorption and emission of radiation by dust particles that are mixed with the gas in the photoionised region. Moreover, the accurate determination of dust temperatures and spectral energy distributions (SEDs) can only be achieved by treating discrete grain sizes and different species separately. All of the above has already been implemented in the new version of MOCASSIN, which is currently undergoing benchmarking tests, as described below.

2 Benchmarking Tests

The dust and gas version of MOCASSIN is benchmarked against the popular 1D code, DUSTY, for the spherically symmetric, homogeneous test models described

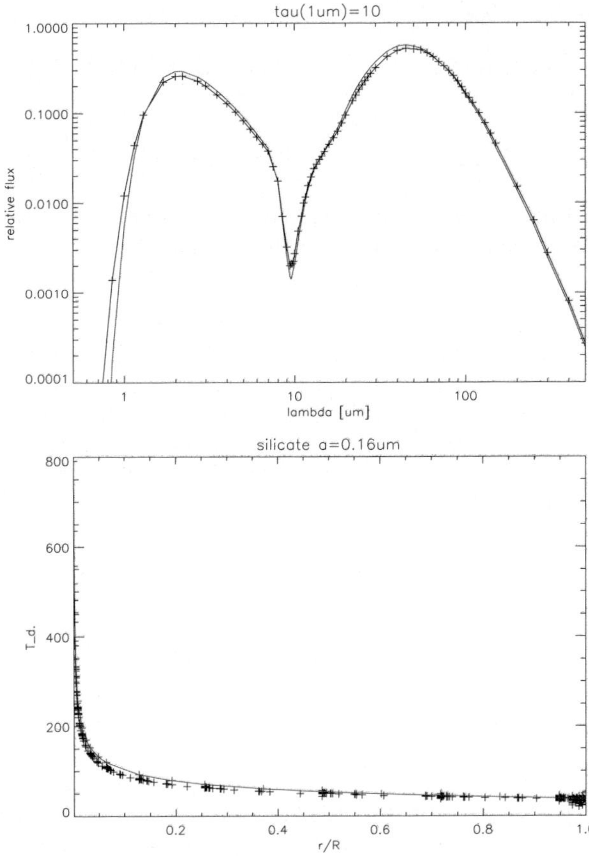

Fig. 1. Spherically symmetric, homogeneous benchmark model. See text for details.

by Ivezic et al. (1997). The input parameters for these pure-silicate models are given in their paper and the grain optical constants used are those included in DUSTY's library, calculated from the data of Draine & Lee (1984).

Figure 1 shows the comparison between MOCASSIN – crosses – and DUSTY (Ivezic et al., 1997) – dashed line – for the p=0 case (see Ivezic et al., 1997) and $\tau_{1\mu m} = 10$. The emerging SEDs in relative flux units are shown on the left panel and the radial grain temperature distribution on the right panel. Taking into account that a completely different approach is used by the two codes, the results obtained by the benchmarking are in good agreement. 2D disk benchmarks are currently being carried out for the pure dust case, whilst a dust and gas model of the PN NGC 3918 is also being constructed (Ercolano et al. in preparation).

While a number of Monte Carlo dust radiative transfer codes already exist (e.g. Bjorkman & Wood, 2001), MOCASSIN is currently the only 3D code capable of self-consistently treating dust RT within the photoionised region, providing the ideal tool for the modelling of non-spherical, density and/or chemically inhomogeneous dusty photoionised environments.

References

1. J.E. Bjorkman, K. Wood: ApJ 554, 615 (2001)
2. D.T. Draine, H.M. Lee: ApJ 285, 89 (1984)
3. B. Ercolano, M.J. Barlow, P.J. Storey, X.-W. Liu: MNRAS 340, 11 36 (2003a)
4. B. Ercolano, C. Morisset, M.J. Barlow, P.J. Storey, X.-W. Liu: MNRAS 340, 1153 (2003b)
5. B. Ercolano, M.J. Barlow, P.J. Storey, X.-W. Liu, T. Rauch, K. Werner: MNRAS 344, 1145 (2003c)
6. B. Ercolano, R. Wesson, Y. Zhang, M.J. Barlow, O. De Marco, T. Rauch, X.-W. Liu: MNRAS submitted (2004)
7. Z. Ivezic, M.A.T. Groenewegen, A. Men'shchikov, R., Szczerba: M NRAS 291, 121 (1997)

C III] Imagery of Planetary Nebulae and Extragalactic H II Regions

R.J. Dufour[1], R.B.C. Henry[2], K.B. Kwitter[3], B.A. Buckalew[4], B.D. Moore[1], J. Bohigas[5], and C. Esteban[6]

[1] Rice University, Houston, Texas, 77005, USA (rjd@rice.edu)
[2] University of Oklahoma, Norman, Oklahoma, USA
[3] Williams College, Williamstown, Massachusetts, USA
[4] University of Wyoming, Laramie, Wyoming, USA
[5] Instituto de Astronomia, UNAM, Ensenada, Mexico
[6] Instituto Astrofisica Canaris, Tenerife, Spain

Abstract. We report the first results of a Cycle 12 *Hubble Space Telescope* SNAP program imaging the C III] 1909Å lines in planetary nebulae (PNe) and extragalactic H II regions using WFPC2 with the F185W filter. To date (2004 July), four PNe (NGC 6210, NGC 6720, NGC 6826 & NGC 7662) and four H II regions (LMC N160A, SMC N81, NGC 2363 and I Zwicky 18) have been observed. Using continuum imagery of the nebulae made with HST WFPC2 and the knowledge of the UV spectra of regions in the nebulae from archival HST FOS/STIS or IUE data, the UV continuum contribution to the F185W images is estimated and subtracted out to give images of the nebulae in C III] 1909Å. The objectives of the program include: *(a)* studying the effects that spatial variations in T_e, N_e, and dust have on the observed F(1909Å)/F(5007Å) line ratio across the nebulae; *(b)* studying the co-spatiality of the C^{+2} and O^{+2} ionization structure; and *(c)* evaluation of the accuracy of using observations of the integrated F(1909Å)/F(5007Å) line ratio in nebulae for determining the C^{+2}/O^{+2} ratio and the total C/O abundance ratio in PNe and H II regions.

1 Introduction and Poster Access

We present the results of a Cycle 12 *Hubble Space Telescope* SNAP program (GO 09740) imaging several planetary nebulae (PNe) and extragalactic H II regions with the F185W WFPC2 filter that includes the important C III] 1909Å lines used in determining the carbon abundances in such nebulae. A copy of the poster is available on the internet at the URL: `http://galaxy.rice.edu/~rjd/esoposter.pdf`. A few notes on some of our preliminary results are summarized below.

2 The Planetary Nebula NGC 6210

NGC 6210 has numerous WFPC2 observations in the archives (through both the PC1 and WF3 chips), but relatively little about its ionization structure from these data has been published to date. In a three-color composite of its ionization structure in [O II] 3727Å(coded red), Hα(coded green), and [O III] 5007Å(coded blue), one can see a 'crescent' of low excitation material above the central star on the images. We aligned and compared our GO 9740 WF3 images with those taken with other filters in the Cycle 6 GO Program 6792. The C III]/[O III] ratio map shows that the 1909/5007 ratio drops to about 0.6-0.8 (relative to a normalized mean for the ratio of 1.00) in this region above the central star in a clumpy fashion that may be correlated with the strong [O II] emitting material see in the color composite image. However, the [O III]/Hβ map shows that this ionization ratio is remarkably constant across the bright inner nebula. At first, the drop in C III]/[O III] above the central star was thought due to localized reddening, however, an Hβ/Hα map shows very little variation in reddening across the nebula, so the cause of the drop in the C III]/[O III] ratio remains a mystery. We plan to use the reddening map to correct the observed C III]/[O III] map in the future and see if this will have any effect in removing this apparently significant variation in the C III]/[O III] ratio (the reddening fluctuations in this region seem to be far too small for it to produce this effect).

3 The Planetary Nebula NGC 7662

NGC 7662 has an exceptional amount of WFPC2 imagery in the high resolution planetary camera (PC1) mode. This is from HST studies of its expansion (GO 8390) and Fliers (GO 6117). The physical analysis of these images have been published by Balick et al. 1998, *ApJ*, **116**, 360. Unfortunately however, no F487N (Hβ) images have been taken. The multiple ring morphology and ionization structure of nebula, and particularly the low ionization Fliers, are evident on a color composite of the nebula in [N II] (red), Hα (green), and [O III] (blue). From our [O III]/Hα map, it is evident that the nebula has significant ionization structure, with bipolar-looking low ionization structures in the inner bright ring. However, unlike NGC 6210, the C III]/[O III] ratio image is rather constant across the bright ring, though apparently lower outside the ring, dropping to about half the value that it is in the ring.

4 C III] in H II Regions

In the poster we also presented results for the small but bright SMC H II region Henize 81 (Heydari-Malayeri et al. 1999, *A&A*, **344**, 848) for which we found strong C III] emission around stars in its center. In fact, we find that the C III] emission is *much* more concentrated around two O stars than the [O III] emission that is much more spread out in the nebula. We are currently engaged in efforts to improve the continuum subtraction of the F185W imagery and calibrations of all of the WFPC2 imagery through various emission-line filters. Once done, we plan to produce dereddened ratio maps of C III]/[O III] across the nebulae and model these with the photoionization code *CLOUDY*.

Acknowledgements RJD, RCBH, and KBK acknowledges the support of this research through AURA/STScI grants to our respective institutions. RJD also notes the Shell Foundation of Houston supported a summer undergraduate, Andrew Biddle, to assist with the imagery processing.

CRIRES and VISIR, ESO's Latest Tools for Infrared Studies of PNe beyond the Milky Way

Hans Ulrich Käufl

European Southern Observatory, D-85748 Garching bei München, Germany

Abstract. CRIRES and VISIR[1] are part of the first generation of VLT instruments. CRIRES is a cryogenic pre-dispersed long-slit ($\approx$ 40") spectrograph with a nominal resolution $\lambda/\Delta\lambda \approx 10^5$ (i.e. $\approx$ 3kms^{-1}). CRIRES covers the 1-5.5μm wavelength regime. In combination with adaptive optics, CRIRES has a spatial resolution set by the slit size of 0.2". In the absence of guide stars, or if the maximum spectral resolution is not mandatory, the slit can be opened and adjusted, either to the seeing or to the astrophysical requirements. The CRIRES focal plane is equipped with a mosaic of four 1024^2 pixel InSb arrays. This will allow for a quasi-instantaneous access to 2% of the spectrum in one exposure while Nyquist sampling. VISIR is a multi-mode imager/spectrometer for $\lambda \approx 8 - 24\mu m$. The camera has a field of view of typically 30" $*$ 30". It is fully diffraction limited, i.e. at $\lambda \approx 10\mu m$ the spatial resolution is $\approx$ 0.30". In spectroscopy mode VISIR offers three ranges of spectral resolution up to a typical resolution $\lambda/\Delta\lambda \approx 3 * 10^4$ (i.e. $\approx$ 10kms^{-1}). Both VISIR arms are equipped with one 256^2 pixel As:Si BIB array each. VISIR is presently being commissioned on VLT-UT3 while CRIRES shall arrive on Paranal in the third quarter of 2005. Both instruments, in combination with the VLT 8m UTs will allow for detailed infrared studies of Planetary Nebulae and their precursors way beyond the Magellanic Clouds.

1 Introduction

Infrared imaging and spectroscopy have proven to be an indispensable tools for studying PNe. Ground based observations in this context can provide:

- systematic surveys for "dusty" PNe or proto-PNe
- in-detail studies of physical properties (e.g. spatial structure, excitation conditions, kinematics or gas-to-dust ratio) with sub-arcsec spatial and few kms^{-1} equivalent spectral resolution
- studies of PNe progenitors (AGB-stars, proto-PNe) from very basic statistical properties (census, luminosity function, color-color diagrams, reddening) to in-detail studies of the physics of mass-loss (by infrared spectroscopy of stellar atmospheres) or even studies of the mass-loss history (by high spatial resolution imaging spectroscopy)

To make optimum use of the scientific potential of groundbased infrared observations, the VLT has been equipped with a suite of highly specialized optical

[1] CRIRES stands for CRyogenic Infrared Echelle Spectrograph whereas VISIR stands for VLT Imager and Spectrometer for the mid-Infra Red.

and infrared instruments (for a current overview see [12]). CRIRES and VISIR are filling in the $(\lambda, \lambda/\Delta\lambda)$-map the space between 1 and 24μm while providing the maximum spectral (i.e. 3-10kms^{-1}) and spatial resolution (i.e. 0.3-0.2 ").

2 The VISIR Instrument, Results from Commissioning

2.1 Short Description of VISIR

VISIR has been built for ESO by CEA/SAP/DAPNIA in Saclay, France and ASTRON/NFRA in Dwingeloo and Groningen, The Netherlands. The instrument and the laboratory testing including the qualification of the detectors has been reported in great detail elsewhere [11,5,4]. VISIR consists of 2 largely independent sub-units, the imager and the spectrometer, cohabitating in a common cryostat. The spectrometer has a pair of collimators with 6 gratings, 2 of them 40cm-Echelles and 2 grisms (KRS5) for cross-dispersion. The imager has various filters and three magnifications (0.075"/pix, 0.127"/pix and 0.2"/pix). The actual sensitivities and other performance figures can best be accessed via the standard ESO-WWW-documentation (VISIR is being included in the recent call for proposals of ESO).

2.2 [Ne II] Observations of Menzel 3 at $\lambda = 12.8\mu m$

Figure 1 shows a preview of what data can be expected in imaging of Galactic PNe. Menzel 3 (aka the Ant Nebula = Mz 3 = He 2-154 = PK 331-01 1) was picked for its complex morphology in HST-images (c.f. Fig. 1). Sophisticated models exist (c.f. [2,6]) and Mz 3 also exhibits potentially a complex structure in X-rays[10]. Figure 2 shows an exemplary long-slit Echelle spectrum in [Ne II] of Mz 3, allowing to study kinematics with few kms^{-1} velocity resolution while providing 0.4" spatial resolution along the slit. Similar observations for [Ar III], $\lambda \approx 8.99138\mu m$ and [S IV], $\lambda \approx 10.51049\mu m$ will also become possible. Acquisition is done by taking a narrow-band image first, then switching to spectroscopy. Thus the registration of the long-slit on the target is well established.

2.3 Outlook and Sensitivities

VISIR (and to a certain extent also CRIRES) work in the Background Noise LImited Performance (BLIP) regime. This implies, that for a given line the sensitivity increases (!) with $\sqrt{\lambda/\Delta\lambda}$. As long as the line is not excessively over-resolved the line flux per pixel stays constant while the thermal background gets diluted in dispersion direction. This is very favorable for observing PNe. Detection is merely a question of surface brightness. This in turn means, that observations such as presented here are feasible for all PNe as long as they have an apparent diameter of ≥ 0.4 ". As Mz 3 is clearly not at the detection limit, it could be detected in imaging and observed in high resolution Echelle spectroscopy at least up to a distance of 100 kpc.

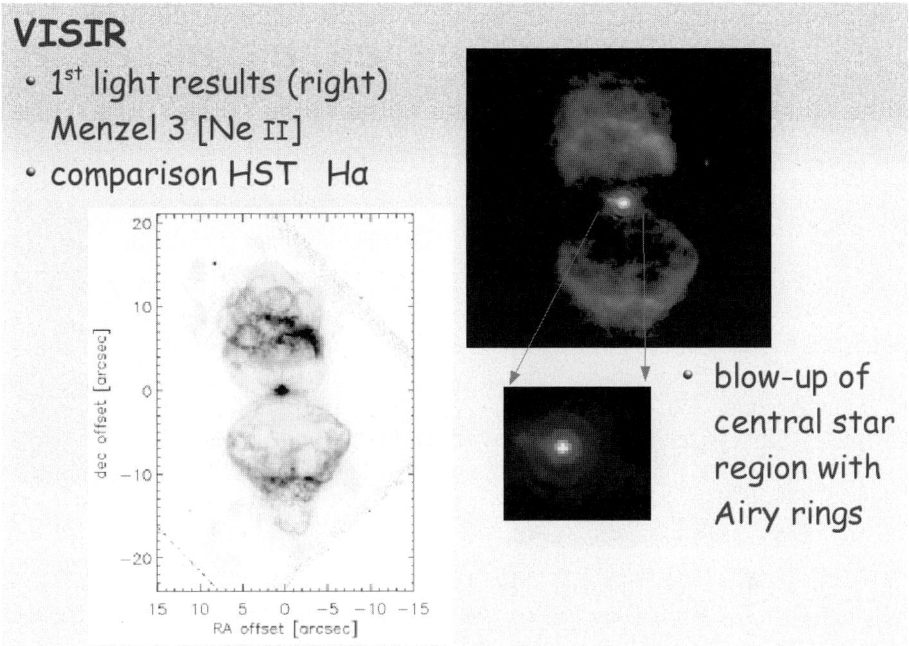

Fig. 1. VISIR narrow-band imaging [Ne II] observations of Mz 3 at $\lambda = 12.8\mu m$ are shown (right) in comparison with an H_α-image obtained with the HST. Both images are scaled identically. Most of the complex morphology in the H_α-image can be retraced in the [Ne II] image. This is by itself already a quite interesting result, as it basically proves that the entire nebula is optical thin for H_α. **Blow Up:** The PSF of VISIR/VLT is close to theoretical expectation: diffraction limited core with a FWHM of 0.39" surrounded by the first and second Airy ring (as this is a narrow-band filter image, the rings have the theoretical contrast). The structure next to the central star is most likely an artifact due to synchronization problems in the first commissioning nights.

3 Description and Status of CRIRES

CRIRES is a cryogenic, pre-dispersed, infrared echelle spectrograph designed to provide a resolving power $\lambda/\Delta\lambda$ of 10^5 between 1 and $5\mu m$ at the Nasmyth focus B of the 8m VLT unit telescope #**1** (Antu). A curvature sensing adaptive optics system feed is used to minimize slit losses and to provide diffraction limited spatial resolution along the slit. A mosaic of 4 Aladdin III InSb-arrays packaged on custom-fabricated ceramics boards has been developed. This provides for an effective 4096x512 pixel focal plane array, to maximize the free spectral range covered in each exposure. Stability, overall precision and calibration of CRIRES aim at a precision of $75ms^{-1}$. Insertion of gas cells to measure radial velocities with higher precision is foreseen. For measurement of circular polarization in lines a Fresnel rhomb in combination with a Wollaston prism for magnetic

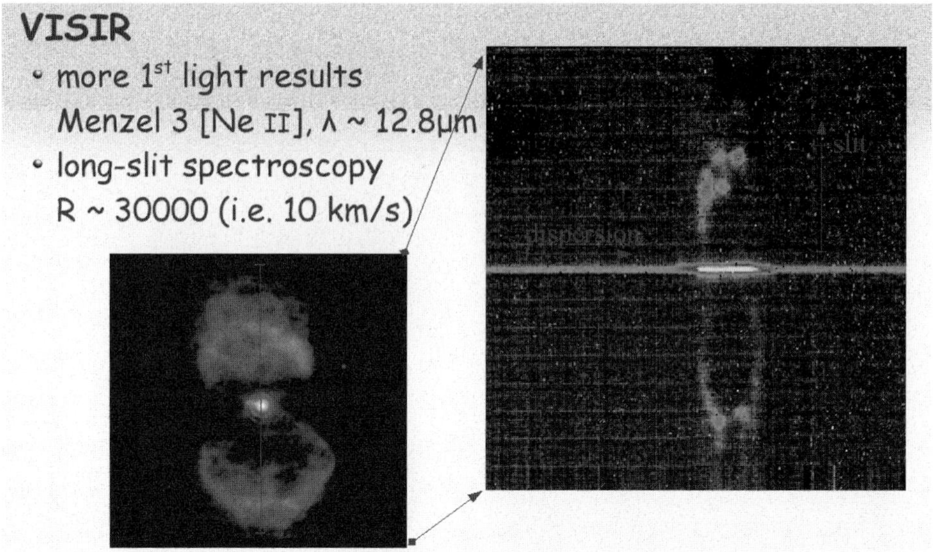

Fig. 2. left panel: [Ne II] image, used for acquisition with the slit in N/S-direction superimposed. The **right panel** shows the actual spectrum. In the dispersion direction the image covers $\approx$ 840 kms^{-1}, while one pixel corresponds to $\approx$ 3.3 kms^{-1}. In the middle of the 2-dimensional spectrum one sees the continuum originating from the compact central source. It is quite obvious that the velocity structure of the Northern lobe is quite different from that of the Southern. Moreover, the velocity structure is quite knotty, which is not accounted for in the theoretical models [6]. The highest velocities are found at the central star.

Doppler imaging is foreseen. The implementation of full spectropolarimetry is under study. This is one result of a scientific workshop held at ESO in late 2003 to refine the science-case of CRIRES [7]. Installation at the VLT is scheduled during the second half of 2005. An detailed description of the instrument including sensitivities is given elsewhere [8,13].

4 High Spectral Resolution in the IR for Proto-PNe

Important atomic lines exist also in the wavelength range of CRIRES[3]: e.g. various recombination lines of H and He: e.g. Pa$_\beta$, Br$_\alpha$, Br$_\gamma$ or Pfβ, forbidden lines (e.g. in the J-band [Fe II], in the K-band [K III] or in the L-band [Zn IV]). In the K-band H_2 lines from hot shocked molecular gas (quadrupole transitions) can be abundant. However, the real interest in this λ-range is to study fully resolved rotational-vibrational molecular transitions from CO or SiO in the atmospheres of AGB-stars and proto-PNs. These lines allow to study and understand massloss[1], dust-formation and s-process element synthesis on the AGB[9]. The latter will be achieved by studying isotopic ratios in Oxygen, Carbon and Silicon (note that isotopic shifts in molecules are of order of 1%). The combination of the high throughput of CRIRES with the VLT will allow these kind of studies to be performed on samples of objects which are selected on astrophysical criteria other than apparent brightness.

5 Conclusions

While ground-based astronomy offers extremely high resolution both spatially and spectrally, space mission are extremely complementary for two reasons: a) the sensitivity in the infrared is orders of magnitude higher, at least in imaging; and b) the interference by telluric absorption features can be a show-stopper for many programmes. In that sense a careful combination of ground- and space-based observations is the key to success.

Acknowledgments: Special thanks go to the VISIR team for making some spectacular early commissioning data available for this presentation.

References

1. Aringer, B., Höfner, S., Wiedemann, G. et al. 1999 A&A342, p. 799
2. Balick, B. & Frank, A. 2002, Ann. Rev. Astron. Astroph., **40**, p. 48
3. Barlow, M. 2004 "Post-AGB Objects and PNe" in Proceedings of ESO-workshop "High Resolution Infrared Spectroscopy in Astronomy", Springer, in press
4. Claret, A.J., Lagage, P.-O., Pantin, E., et al 2003 "Calibration of VISIR, the VLT mid-infrared imager/spectrometer" in Proceedings SPIE **4841**, pp. 252-262
5. Galdemard, P., Garnier, F., Mulet, P., & Reynolds, D. 2003 "Characterization of DRS Technologies' 256x256 mid-IR arrays for VISIR " in Proceedings SPIE **4841**, p. 129
6. García-Segura, G., Langer, N., Różyczca, M. & Franco, J. 1999 ApJ, 517, p. 767-781
7. Käufl, H.U., Moorwood, A.F.M. & Siebenmorgen, R. 2004 Proc. of ESO-workshop "High Resolution Infrared Spectroscopy in Astronomy", Springer, in press
8. Käufl, H.U., Ballester, B., Biereichel, P. et al 2004 "CRIRES: A High Resolution Infrared Spectrograph for ESO's VLT" in Proceedings SPIE **5492** , in press
9. Käufl, H.U., Aringer, B., Uttenthaler, S. & Hron, J. 2005 "Isotopic Abundances in AGB-Stars " in Proc. of ESO/Arcetri-workshop on "Chemical Abundances and Mixing in Stars in the Milky Way and its Satellites", Springer, in press
10. Kastner, J.H., Balick, B., Blackman, E.G. et al. 2003, ApJ591, l-37-L40
11. Lagage, P.-O., Pel, J.W., Claret, A., et al 2003 "Characterization in the laboratory of VISIR, the mid-infrared imager and spectrometer for the VLT " in Proceedings SPIE **4841**, pp. 923-931
12. Moorwood, A.F.M. 2004 "Instrumentation at the ESO VLT" in Proceedings SPIE **5492** , in press
13. Moorwood, A.F.M., 2004 "CRIRES: Context and Status" Proc. of ESO-workshop "High Resolution Infrared Spectroscopy in Astronomy", Springer, in press

Part VI

PNe as Chemical Probes of Galaxies

Chemical Composition of Planetary Nebulae: The Galaxy and the Magellanic Clouds

Walter J. Maciel, Roberto D.D. Costa, and Thais E.P. Idiart

Instituto de Astronomia, Geofísica e Ciências Atmosféricas (IAG), Universidade de São Paulo, São Paulo SP, Brazil

Abstract. An analysis is made of the abundances of planetary nebulae (PN) in the Galaxy (MW), the Large Magellanic Cloud (LMC) and the Small Magellanic Cloud (SMC). The data have been gathered by the IAG/USP group in a homogeneous procedure, involving observational techniques, data acquisition and reduction. Emphasis is placed on distance-independent correlations, which are particularly important for PN in the Galactic disk. It is shown that: (i) our abundances are comparable to the accurate abundances of nebulae with ISO data; (ii) most abundance correlations hold for all three systems, taking into account the metallicity differences between the Galaxy and the Magellanic Clouds; and (iii) there is a good general agreement of our data with the predictions of recent theoretical models for PN progenitors.

1 Introduction

Planetary nebulae (PN) are especially important for studies of the chemical evolution of the Galaxy and the Magellanic Clouds. They are the offspring of stars in a mass range of roughly 0.8 to 8 $M_{\odot}$, so that a relatively wide distribution of ages and population types can in principle be observed.

Chemical abundances of PN are derived for several elements, so that empirical distance-independent correlations can be determined with a reasonable degree of accuracy. These correlations can then be compared with the predictions of the nucleosynthetical processes and chemical evolution models of intermediate mass stars (see for example [15], [7], [12]) and act as constraints on these models.

In particular, it is known that there is a general metallicity difference between the Galaxy and the Magellanic Clouds, and between the LMC and SMC themselves. On the other hand, it is not clear whether the chemical evolution has proceeded differently in the Clouds as compared to the Galaxy, or the evolutionary steps were essentially the same, albeit starting from a more metal-poor gas. Some information in this respect can be obtained by the analysis of the abundances of a relatively large sample of PN in these systems.

The total number of PN in the Clouds with accurate abundances is relatively small, as compared with the population of PN in the Galactic disk and bulge (see for example [18] and [13]). Since the early nineties, the IAG/USP group launched a long term project to derive accurate abundances for PN in the Galaxy and the Magellanic Clouds, based on detailed spectroscopic observations secured both in Brazil and in Chile (see [2] for some references). In this paper, we will present a brief account of this work.

2 The Data

In a series of papers by the IAG/USP group, the chemical compositions of over two hundred planetary nebulae have been derived, including objects in four different regions: (i) the Galactic disk; (ii) the Galactic bulge; (iii) the LMC; and (iv) the SMC. The main references are [3], [6], [5], [2], [4], [1], [8], [9], [10], and [11]. All observations have been obtained using either the 1.6 m LNA telescope at Brasópolis, Brazil, or the 1.52 m ESO telescope at La Silla, Chile. At ESO a Boller and Chivens cassegrain spectrograph was used with a CCD detector at a reciprocal dispersion of about 2.5 Å/pixel. At the LNA the dispersion was of about 4.4 Å/pixel. The reductions have been made using the IRAF package. Basically, the electron densities have been determined from the [SII] 6716/6731Å lines, while for the electron temperatures we have used the [OIII] 4363/5007Å and [NII] 5754/6584Å lines. The observational techniques employed, data acquisition and reduction procedures have all been similar, as warranted by the high degree of homogeneity desirable in the analysis of the chemical composition of planetary nebulae.

3 Distance-Independent Correlations

Planetary Nebulae and HII Regions Comparing the average abundances of PN with the corresponding values for HII regions, we find that the abundances of the elements S, Ne, Ar, and O are similar both in the PN and HII regions of each system. Average values for oxygen in PN are $\epsilon(\mathrm{O}) = \log \mathrm{O/H} + 12 \simeq 8.65$, 8.40 and 8.16 dex for the MW, LMC and SMC, respectively. The differences for O, S, Ar and Ne are relatively small, generally under 0.2 dex, similar or smaller than the average abundance uncertainties. The main differences appear for helium and especially nitrogen, which are enriched in PN, and are dredged up in the progenitor stars. The differences amount to about 40% for He/H and 0.5 to 0.7 dex for N/H.

Correlations Involving S, Ar, Ne and O The elements S, Ar and Ne are not expected to be produced by the PN progenitor stars, as they are manufactured in the late evolutionary stages of massive stars. Therefore, S, Ar and Ne abundances as measured in planetary nebulae should reflect the interstellar composition at the time the progenitor stars were formed. Regarding the oxygen abundances, there is some evidence of a reduction in the O/H ratio in some PN due to ON cycling in their progenitor stars, so that there may be some small variations in the O/H abundances in the nebular gas relative to the pristine abundances.

The variation of the ratios S/H, Ar/H and Ne/H with O/H show a good positive correlation for all systems, with similar slopes. Our sample is not complete, so that the average abundances do not necessarily reflect the metallicities of the galaxies considered. However, the metallicity differences can be observed judging from the observed metallicity range. The Galactic nebulae extend to higher metallicities (up to $\epsilon(\mathrm{O}) \simeq 9.2$), while the LMC objects reach $\epsilon(\mathrm{O}) \simeq 8.8$, and the SMC extends to even lower metallicities, namely $\epsilon(\mathrm{O}) \simeq 7.0$. Taking into account the Galactic and MC nebulae by Stasińska et al. [18], these results are confirmed and extended, especially at lower metallicities, $\epsilon(\mathrm{O}) < 8.0$, for which their sample is richer than ours.

Correlations Involving N An anticorrelation between N/O and O/H has been discussed by a number of people (see for example [2], [18] and Perinotto et al., this conference) and may be an evidence for the conversion of O into N in the PN progenitor stars. This scenario is basically supported by our data. The Galactic disk nebulae show a mild correlation, taking into account the majority of objects with $\epsilon(\mathrm{O}) > 8.0$. For the Magellanic Clouds, the anticorrelation appears better defined, particularly for the SMC. In this case, our sample fits nicely the previous sample [18], especially at higher O/H ratios, that is, $\epsilon(\mathrm{O}) > 8.0$.

Assuming that the nitrogen abundance is directly proportional to the oxygen abundance, as is the case for S, Ar and Ne, then for lower O/H ratios the N/O ratio is expected to be essentially constant. Since this ratio increases on average for lower O/H values, we have to conclude that the nitrogen abundance is enhanced, possibly at the expense of oxygen. Since the N/O $\times$ O/H anticorrelation is better defined for the Magellanic Clouds, it seems that the lower metallicities of these galaxies favour the ON-cycling as observed in planetary nebulae.

A well defined N/O *v.* N/H correlation is shown in Fig. 1 for Galactic (crosses), LMC (empty circles) and SMC (filled circles) nebulae. Average error bars are provided at the lower right corner. The dashed line shows the theoretical predictions of the final model for the LMC by Groenewegen and de Jong [12]. The inflection point of the line corresponds to initial stellar masses of about 2 $M_\odot$, lower masses being associated with the left side of the plot and higher masses to the right side. As pointed out in [12], the location of PN in this diagram is a good indicator of main sequence mass of the progenitor star. We can see that our SMC sample includes basically low mass progenitors.

Correlations Involving He Another interesting correlation is a plot of N/O as a function of He/H (figure 2). For the Galaxy (crosses), a good correlation is observed. The inclusion of the LMC and SMC objects generally supports this conclusion, although the scattering increases, probably due to the different metallicities of the galaxies. The main differences occur for the SMC, for which the He/H ratio has approximately the same range as the LMC and even the Galaxy, but N/O is generally lower, at least in our sample. Since the O/H ratio is also smaller in the SMC, due to the metallicity difference, we apparently observe a smaller nitrogen enhancement in the SMC, so that the progenitor star masses are somewhat lower in this galaxy, as also suggested by Fig. 1. In other words, type I PN are probably less frequent in the SMC, in agreement with the conclusion by Stanghellini et al. [17], based on HST data, that bipolar PN are much rarer in the SMC than in the LMC or in the Galaxy (see also Shaw and Villaver, this conference). It is argued that the lower metallicity environment of

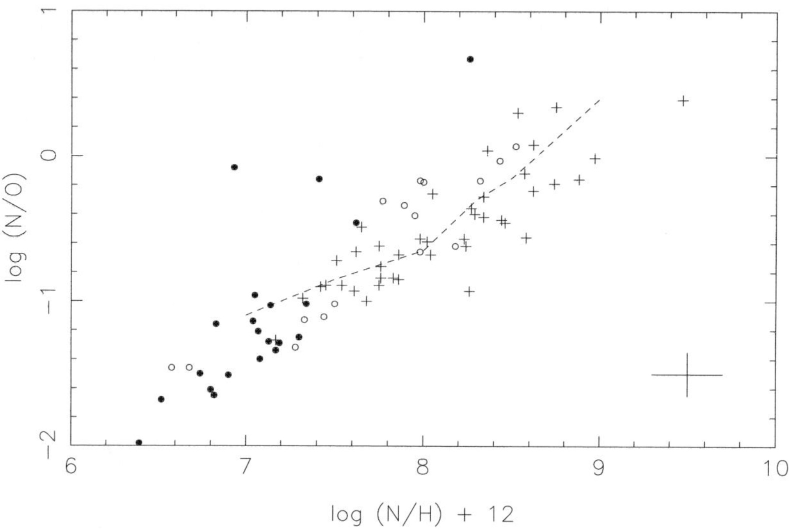

Fig. 1. Nitrogen abundances of PN in the Galaxy (crosses), LMC (empty circles) and SMC (filled circles). The dashed line shows predictions of the final model by Groenewegen and de Jong [12]. Average error bars are shown at the lower right corner.

the SMC does not favour the formation of the higher mass progenitors that give rise to Type I bipolar (or asymmetric) PN.

Recently, Marigo et al. [15] have presented abundances of Galactic PN for which ISO and IUE spectra are available, so that their abundances are presumably well determined, as all important ionization stages have been considered, making it practically unnecessary to use ionization correction factors. Also, infrared data are less dependent on the electron temperatures and their intrinsic uncertainties. These objects are included in Fig. 2 (triangles connected by dashed lines), and show a good agreement with our data, at least for the Milky Way. Another comparison of our Galactic PN with nebulae having ISO abundances can be seen in Fig. 3, where we plot our PN along with nebulae having ISO abundances from Pottasch et al. [16] (triangles connected by dashed lines). Also shown in the figure are the Sun, Orion (lozenge) and predictions of theoretical models by Marigo et al. [15] (continuous line), which consist of synthetic evolutionary models for the thermally pulsing asymptotic giant branch stars.

4 Final Remarks

Recently, Costa et al. [3] have analyzed a sample of PN located in the direction of the Galactic anticenter. It was shown that the observed radial O/H gradient tends to flatten out for large galactocentric distances, roughly $R > 10$ kpc, in agreement with previous results by Maciel and Quireza [14]. Therefore, it is interesting to investigate whether anticenter PN – which are in principle more

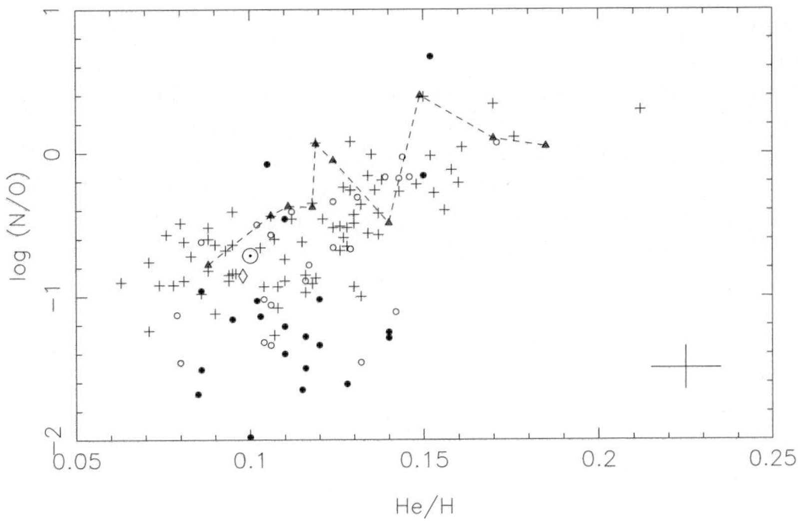

Fig. 2. The N/O *v.* He/H correlation for PN. Symbols are as in Fig. 1. Also included are the Sun, Orion (lozenge) and Galactic PN with ISO data from Marigo et al. [15] (triangles connected by dashed lines).

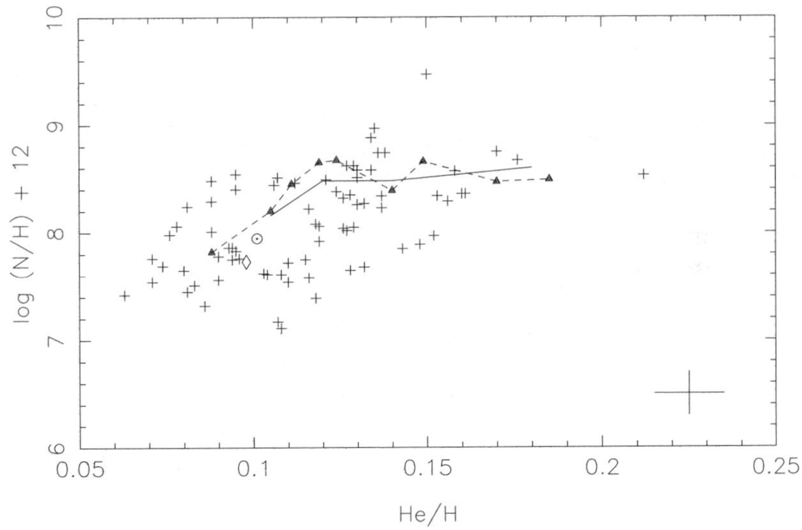

Fig. 3. The N/H *v.* He/H correlation for Galactic PN. Also included are the Sun, Orion (lozenge), PN with ISO data from Pottasch et al. [16] (triangles connected by dashed lines) and predictions of theoretical models by Marigo et al. [15] (continuous line).

metal poor than their inner counterparts – have abundances closer to metal poor systems, namely the LMC and SMC.

An analysis of our sample shows that all Galactic nebulae, comprising inner and outer objects, are evenly distributed along the disk, so that there is apparently no association of the outer nebulae with the lower metallicity objects of the Magellanic Clouds. However, taking into account the nebulae of the larger sample of Lago and Maciel (in preparation), which includes objects from the literature at low galactocentric distances, we conclude that a tendency can be observed, in the sense that the higher metallicity objects are preferentially located closer to the Galactic center.

Finally, it is interesting to notice that the same correlations found for PN in the Galactic disk and in the Magellanic Clouds can also be observed in the objects of the Galactic bulge, as recently shown by Escudero et al. [6].

Acknowledgements. We thank J. A. de Freitas Pacheco, M. M. M. Uchida, L. G. Lago and A. V. Escudero for some helpful discussions. This work was partially supported by CNPq and FAPESP.

References

1. R. D. D. Costa, C. Chiappini, W. J. Maciel, J. A. de Freitas Pacheco: Astron. Astrophys. Suppl. **116**, 249 (1996)
2. R. D. D. Costa, J. A. de Freitas Pacheco, T. E. P. Idiart: Astron. Astrophys. Suppl. **145**, 467 (2000)
3. R. D. D. Costa, M. M. M. Uchida, W. J. Maciel: Astron. Astrophys. (in press)
4. F. Cuisinier, W. J. Maciel, J. Köppen, A. Acker, B. Stenholm:, B. Astron. Astrophys. **353**, 543 (2000)
5. A. V. Escudero, R. D. D. Costa: Astron. Astrophys. **380**, 300 (2001)
6. A. V. Escudero, R. D. D. Costa, W. J. Maciel: Astron. Astrophys. **414**, 211 (2004)
7. M. Forestini, C. Charbonnel: Astron. Astrophys. Suppl. **123**, 241 (1997)
8. J. A. de Freitas Pacheco, B., Barbuy, R. D. D. Costa, T. E. P. Idiart: Astron. Astrophys. **271**, 429 (1993a)
9. J. A. de Freitas Pacheco, R. D. D. Costa, W. J. Maciel: Astron. Astrophys. **279**, 567 (1993b)
10. J. A. de Freitas Pacheco, W. J. Maciel, R. D. D. Costa: Astron. Astrophys. **261**, 579 (1992)
11. J. A. de Freitas Pacheco, W. J. Maciel, R. D. D. Costa, B. Barbuy: Astron. Astrophys. **250**, 159 (1991)
12. M. A. T. Groenewegen, T. de Jong: Astron. Astrophys. **282**, 127 (1994)
13. P. Leisy, M. Dennefeld: Astron. Astrophys. Suppl. **116**, 95 (1996)
14. W. J. Maciel, C. Quireza: Astron. Astrophys. **345**, 629 (1999)
15. P. Marigo, J. Bernard Salas, S. R. Pottasch, A. G. G. Tielens, P. R. Wesselius: Astron. Astrophys. **409**, 619 (2003)
16. S. R. Pottasch, D. A. Beintema, J. Bernard Salas, W. A. Feibelman: *IAU Symposium 209*, ed. S. Kwok, M. Dopita, R. Sutherland, ASP, 353 (2003)
17. L. Stanghellini, R. A. Shaw, B. Balick, M. Mutchler, J. C. Blades, E. Villaver: Astrophys. J. **596**, 1014 (2003)
18. G. Stasińska, M. G. Richer, M. L. McCall: Astron. Astrophys. **336**, 667 (1998)

Planetary Nebulae in the Sagittarius Dwarf Spheroidal Galaxy

Albert A. Zijlstra[1], Krzysztof Gesicki[2], Jeremy Walsh[3], and Daniel Péquignot[4]

[1] Department of Physics and Astronomy, The University of Manchester, P.O. Box 88, Sackville Street, Manchester M60 1QD, United Kingdom
[2] Centrum Astronomii UMK, ul.Gagarina 11, PL-87-100 Torun, Poland
[3] ST-ECF, ESO, Karl-Schwarzschild-Strasse 2, 85748 Garching, Germany
[4] Observatoire de Paris-Meudon, F-92195 Meudon Cédex, France

1 The Three PN

The Sagittarius dwarf galaxy (Sgr) is a satellite of the Galaxy, located behind the Galactic Bulge. It is the nearest extra-galactic system, at a distance of 25 kpc. Sgr contains three known planetary nebulae (PNe) (two from Zijlstra & Walsh 1996, and one new identification). Together with the single PN in Fornax, these are the only PNe known in dwarf spheroidal galaxies.

We have observed all three Sgr PNe with the HST: [O III], V-band and Hα images are shown in Fig. 1. The maximum diameters are 0.6 arcsec, 1.3 arcsec and 2.7 arcsec. All three show a very similar, bipolar morphology.

Deep spectra were used to derive abundances. Echelle spectra were used with the Torun models (Gesicki & Zijlstra 2003) to derive velocity fields, and infer kinematic ages and core masses. Results are listed in Table 1.

Table 1. Model results

	He 2-436	Wray 16-423	StWr 2-21
Central star	[WC6]	[WC 4-6]	[WO2]
log (Hβ flux)	−11.4	−11.89	−12.6
$T_{\rm eff}$	59100	100000	120000
[O/H]	−0.52	−0.51	−0.35
$\log L/L_\odot$	3.96	3.7	2.84
$V_{\rm exp}$ [km/s]	13	33	31
Turbulence [km/s]	10	0	0
Age [yr]	2600	2800	5000
Core mass ($M_\odot$)	0.61	0.61	0.61

2 Discussion

Origin: He 2-436 and Wray 16-423 have almost identical abundances, and are believed to date from the same star formation event, approximately 5 Gyr ago

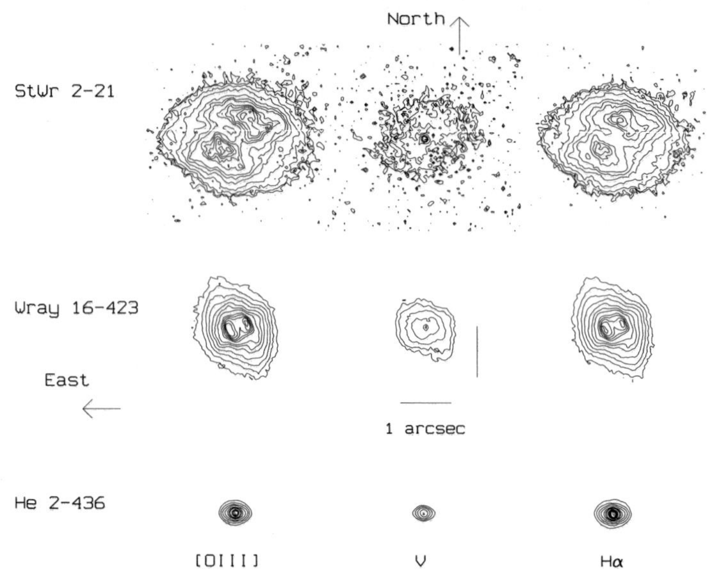

Fig. 1. Contour plots in [O III], V-band and Hα of the three Sgr PNe. The peak fluxes are scaled to give approximately the same peak value for all [O III] images. Relative contour levels are at 0.5,1,2,4,8,12,16,24,28,32,36,40.

(Dudziak et al. 2000). StWr 2-21 is found to have higher abundances: it is likely related to the last star formation episode in Sgr, ~ 1 Gyr ago.
Morphology: In Galactic PNe, bipolar morphologies are more common in type-I, disk PN. The fact that the Sgr PN also show such morphologies favours progenitor mass, rather than metallicity, as the driving cause.
[WC] stars: WR-type features are seen in all three spectra. In the Galaxy, about 10% of PNe have [WC] central stars. This is the first indication that progenitor mass/age and metallicity determines the emergence of a [WC] star.
Turbulence: Galactic PNe with WC stars show a turbulent component in the velocity fields. The Sgr PNe concur: He 2-436, with the strongest [WC] wind, shows significant turbulence, the others do not. The turbulent velocity is typical for its subclass.
Thickness: All three PN show very small cavities compared to the outer radius. This effect also occurs in the SMC (Gesicki & Zijlstra 2003), and could be a residual from slow expansion during the post-AGB evolution.

References

1. Dudziak G., Péquignot D., Zijlstra A.A., Walsh J.R., 2000, A&A, 363, 717
2. Gesicki K., Zijlstra A.A., 2003, MNRAS, 338, 347
3. Zijlstra A.A., Walsh J.R., 1996, A&A, 312, L21

What PNe Could Tell About Extinction Inside External Galaxies

Joachim Köppen

Observatoire de Strasbourg, 11 Rue de l'Université, F-67000 Strasbourg, France

Abstract. With the extinctions measured in PNe of the Galactic Bulge we had verified that the classical small clouds model for extinction in the Solar Neighbourhood is easily generalized to the entire Galaxy. I use this model to explore the behaviour of extinction of PNe observed in an external galaxy. With a reasonable number one can derive the opacity of the extinction disk. A larger number would be required to constrain the ratio of the scale heights of PNe and extinction. The measurement of the Hα/Hβ ratio of PNs in disk galaxies provides a most useful probe to learn about the interstellar medium in external galaxies.

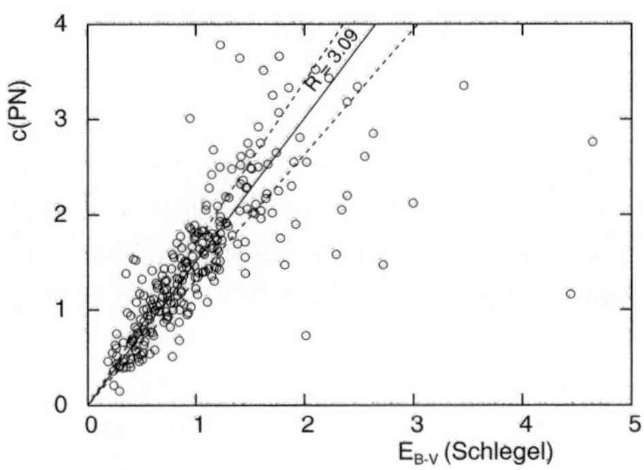

Fig. 1. The extinctions c measured in 271 Galactic Bulge PNe compared to the reddening at their position estimated from the map of Schlegel et al. (1998)

The extinctions measured in planetary nebulae of the Galactic Bulge show a systematic decrease with the nebula's galactic latitude. This relation is well reproduced with an extension of the classical statistical model of Galactic extinction, which assumes that small clouds of equal extinction are randomly distributed in a thin exponential disk (Köppen & Vergely, 1998). The observed extinctions also compare well with the extinctions one obtains from the map of Schlegel et al. (1998) for the positions of each nebula. Not only the average relation is as expected, but also the scatter agrees with the Poissonian model.

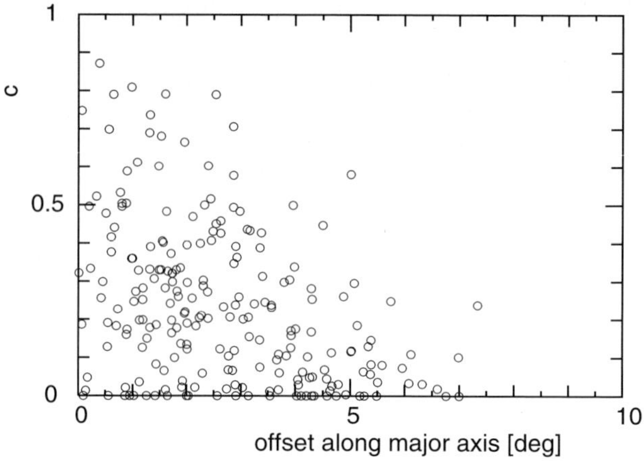

Fig. 2. The expected extinction of PNe in the disk of an external galaxy as a function of offset along the major axis (10° correspond to 15.7 kpc)

What would one observe, if one could look from the outside at the population of Galactic disk PNe? We assume an inclination of 27° and an observer's distance of 89 kpc. The PNe are distributed in an exponential disk of radial scale of 3 kpc and a scale height of 250 pc. If one selects PNe in a 2° wide strip along the major axis, the model predicts the relation between the extinction measure c and the offset from the galactic centre, shown in Fig. 2. At each position there is a range of extinctions possible, as one samples the nebulae in front of the dust disk (low extinction) to those behind (high extinction). The range decreases in the outer zones of the galaxy. Most objects are in the galactic plane, hence they are found as a ridge of slightly enhanced density in the middle of the triangular-shaped distribution. The upper envelope of the objects in the plot yields the opacity of the galaxy, i.e. the average opacity of the clouds and the number density of the clouds with respect to the gas density. The simulations show that this can be achieved with less than about 300 objects.

Information on the relative distribution of planetary nebulae and dust is contained in the details of the distribution of objects. The larger the PN scale height is in comparison with that of the dust disk, the more evenly are the objects spread out in the diagram. However, the simulations show that one needs a rather large number of objects to reliably extract this interesting information about the PN population in that galaxy.

Acknowledgement

I thank Rodrigo Ibata for extracting the extinctions from the Schlegel map.

References

1. J. Köppen, J.-L.Vergely: MNRAS **299**, 567 (1998)
2. D.J.Schlegel, D.P.Finkbeiner, M.Davis: ApJ **500**, 525 (1998)

Planetary Nebulae as Probes of the Chemical Evolution of Dynamically Hot Systems

Michael G. Richer[1] and Marshall L. McCall[2]

[1] Observatorio Astronómico Nacional, Instituto de Astronomía, UNAM, P.O. Box 439027, San Diego, CA 92143

[2] Department of Physics and Astronomy, York University, 4700 Keele St., Toronto, Ontario, Canada M3J 1P3

Abstract. The measurement of chemical abundances in planetary nebulae in nearby galaxies is now relatively straightforward. The challenge is to use these chemical abundances to infer the chemical evolution of their host galaxies. At this point, our understanding of chemical evolution based upon planetary nebulae in galaxies without star formation is strongly coupled to our understanding of the relationship between the chemical abundances in stars and planetary nebulae in the Milky Way bulge. Supposing that the same relation holds in all systems where star formation ceased long ago, these systems follow a metallicity-luminosity relation that is displaced to higher chemical abundances compared to that found for dwarf irregular galaxies. A more efficient enrichment process appears to be required as part of the explanation for this shift, in addition to the inevitable fading of these galaxies.

1 Introduction

As a result of its evolution, a galaxy consumes gas to form stars. Given a stellar initial mass function, the stars born in any given generation will have a variety of lifetimes. The most massive stars live only several 10^6 years before exploding as supernovae (SNe), their lifetimes being comparable to the time scale for star formation. When they explode as supernovae, these stars return some fraction of their initial mass to the interstellar medium (ISM). Of this returned mass, some of it will consist of elements synthesized during the star's evolution and so the composition of the ISM will change over time as these contributions accumulate. One consequence is that the stars forming at later times contain higher abundances of these newly-synthesized elements. Most stars contribute to this process of enriching the ISM, though the time scales for the majority of stars to contribute is of order several 10^9 years. The study of the chemical evolution of galaxies attempts to characterize this process of chemical enrichment as a function of galaxy type, mass, and environment to better constrain theories of galaxy formation and evolution.

Dynamically hot systems (DHSs) are stellar systems whose kinematics are dominated by random motions, such as elliptical and dwarf spheroidal galaxies and the bulges of spirals [1]. Generally, DHSs have little gas and their principal phase of star formation ceased several 10^9 years ago. Traditionally, the integrated spectrum of starlight has been used to study the chemical evolution of these systems, though the interpretation of integrated spectra is difficult due to the interplay of the metallicities and ages of the stellar population mix in any real galaxy. Nonetheless, it has long been known that these systems define a trend between the strength of their absorption lines, e.g., the Mg_2 index, and their stellar velocity dispersions, e.g., [7,2]. This trend is believed to arise due to winds that develop as a result of the energy that supernovae inject into a galaxy's ISM while star formation is still underway [21].

The lack of ongoing star formation in DHSs requires that another probe besides H II regions, the usual tool, be used to study their chemical evolution. This is unfortunate, since, in H II regions, it is possible to measure the present abundances of certain elements whose production appears easy to understand. It is possible to study the chemical evolution of DHSs using stars or planetary nebulae (PNe) as probes, e.g., [41], provided that it is possible to observe elements whose abundances are unchanged as a result of the probe's evolution. These probes have the advantage over integrated spectra that their elemental abundances may be obtained directly from their spectra. The principal disadvantages of stars and PNe, as compared to H II regions, are determining their ages, and therefore the epoch in a galaxy's evolution to which their chemical composition corresponds, whether their chemical abundances have been altered by stellar evolution, and at what enrichment level the process of chemical evolution ceased in a given galaxy. A wider range of elemental abundances may be determined from the spectra of stars than from those of PNe, but stellar spectra are more difficult to measure to the required precision unless the stars are much brighter than the galaxy background or unless the latter is faint. In the central parts of galaxies, it is easier to obtain spectra of PNe, which stimulates interest in their use for studying the chemical evolution of DHSs.

Recently, [26] reviewed this topic. That review focussed more upon some of the theoretical underpinnings of studying the chemical evolution of DHSs. Here we shall focus more on issues related to observations.

2 Spectroscopy of Extragalactic Planetary Nebulae

To date, published data have come from 4m-class telescopes. 8m-class telescopes should allow the derivation of oxygen abundances from measured electron temperatures for galaxies out to at least 5 Mpc, and further if high spectral and spatial resolution are used for measuring [O III]λ4363.

A variety of observational difficulties must be confronted when obtaining spectroscopy of extragalactic PNe. The first arises from differential atmospheric refraction over the wide wavelength interval that must be observed [17,47]. Multi-object spectrographs are an attractive option to increase the efficiency of the observations by observing many PNe in a galaxy at once, but many of these instruments lack atmospheric dispersion compensators and, unlike in the case of long slit spectroscopy, it is not possible to orient multi-object spectrographs so that the slitlets align with the parallactic angle. The second difficulty is subtracting the light from the background galaxy, which is often much brighter than the sky. The stellar absorption lines coincide with the nebular Balmer lines and can vary in intensity on small spatial scales. If there is an underlying galaxy disk, the diffuse interstellar medium and its small-scale variation further complicate the background subtraction by emitting a nebular spectrum whose line intensities can be dramatically different from those in the PNe under study. Emission lines from such gas preferentially comes from low ionization states, e.g., H I, [O II], [N II], and [S II], but [O III] is also detected, though He I frequently is not [12]. Emission from the night sky near He Iλ5876 and Hα-[N II] is detrimental too, though it is spatially uniform, in contrast to the galaxy background. [36] provide an instructive discussion of this issue and argue convincingly of the advantages of 3D spectroscopy, especially the advantage of having a 2D spatial map of the background emission to better subtract this emission. A third difficulty is the weakness of the [O III]λ4363 line, which is about 100 times fainter than [O III]λ5007 [35], and its proximity to the Hg Iλ4358 line at observatories whose night sky emission is contaminated by street lights. [O III]λ4363 must be observed if the electron temperature is to be measured and used to obtain a reliable estimate of the oxygen abundance.

It is difficult to characterize the absolute precision of the line intensities measured to date in extragalactic PNe and the chemical abundances derived from them. The most extensive comparison may be made using the PNe studied in M32 by [35,32], who used the same equipment and refinements of the same observational technique. For lines stronger than Hβ, the uncertainty appears to be about $\pm$25%. For lines whose strength is 50% that of Hβ, the uncertainty appears to be about a factor of two. However, in all cases, the abundances derived in both studies agree within uncertainties. These "internal" uncertainties can be as low as $\pm$15% and $\pm$0.1 dex for He and O abundances, respectively, but may underestimate the true uncertainties if they do not include all systematic uncertainties.

There are few observations of a given object by different groups. In M31, PN29 is common to three studies. The values found for the oxygen abundance, the Ne/O ratio, and the N/O ratio all span a range of about 0.3 dex [35,17,36]. The helium abundance differs by almost 40% between two of the studies [17,36]. PN23 has been observed twice and the oxygen abundances and Ne/O ratios agree within uncertainties [35,17]. PN27 has also been observed twice [17,36], with the resulting oxygen abundances in excellent agreement, but the helium abundances differing by 35%. Although it is an indirect indicator, there are several detections of He Iλ5876 in PNe in M31 that are either anomalously strong or weak, perhaps as a result of significant background contamination [35,17]. In M32, the line intensities for PN1 are in generally good agreement among the available measurements [18,35,32]. Considering all of the data available, the case of PN29 in M31 appears unusual. Were the oxygen abundances generally as uncertain as those measurements imply, dispersions in abundances within a given galaxy should be at least 0.3 dex. In M32, NGC 185, and NGC 205, the dispersions in the oxygen abundances for PNe with measured electron temperatures are 0.19 dex, 0.27 dex, and 0.37 dex, respectively [35,32]. A dispersion as low as that found for M32 (13 PNe) implies that the typical uncertainty in oxygen abundances should be less than 0.2 dex, which is comparable to the uncertainties quoted for individual abundances (when quoted).

It appears that background subtraction may be the dominant source of uncertainty in the line intensities at the moment [36]. Currently, there are too few measurements to determine whether the differences among studies are systematic. Helium abundances appear to be uncertain by up to about 40%, while oxygen abundances may be uncertain by up to 60%. Abundance ratios of neon and nitrogen relative to oxygen may be less uncertain, but one can anticipate difficulties with the background subtraction for nitrogen lines (and eventually those of sulphur). In principle, observations with higher spectral resolution should help, since the PNe should have velocities that scatter about the velocity of the diffuse ionized gas in the ISM, allowing separation of their components.

What matters, however, for studies of the chemical evolution of galaxies are the mean abundances and their dispersions for a population of PNe. Provided that the oxygen abundances for individual PNe in current studies do not deviate systematically from their true values, the mean value for a sample of PNe should be robust. On the other hand, the dispersions in abundances may be more seriously affected if the true uncertainties are severely underestimated.

3 Interpreting the Chemical Abundances in Bright Planetary Nebulae

The first issue that has an important bearing upon the interpretation of the abundances measured in PNe is the object selection process. To derive the most secure oxygen abundances, the electron temperature must be measured, requiring that [O III]λ4363 be detected. Since this line is about 100 times fainter than [O III]λ5007, the sample will normally be restricted to PNe that are bright in [O III]λ5007. "Bright PNe" is adopted to denote PNe bright in [O III]λ5007, usually within 2 mag of the peak of the PN luminosity function. To use this population of PNe to infer the chemical evolution of their host galaxies, it is necessary to understand the biases intrinsic to this population of PNe.

Second, planetary nebulae are the descendants of evolved stars. It is necessary to know how the nucleosynthesis associated with the evolution of the progenitor star has modified its original chemical composition, for only elements whose abundance survives this evolution unchanged are useful for studying the chemical evolution of the host galaxies. It is well-established that the stellar progenitors of PNe modify their original abundances of He, C, N, and s-process elements, e.g., [25,8]. Fig. 1 implies that the processes of nucleosynthesis and dredge-up in PN progenitors appear to be approximately equally efficient in all galaxies studied so far, since the N/O ratios in their bright PNe span the same range of values. There is still considerable debate regarding whether O and Ne are modified. Models indicate that PN progenitors may modify their initial abundance of O and Ne slightly, though the magnitude and even the sign of the effect is unclear [25,8]. Observationally, the abundances of O and Ne in bright PNe correlate in the same

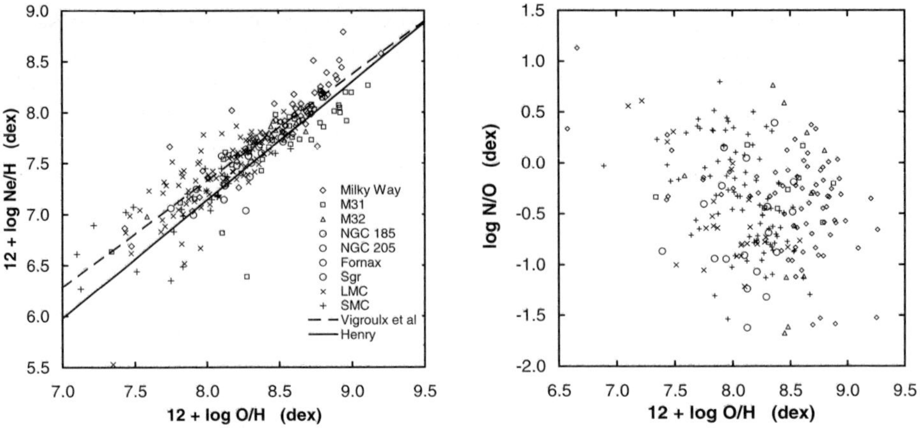

Fig. 1. Left: The oxygen and neon abundances observed in bright PNe follow the same relation as found for the ISM in star-forming galaxies [45,13]. The simplest interpretation is that the progenitors of bright PNe do not significantly modify either of these abundances. Right: The N/O ratios for PNe in all galaxies scatter over the same range of values, indicating that nucleosynthesis and mixing processes in the progenitor stars of bright PNe have approximately the same effect in all galaxies. The symbols have the same meaning in both figures. The data are adopted from [40,17,36].

way as in giant extragalactic H II regions (Fig. 1) [45,13,40]. Type II supernovae are believed to be primarily responsible for setting this abundance ratio in the ISM, so observing the same correlation in bright PNe and H II regions implies that the progenitors of bright PNe probably do not significantly modify either their O or Ne abundances. Both O and Ne abundances in bright PNe should be useful probes for studying the chemical evolution of their host galaxies.

Finally, it is necessary to understand what the mean oxygen abundances in bright PNe represent. In the Magellanic Clouds, bright PNe, H II regions, and young stars all have similar oxygen abundances [30,14,43]. The same situation is found in NGC 6822 [44]. The agreement between stellar and nebular abundances implies that there are no serious problems with the nebular abundance scale based upon forbidden lines, since the stellar abundances are based upon permitted lines. Bright PNe and H II regions in NGC 3109, Sextans A, and Sextans B likewise have similar oxygen abundances [24,19]. Therefore, in star-forming systems, in general, bright PNe have oxygen abundances very similar to those in the ISM. Fewer systems without star formation are available for similar study. The best case study is the bulge of the Milky Way. There, bright PNe have a mean oxygen abundance similar to the mean oxygen abundance in the stars [27,40]. This result may hold more generally for the entire PN population [6]. In the Fornax and Sagittarius dwarf spheroidals, the oxygen abundances of their PNe are typical of the younger, more metal-rich stellar populations. In Sagittarius, this population appears to dominate the mass [5]. In Fornax, the stellar population of intermediate metallicity dominates the mass [37], but

Fornax's total stellar mass is so low that it is fortunate to observe a single bright PN. Therefore, in systems without star formation, the oxygen abundance in the bright PNe appears to be representative of the mean oxygen abundance in the stars.

It is worthwhile trying to understand why this abundance dichotomy arises in systems with and without star formation. Many processes in stellar and galactic evolution can affect the population of bright PNe. The means by which mass is lost by a PN precursor on and immediately after the asymptotic giant branch (AGB) should affect the eventual PN morphology. In the Milky Way, the morphology of PNe correlates with their scale height above the galactic plane [9], so morphology should correlate with progenitor mass. Morphology could significantly influence PN luminosity, since it will affect the angular distribution of the nebular optical depth. There is a sensitive interplay between the time scale to evolve from the AGB to the PN phase and the nebular expansion velocity, since both affect the density of the nebular shell and its consequent optical depth and luminosity [40]. A bright PN will only be observed if the central star is hot and bright while the nebular shell is sufficiently dense. The time scale of the phase during which the central star is bright is a strong function of the mass of the central star, e.g., [3], which, in turn, is a function of the progenitor mass. The time scale during which the central star is bright also depends upon whether the the star leaves the AGB burning H or He, e.g., [42]. Finally, the time required for a star to evolve from the AGB to the PN phase further depends upon the envelope mass remaining when the star evolves off the AGB [39], since this mass must be disposed of through either a stellar wind or nuclear burning. All of these effects may well depend upon the progenitor mass and metallicity. A key unknown is whether the evolution of the nebular shell is correlated with that of the central star [46,33].

In a galactic context, chemical evolution usually ensures a correlation between the masses and metallicities of the PN progenitors. This arises through the history of star formation, which defines the mass range spanned by the PN progenitors and their number distribution [33]. In star-forming galaxies, the PN progenitors span the entire range of masses allowed. Since the death rate in a stellar population is a strong function of the population's age [29], the youngest populations can easily dominate the production of PNe, if the star formation rate has been roughly constant (often a reasonable approximation). A high death rate alone is no guarantee that the youngest stellar populations dominate the production of bright PNe, as the highest-mass central stars evolve so fast that it is statistically unlikely to observe such PNe [16,28]. In the LMC, the youngest stellar populations apparently produce PNe fainter than do somewhat older stellar populations [38]. In DHSs, where star formation ceased several 10^9 years ago, the PN progenitors will span a small range of masses, so the death rates among the different stellar populations will be similar and each population should contribute PNe in proportion to its mass, yielding mean chemical abundances in the PN population very similar to those in the stars. These arguments are in qualitative agreement with attempts to model this process [33] and illustrate why the

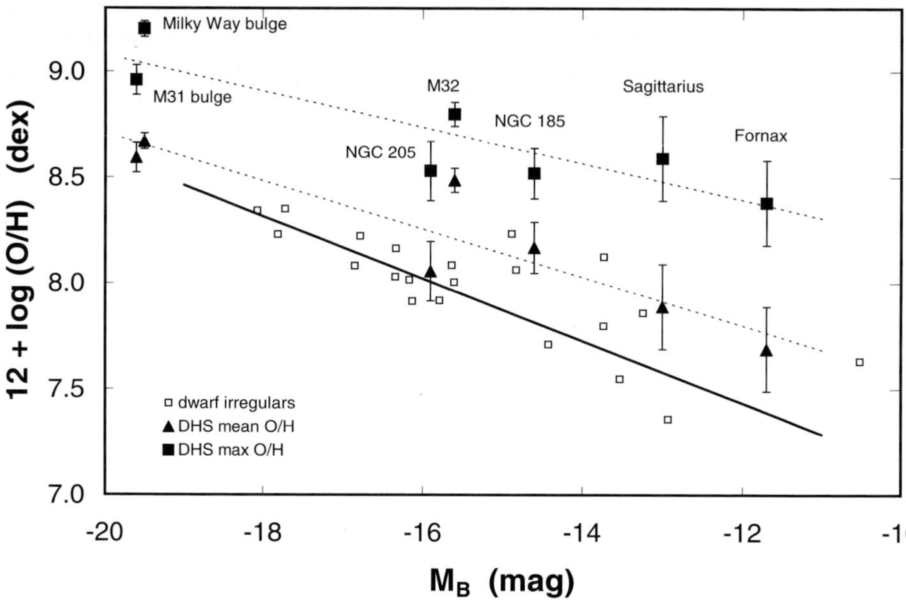

Fig. 2. This figure presents the relation between oxygen abundance and luminosity in the *B*-band for a sample of DHSs. The two dotted lines show the mean and maximum oxygen abundances found for PNe in DHSs. For reference, the same relation is shown for dwarf irregular galaxies, which is based upon oxygen abundances measured in H II regions, and thus should be compared with the maximum abundances observed in DHSs [31].

relations between abundances in bright PNe and other tracers in galaxies with and without star formation might arise.

4 The Chemical Evolution of DHSs

A fundamental assumption in the analysis that follows is that the bright PNe in all DHSs have mean oxygen abundances that are representative of the mean oxygen abundance in their stars, as occurs in the bulge of the Milky Way. The dwarf spheroidals Fornax and Sagittarius are the only exceptions, where we assume that their PN populations represent only their more metal-rich stellar populations and the mean oxygen abundance for their entire stellar content is reduced appropriately. The results depend upon the validity of this assumption.

DHSs follow a well-defined relation between their oxygen abundances and luminosities, as presented in Fig. 2. The relation is plotted using both the mean oxygen abundance observed in the PNe in each DHS as well as using the maximum abundance observed. The latter relation should approximate the maximum oxygen abundances found in the ISM before star formation ceased in these DHSs.

These maximum abundances should be approximately analogous to the oxygen abundances observed in H II regions in star-forming galaxies [31]. In star-forming galaxies, this relation is thought to arise because galaxies of different mass have converted different fractions of their gas into stars, and so the remaining ISM has been enriched more in those galaxies with the least gas remaining [23]. This same process is believed to have occurred in DHSs, but that, at some point, the DHSs lost their gas.

DHSs must have faded since they ceased forming stars. As a result, their abundance-luminosity relation has shifted to lower luminosities over time. It is easy to illustrate that fading a dwarf irregular is insufficient of itself to reproduce the abundance-luminosity relation followed by DHSs. If a typical dwarf irregular has a colour $(B - V) = 0.5\,\mathrm{mag}$, then after 10^{10} years this galaxy will have faded by less than 2.5 mag [11] if we approximate it as a single stellar population with a metallicity $Z = 0.008$, i.e., similar to the metallicity of the LMC. The approximation of a single stellar population will exaggerate the amount of fading, since much of the light is contributed by stellar populations with redder colours that will fade less. However, 5 mag separate the abundance-luminosity relation for dwarf irregulars and the maximum abundance relation for DHSs. Therefore, fading alone cannot explain the shift of the abundance-luminosity relations for DHSs.

The abundance-luminosity relation for DHSs may also be offset if their ISM was enriched more efficiently in newly synthesized elements. Preliminary attempts to model the mean oxygen abundances and their dispersions find that it is difficult to explain simultaneously the mean abundances and their dispersions in DHSs unless DHSs were losing gas even while still forming stars [26]. As a result, the enrichment of their ISM proceeded at a faster rate than in today's dwarf irregulars because of reduced dilution of the nucleosynthetic products ejected by supernovae. The effective yield of oxygen derived for DHSs exceeds that derived for dwarf irregulars [26,23], and even that derived recently for the inner parts of spiral galaxy disks [10,4], but well within the range expected from stellar nucleosynthesis calculations [20].

The helium abundances of the PNe in DHSs offer additional evidence in favour of efficient enrichment of their ISM. In Fig. 3, the helium abundances in PNe in different galaxies are plotted as a function of their oxygen abundances. At a given oxygen abundance, the PNe in M32 and the dwarf spheroidals, Fornax, Sagittarius, NGC 185 and NGC 205 have higher helium abundances than the PNe in the Magellanic Clouds. This effect is not so obvious for the bulges of the Milky Way or M31, but these are also the DHSs that fall closest to the abundance-luminosity relation for dwarf irregulars in Fig. 2. If the similar scatter in N/O is taken as evidence that mixing and nucleosynthesis in the PN progenitors was similar in all galaxies (Fig. 1), the cause of the high helium abundances in the DHSs should then be the result of the evolution of their host galaxies, and not the PN progenitors. Efficient enrichment of the ISM is such a process. In fact, gas outflow should affect helium enrichment more than oxygen enrichment because the production of helium is dominated by stars of lower mass (and which

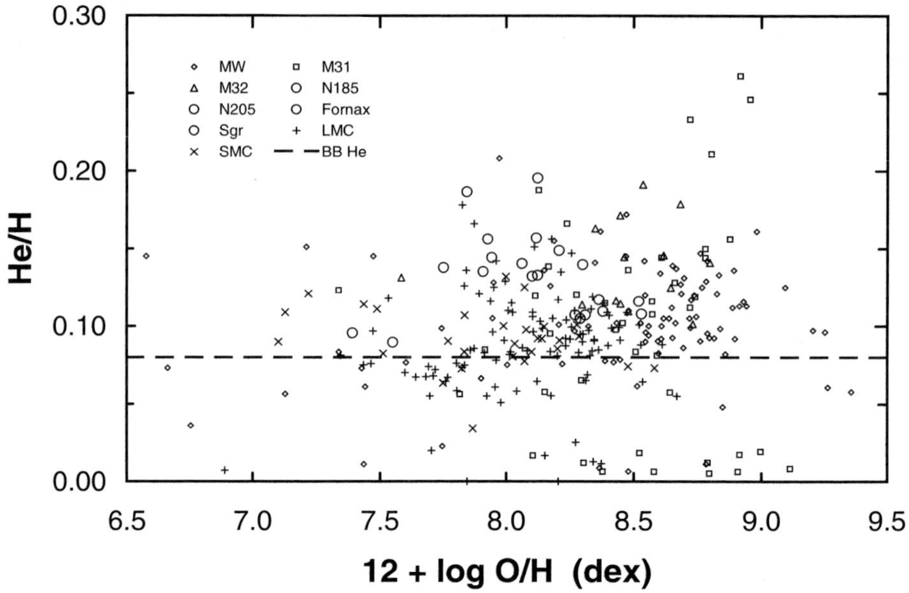

Fig. 3. Helium abundances are plotted as a function of the oxygen abundance for the PNe in various galaxies. At a given oxygen abundance, the helium abundance is higher in the PNe in M32 and the dwarf spheroidals than it is in the PNe in the Magellanic Clouds. Given that mixing and nucleosynthesis have the same effects upon the PN progenitors in all galaxies (Fig. 1), processes related to the evolution of the host galaxies, such as the enrichment of the ISM, should be responsible for the higher helium abundances in the DHSs. The very low helium abundances (near zero) are the result of missing ionization stages. The data are adopted from [40,17,36].

live longer) than is the case for oxygen [22], and so the ISM has more time to evolve (lose mass) during the production of helium.

Supernova-driven winds are commonly believed to have removed the gas from DHSs [21]. The basis for this belief is the correlation between metallicity indices and the velocity dispersion, e.g., the $Mg_2 - \sigma$ relation [2]. SNe are responsible for injecting newly-synthesized elements into the ISM, but they likewise inject large quantities of energy into the ISM. If the energy injection rate exceeds the dissipation rate, the thermal energy of the ISM increases and it may eventually exceed its gravitational binding energy. At that point, the ISM begins to flow out of the galaxy in a wind, presumably halting star formation at the same time. Massive galaxies are enriched more than low-mass galaxies since more SNe are required to initiate a wind. In this scenario, the chemical enrichment of a galaxy is related to the velocity dispersion of its stars only because both are controlled by the gravitational potential. Since oxygen is produced by core collapse SNe, the oxygen abundance is a faithful tracer of the energy they injected

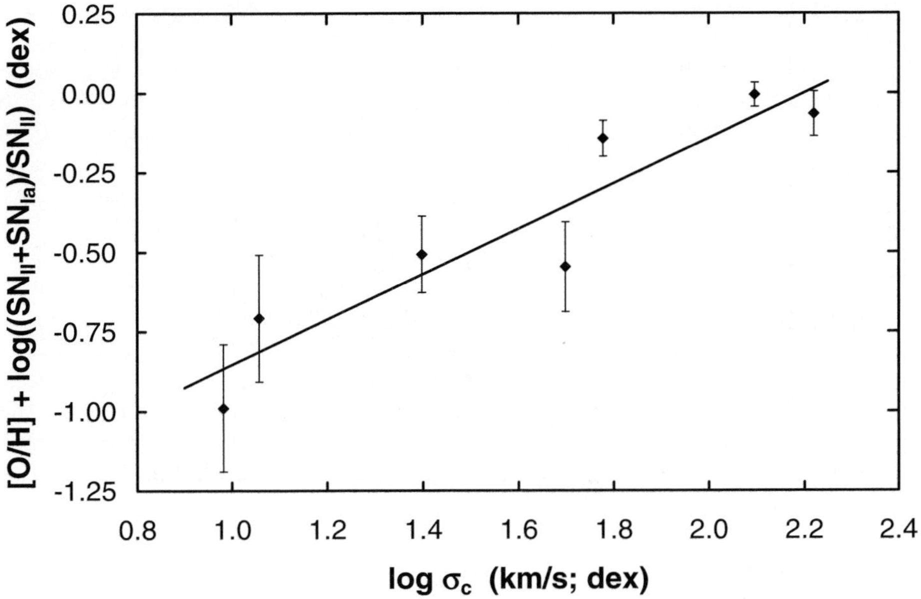

Fig. 4. Oxygen abundances are plotted versus velocity dispersion in DHSs. The oxygen abundances have been corrected to account for the energy injection from type Ia SNe. This relation arises naturally if supernova-driven winds are the mechanism that terminates chemical evolution in DHSs.

into the ISM. However, star formation in DHSs may have extended over a period long enough that type Ia supernovae also contributed energy and matter to the ISM. One may correct the oxygen abundances to account for the contribution of type Ia SNe using the O/Fe ratio observed in stars [34], since type Ia SNe contribute significantly to iron enrichment, but not to oxygen enrichment. Fig. 4 presents these corrected oxygen abundances plotted as a function of the velocity dispersion in their host galaxy. A correlation is found that is consistent with expectations if the chemical evolution of DHSs was terminated by the action of supernova-driven winds.

The role of mergers in the formation of DHSs is difficult to discern from these results. For gas-rich, major mergers (components with similar masses), the chemical evolution is changed very little compared to a monolithic formation scenario, since a large fraction of the star formation occurs after assembling the baryonic mass. Major mergers that are gas-poor would flatten the abundance-luminosity relations in Fig. 2 and the abundance-velocity dispersion relation in Fig. 4, for most of the stars would form in a shallower gravitational potential than that in which they now reside. The slope and not the intercept of the relations is affected because the smallest DHSs considered here are too small to be the products of mergers themselves. Minor mergers (components with very different masses) would not be distinguishable in Figs. 2 and 4 regardless of their gas content. Clearly, the only scenario ruled out is the assembly of large DHSs entirely from a large number of small, gas-poor systems. This would produce a DHS that falls below the relations in Figs. 2 and 4, which has not been observed so far. Basically, one would find the same oxygen abundances in all DHSs, regardless of mass.

5 Conclusions

Two observations indicate that bright PNe in DHSs should be useful probes to infer the chemical evolution of these galaxies. First, the precursors of bright PNe do not significantly modify their original oxygen abundances. Second, the mean oxygen abundance measured in bright PNe appears to be representative of the average oxygen abundance for the entire stellar population in DHSs. Nonetheless, there are numerous uncertainties concerning the production of populations of bright PNe and any advance in quantifying the processes involved, at either the stellar or galactic level, will improve the utility of PNe for studying the chemical evolution of DHSs.

DHSs are found to follow a relation between oxygen abundance and luminosity. This relation is shifted to higher abundances than that followed by dwarf irregular galaxies, independently of whether the mean or maximum oxygen abundances for the DHSs are considered. Fading appears to be insufficient to explain the shift. DHSs appear to have incorporated their nucleosynthetic production into subsequent stellar generations more efficiently than did dwarf irregulars.

References

1. Bender, R., Burstein, D., & Faber, S. M. 1992, ApJ, 399, 462
2. Bender, R., Burstein, D., & Faber, S. M. 1993, ApJ, 411, 153
3. Blöcker, T. 1995, A&A, 299, 755
4. Bresolin, F., Garnett, D. R., & Kennicutt, R. C., Jr. 2004, ApJ, in press; also astro-ph/0407065
5. Cole, A. 2001, ApJ, 559, L17
6. Exter, K. M., Barlow, M. J., & Walton, N. A. 2004, MNRAS, 349, 1291
7. Faber, S. M., ApJ, 179, 423
8. Forestini, M., & Charbonnel, C. 1997, A&AS, 123, 241
9. Garcia-Segura, G., Franco, J., López, J. A., Langer, N., & Różyczka, M. 2002, RMAAC, 12, 117
10. Garnett, D. R., Kennicutt, R. C., Jr., & Bresolin, F. 2004, ApJ, 607, L21
11. Girardi, L., Bressan, A., Bertelli, G., & Chiosi, C. 2000, A&AS, 141, 371
12. Greenawalt, B., Walterboss, R. A. M., & Braun, R. 1997, ApJ, 483, 666
13. Henry, R. B. C. 1989, MNRAS, 241, 453
14. Hill, V., Andrievsky, S., & Spite, M. 1995, A&A, 293, 347
15. Izotov, Y. I., & Thuan, T. X. 1998, ApJ, 500, 188
16. Jacoby, G. H. 1989, ApJ, 339, 39
17. Jacoby, G. H., & Ciardullo, R. 1999, ApJ, 515, 169

18. Jenner, D. C., Ford, H. C., & Jacoby, G. H. 1979, ApJ, 227, 391
19. Kniazev, A. Y., Grebel, E. K., Pramskij, A. G., & Pustilnik, S. A. 2004, these proceedings
20. Köppen, J., & Arimoto, N. 1991, A&AS, 87, 109
21. Larson, R. B. 1974, MNRAS, 169, 229
22. Maeder, A. 1992, A&A, 264, 105
23. Lee, H., McCall, M. L., Kingsburgh, R. L., Ross, R., & Stevenson, C. C. 2003, AJ, 125, 146
24. Magrini, L., Perinotto, M., Corradi, R., Leisy, P., & Mampaso, A. 2004, these proceedings
25. Marigo, P., Bressan, A., & Chiosi, C. 1996, A&A, 313, 545
26. McCall, M. L., & Richer, M. G. 2003, in IAU Symp. 209, Planetary Nebulae: Their Evolution and Role in the Universe, eds. S. Kwok, M. Dopita, and R. Sutherland (ASP: San Francisco), 583
27. McWilliam, A., & Rich, R. M. 1994, ApJS, 91, 749
28. Méndez, R. H., & Soffner, T. 1997, A&A, 321, 898
29. Renzini, A., & Buzzoni, A. 1986, in Spectral Evolution of Galaxies, eds. C. Chiosi and A. Renzini (D. Reidel Publishing Co.: Dordrecht), 195
30. Richer, M. G. 1993, ApJ, 415, 240
31. Richer, M. G., & McCall, M. L. 1995, ApJ, 445, 642
32. Richer, M. G., & McCall, M. L. 2004, in preparation
33. Richer, M. G., McCall, M. L., & Arimoto, N. 1997, A&AS, 122, 215
34. Richer, M. G., McCall, M. L., & Stasińska, G. 1998, A&A, 340, 67
35. Richer, M. G., Stasińska, G., & McCall, M. L. 1999, A&AS, 135, 203
36. Roth, M. M., Becker, T., Kelz, A., & Schmoll, J. 2004, ApJ, 603, 531
37. Saviane, I., Held, E. V., & Bertelli, G. 2000, A&A, 355, 56
38. Stanghellini, L. 2004, these proceedings
39. Stanghellini, L., & Renzini, A. 2000, ApJ, 542, 308
40. Stasińska, G., Richer, M. G., & McCall, M. L. 1998, A&A, 336, 667
41. Tolstoy, E., Venn, K. A., Shetrone, M., Primas, F., Hill, V., Kaufer, A., & Szeifert, T. 2003, AJ, 125, 707
42. Vassiliadis, E., & Wood, P. R. 1994, ApJS, 92, 125
43. Venn, K. A. 1999, ApJ, 518, 405
44. Venn, K. A., Lennon, D. J., Kaufer, A., McCarthy, J. K., Przybilla, N., Kudritzki, R. P., Lemke, M., Skillman, E. D., & Smart, S. J. 2001, ApJ, 547, 765
45. Vigroulx, L., Stasińska, G., & Comte, G. 1987, A&A, 172, 15
46. Villaver, E., Manchado, A., & García-Segura, G. 2002, ApJ, 581, 1204
47. Walsh, J. R., Walton, N. A., Jacoby, G. H., & Peletier, R. F. 1999, A&A, 346, 753

A Study of Chemical Abundances of Planetary Nebulae in M33

Mario Perinotto[1], Laura Magrini[1], Antonio Mampaso[2], and Romano L.M. Corradi[3]

[1] Dipartimento di Astronomia e Scienza dello Spazio, Universitá di Firenze, Italy
[2] Instituto de Astrofísica de Canarias, Tenerife, Canarias, Spain
[3] Isaac Newton Group of Telescopes, La Palma, Canarias, Spain

Abstract. Eleven Planetary Nebulae (PNe) in the spiral galaxy M 33 have been studied with the CLOUDY 94.00 photoionization code (Ferland et al. 1998) using observations obtained with the 4.2 m WHT (La Palma, Spain) equipped with AF2/WYFFOS at a resolution of 3.0Å/pixel in the spectral range 4300 to 7380 Å. A galactocentric gradient in O/H is suggested. The average [O/H] is very close to the same quantity from H II regions (Vílchez et al. 1988).

1 Introduction

The study of chemical abundances in extragalactic PNe has so far been conducted in only a few galaxies. In most of them the simplified ICF (Ionization Correction Factor) method has been used with direct determination of T_e, derived from [OIII] $\lambda\lambda$ 4363/5007 Å or, in the worst cases when the λ 4363 Å is not measured, using its upper limit. We have attempted to determine relatively reliable abundances in cases in which T_e could not be determined observationally, by use of a detailed photoionization model.

2 The Model

Each object has been modeled with the following basic conditions: *i)* the central star was considered a blackbody whose temperature was constrained by He I and He II or [OIII] and He II line ratios; *ii)* nebular radii between 0.0032 up to 0.32 pc were considered, constraining it with the [OIII]/[OII] line ratio when possible; *iii)* electron density from the observed [SII] doubled was adopted along with the approximation of constant density or of r^{-2} dependence. Tests with well-observed Galactic PNe were done (Perinotto et al. 2004).

3 Results

Chemical abundances of He/H, N/H, O/H, Ar/H, and S/H were derived with estimated errors of the order of 0.15 dex for He/H and 0.2 dex for O/H and between 0.3 and 0.4 dex for the other elements. Within these errors a galactocentric gradient is barely seen in O/H. Average values of chemical abundances

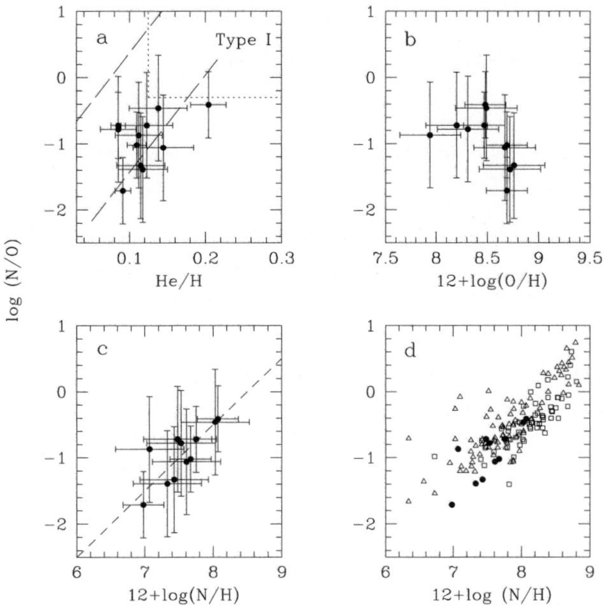

Fig. 1. Relationships among chemical abundances.

are compared with those from H II regions of M 33. A significant result is that O/H is about the same as in H II regions. This indicates that the metal abundance of the interstellar medium was unchanged by the time of the formation of the progenitors of PNe up to the present time. Some relationships between chemical abundances are shown in Fig. 1:

a) He/H *vs.* N/O. The area included inside the dashed lines is the main locus of Galactic PNe derived from Perinotto et al. (2004). The dotted lines show the limits of type I PNe area;

b) O/H *vs.* N/O: an anti-correlation in shown as seen by Jacoby & Ciardullo (1999) in M 31;

c) N/H *vs.* N/O. The dashed line shows the Galactic relation derived by Henry (1990). In spite of some dispersion, the Galactic relation between N/H and N/O appears to apply as well to M 33;

d) N/H *vs.* N/O. Filled circles are PNe of M 33 (this work), triangles are PNe of LMC (Leisy et al. 2004) and squares are Galactic PNe (Perinotto et al. 2004).

References

1. Henry R. B. C., 1990, ApJ, 356, 229
2. Leisy P.& Dennefeld, M., 2004, in preparation
3. Perinotto M., Morbidelli L., & Scatarzi A., 2004, MNRAS, 349, 793
4. Vílchez J.M., Pagel B.E.J., Diaz A.I., Terlevich E., Edmunds M.G. 1988, MNRAS 235, 633

The Inner Abundance Gradient of M33 from Bright Planetary Nebulae

Grażyna Stasińska[1], José M. Vílchez[2], Enrique Pérez[2], Rosa M. Gonzalez Delgado[2], Romano L.M. Corradi[3], Antonio Mampaso[4], and Laura Magrini[5]

[1] LUTH, Observatoire de Meudon, 92195 Meudon Cedex, France
[2] Instituto de Astrofísica de Andalucía, Apartado 3004, 18080 Granada, Spain
[3] Isaac Newton Group of Telescopes, Ap. de Correos 321, 38700 Sta. Cruz de la Palma, Spain
[4] Instituto de Astrofísica de Canarias, Tenerife, Spain
[5] Dipartimento di Astronomia e Scienza dello Spazio, Universitá di Firenze, L.go E. Fermi 2, 50125 Firenze, Italy

Abstract. We show why planetary nebulae should be better metallicity indicators than HII regions for metal-rich zones in galaxies. We present preliminary results on O and Ne abundances obtained by an electron-temperature based method in 4 planetary nebulae that we observed in the inner region of M33. We find them to be definitely subsolar, indicating that the metallicity gradient levels off in the inner part of this galaxy.

1 Introduction

The easiest and most popular way to determine abundance gradients in spiral galaxies is to use giant HII regions. Giant HII regions are bright, easily observed, and their chemical composition is easy to derive, at least regarding the most abundant elements (He, N, O, Ne ...). If the electron temperature can be determined from the [O III]4363/5007Å line ratio, the ionic abundances are obtained directly from observed emission line ratios, using the emissivities at the corresponding temperature. If the electron temperature is not measured, the metallicity is derived from empirical methods, based on strong lines only, and calibrated on photoionization models of HII regions. In the case of M33, combining their own observations with those of Smith (1975) and Kwitter & Aller (1981), Vílchez et al. (1988) found an oxygen abundance gradient across the galaxy of Δ log O/H / ΔR =0.12 dex kpc^{-1}, with an indication of a steeper gradient in the inner galaxy ($\rho/\rho_0 < 0.3$).

Other methods, often more complex and sometimes less reliable (abundance determinations in supernova remnants, Smith et al. 1993, analysis of stellar clusters, Ma et al. 2004, abundance determinations in the atmospheres of B supergiants Monteverde et al. 1997, 2000), have also been used to study abundance gradients in spiral galaxies, with sometimes contradictory results.

So far, abundance gradients from planetary nebulae (PNe) have been derived only in the Milky Way (Maciel & Köppen 1994, Maciel & Quireza 1999), where one of the largest problems is the uncertainty in the distances of these objects.

In the inner zones of spiral galaxies the metallicity is supposedly higher than solar (e.g. Vílchez et al. 1988, Zaritsky et al. 1994, Henry & Howard 1995). However, recent studies based on stars (Smartt et al. 2001) or giant HII regions (Kennicutt et al. 2003, Garnett et al. 2004, Bresolin et al. 2004) indicate that abundance gradients in these central parts might be flatter than previously thought.

Here, we describe preliminary results on the derivation of the inner metallicity gradient in the spiral galaxy M33 by using bright PNe. We first show why PNe are expected to be superior to HII regions for this purpose.

2 What is Wrong with Abundance Determinations in Metal-Rich HII Regions?

At high metallicities cooling by the metals depresses the electron temperature and the weak [O III]4363Å line is not observed. One has to rely on 'strong line' methods, which are only of statistical value (see e.g. Stasińska 2004a for a discussion). The derived values of the abundances strongly depend on the calibration used (see e.g. Pindao et al. 2002 or Kennicutt et al. 2003). Observations with Very Large Telescopes may be misleading: even if [O III]4363Å is measured the derived O/H will be underestimated since strong electron temperature gradients due to cooling by [O III] 52, 88μm make the [O III]4363/5007Å ratio overestimate the temperature in the zones emitting the optical forbidden lines (see Stasińska 2004b, in preparation).

3 The Advantage of Using Extragalactic Planetary Nebulae

Useful spectroscopy for accurate abundance determinations in extragalactic PNe can be done only on luminous objects. Simple models of evolution of expanding PNe around evolving central stars, such as described in Stasińska et al. (1998) show that the majority of PNe that are luminous in [O III]5007Å have a high effective temperature ($>$ 100,000 K) and a high density ($>$ 1000 cm^{-3}) as illustrated in Fig. 1. Both conditions result in a substantial increase of the electron temperature with respect to giant HII regions of same metallicity, as mentioned in Stasińska (2002) and further discussed in Stasińska (2004b, in preparation). This eliminates the strong bias in the determination of the elemental abundances that occurs in the case of metal-rich HII regions.

As shown by Stasińska (2002), even if PNe are less luminous than giant HII regions by severals orders on magnitude, the expected fluxes in the [O III]4363Å line are comparable at high metallicities.

Since PNe are post-AGB objects, their chemical composition might be affected by nucleosynthesis and mixing processes in the progenitor stars. This is certainly the case for at least He, C and N. The oxygen abundance could be affected by hot bottom burning, especially if the initial metallicity is less than 1/4

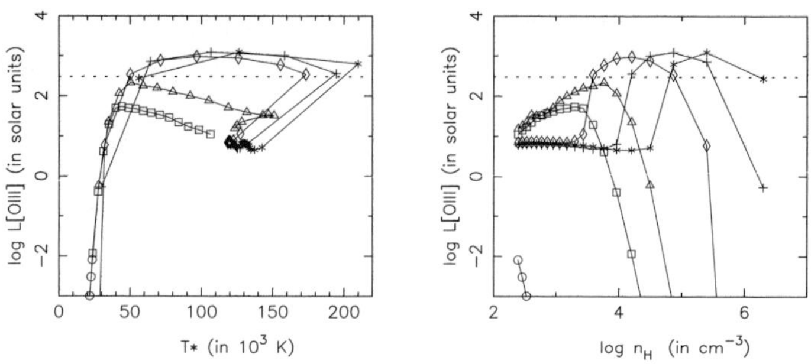

Fig. 1. Evolution tracks in the (log ([O III], $T*$) and the (log ([O III], log n_H) planes of photoionization models of PNe of mass 0.2 $M_\odot$ expanding at 20 km s^{-1}around evolving post-AGB stars of various masses. Symbols mark time in 500 yr intervals. Circles: M_* = 0.56; squares: M_* = 0.58; triangles: M_* = 0.60; diamonds: M_* = 0.62; crosses: M_* = 0.64 M; stars: M_* = 0.66 $M_\odot$.

of solar (Marigo 2001). However both the comparison the oxygen abundances in Galactic and Magellanic Clouds PNe and HII regions and theoretical predictions (Marigo 2001) indicate that this effect is not expected to be important. The neon abundance is not affected by these processes and is expected to trace the neon abundance of the interstellar gas out of which the progenitor was born.

It is important to relate the oxygen and neon abundances measured in PNe to the present day abundances in the interstellar medium. As seen in Fig. 1, the most luminous PNe have central star masses of about 0.62 – 0.66 $M_\odot$ and are therefore descendants from stars with initial masses of about 3 $M_\odot$ (see the discussion about the initial-final mass relation by Weidemann 2000). Their Ne, Ar, S (and probably O) abundances are expected to be the same as those of the interstellar medium out of which the progenitors were formed about 1 Gyr ago.

4 Our Observations of Planetary Nebulae in M33

We planned to observe 10 bright PNe in the inner disk of M33, from the list of Magrini et al. (2001). Due to weather conditions only 4 PNe were observed. The observations were performed using the ISIS spectrograph attached to the WHT 4.2m telescope in La Palma (Canary Islands) during the night 28 Oct 2003. Both blue and red arms were used with a dichroic set to an effective wavelength of 5400 A. The slit width was set to 1.5 arcsec and the seeing oscillating between 1.2 to 1.6 arcsec. The weather was clear but with light clouds passing over the night. The grating R600B was used in the blue arm, centered at 4400 A yielding a reciprocal dispersion of 0.45 A/pix ; the grating R300R was used in the red arm centered at 6130 A yielding a dispersion of 0.84 A/pix. The exposures were of 3 hours per spectrum.

The spectra have been reduced and calibrated with standard procedures. As recently emphasized by Roth et al. (2004), one of the problems for a quantitative analysis of abundances is the subtraction of the diffuse emission and of the

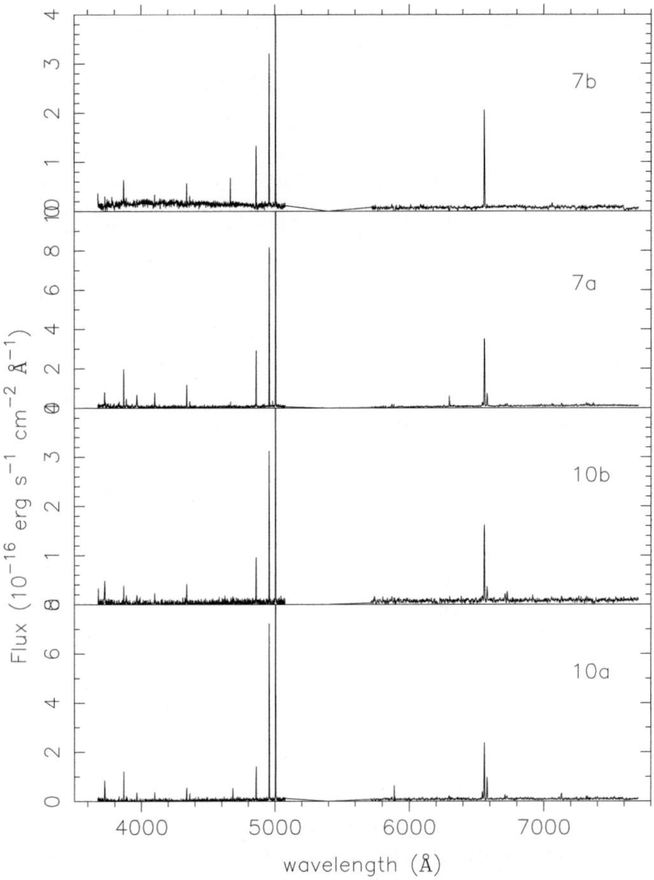

Fig. 2. Calibrated spectra of the 4 PNe we observed in M33

stellar absorption lines, which we did with particular scrutiny. Fig. 2. shows the calibrated spectra of the four PNe observed. We could detect the following lines in at least some of the spectra: Hα, Hβ, Hγ, Hδ, He I 5876, He II 4686Å, [O II]3727, [O III]4959, 5007, [O III]4363, [Ne III]3869, [N II]6585, [S II]6716, 6731, [Ar III] 7135Å.

The emission lines were corrected for extinction using the theoretical value for the Balmer decrement and the Seaton (1979) reddening law. The abundances were then derived using the electron temperature from the [O III] 4363/5007 ratio and the electron density from the [SII] 6716/6731 ratio. The ionization correction factors were taken from Kingsburgh & Barlow (1994). Table 1 presents our preliminary results on the chemical composition of the four PNe we observed.

We see that the (preliminary) abundances of all of our PNe appear to be well below solar! If confirmed, this would indicate that the abundance gradient in M33 levels off and is perhaps even negative in the inner zones of M33. We note

Table 1. Te[O III] and O/H & Ne/H ratios in the 4 observed PNe

	Te[O III] (K)	O/H (10^{-6})	Ne/H (10^{-6})
PN 7a	12000 –13300	140 – 184	4.8 – 3.5
PN 7b	14200 –17900	66 – 118	39 – 73
PN 10a	14000 –15700	323 – 176	8.5 – 11.5
PN 10b	$<$ 5000 – 11500	221 – $>$ 5200	30 – $>$ 1300
Sun		490	120

that our finding appears consistent with the neon abundances estimated from far infra-red line measurements by Willner & Nelson-Patel (2002), although in the latter case some systematics occur from the neglect of the electron temperature dependence of the hydrogen emission.

References

1. Bresolin , F., Garnett, D.R., Kennicutt, R.C. 2004, astro-ph/0407065
2. Garnett, D.R., Kennicutt, R.C., Jr., Bresolin, F. 2004, ApJ, 607,L21
3. Roth, M.M., Becker, T., Kelz, A., Schmoll, J. 2004, ApJ, 603, 531
4. Henry, R.B.C., Howard, J.W. 1995, ApJ, 438, 170
5. Kennicutt, R.C., Jr., Bresolin, F., Garnett, D.R. 2003, ApJ, 591, 801
6. Kingsburgh, R.L., Barlow, M.J. 1994, MNRAS, 271, 257
7. Kwitter, K.B., Aller, L.H. 1981, MNRAS, 195, 939
8. Ma, J., Zhou, X., Chen, J. 2004, A&A, 413, 563
9. Maciel, W.J., Koppen, J. 1994, A&A, 282, 436
10. Maciel, W.J., Quireza, C. 1999, A&A, 345, 629
11. Magrini, L., Cardwell, A., Corradi, R.L.M., et al. 2001, A&A, 367, 498
12. Marigo, P. 2001, A&A, 370, 194
13. Monteverde, M.I., Herrero, A., Lennon, D.J. 2000, ApJ, 545, 813
14. Monteverde, M.I., Herrero, A., Lennon, D.J., Kudritzki, R.-P. 1997, ApJ, 474, L107
15. Pindao, M., Schaerer, D., González Delgado, R.M., Stasińska, G. 2002, A&A, 394, 443
16. Seaton, M.J., 1979 MNRAS, 187, P73
17. Smartt, S.J., Venn, K.A., Dufton, P.L., et al., 2001, A&A, 367, 86
18. Smith, H.E., 1975, ApJ, 199, 591
19. Smith, R.C., Kirshner, R.P., Blair, W.P. et al., 1993, ApJ, 407, 564
20. Stasińska, G., 2002, RevMexAA (Serie de Conferencias), 12, 62
21. Stasińska, G., Richer, M.G., McCall, M.L., 1998, A&A, 336, 667
22. Stasińska, G., 2004a, Cosmochemistry. The melting pot of the elements. eds. C. Esteban, R. J. García López, et al., CUP, p. 115
23. Vílchez, J.M., Pagel, B.E.J., Diaz, A.I. et al., 1988, MNRAS, 235, 633
24. Weidemann, V., 2000, A&A, 363, 647
25. Willner, S.P., Nelson-Patel, K., 2002, ApJ, 568, 679
26. Zaritsky, D., Kennicutt, R.C., Jr., Huchra, J.P, 1994, ApJ, 420, 87Z

Crowded Field 3D Spectrophotometry of Extragalactic Planetary Nebulae

Martin M. Roth[1], Thomas Becker[1], Petra Böhm[1], Detlef Schönberner[1], Matthias Steffen[1], and Katrina Exter[2]

[1] Astrophysikalisches Institut Potsdam, An der Sternwarte 16, D-14482 Potsdam, Germany
[2] Instituto de Astrofísica de Canarias, La Laguna, Tenerife, Spain

Abstract. The spectrophotometry of extragalactic planetary nebulae in high surface brightness regions of nearby galaxies is notoriously difficult because of the problem of accurately subtracting the local background of unresolved stars and gaseous emission of the ISM. Conventional slit spectroscopy is not very well suited for providing a good estimate of the two-dimensional surface brightness distribution in the continuum, absorption lines, and superimposed emission lines from all of these background sources. The emergent technique of integral field (3D) spectroscopy holds the promise to overcome these limitations, similar to the advance of CCD photometry vs. photoelectric aperture photometry. We discuss basic properties of 3D spectrophotometry, results from a pilot study with planetary nebulae in the bulge of M31, and some more recent results from observations in other local group galaxies.

1 Introduction

The first abundance studies with extragalactic planetary nebulae (XPN) in local group galaxies beyond the Magellanic Clouds as advocated by Walsh et al. (2000), which were performed by Jacoby & Ciardullo (1999), Richer et al. (1999), and Walsh et al. (1999), have shown that it is indeed possible to measure faint emission line spectra of XPN with 4m class telescopes and multi-object spectrographs even out to galaxies as far as NGC5128. With modern 8-10m class telescopes, high efficiency spectrographs, and under excellent observing conditions, one should expect considerable progress in this field. However, the former studies have also shown that observations of XPN in high surface brightness regions of their host galaxies tend to suffer significantly from systematic errors related to background subtraction problems. These can be summarized as follows:

- light losses due to finite slit-width
- seeing FWHM = f(t), hence temporal variation of slit losses
- seeing FWHM = f(λ), i.e. wavelength-dependent slit losses
- differential atmospheric refraction
- spatially variable background continuum of unresolved stars
- spatially variable background emission line spectrum of ISM

A thorough study of these effects was presented by Jacoby & Kaler (1993). Obviously, the geometry of the spectrograph slit is the limiting factor in our ability to perform spectrophotometry, and to provide a better prediction of the two-dimensional background light distribution, which must be subtracted accurately enough if one wants to perform a reliable diagnostic analysis of XPN spectra (see also Stásinska, these proceedings).

2 3D Spectrophotometry

In recognizing that the technique of integral field (3D) spectroscopy, which was initially introduced for measuring velocity fields and stellar populations in galaxies, produces spectra in a two-dimensional field-of-view without any slit losses, Roth et al. (1997) proposed the development of *crowded field 3D spectroscopy* as a new method to overcome source confusion in spectroscopy. Using the MPFS (Si'lchenko & Afanasiev 2000) and PMAS instruments (Roth et al. 2000a), Roth et al. (2004) presented a pilot study on XPN in the bulge of M31, demonstrating that the use of the full 2-dimensional spatial information of the 3D datacube in

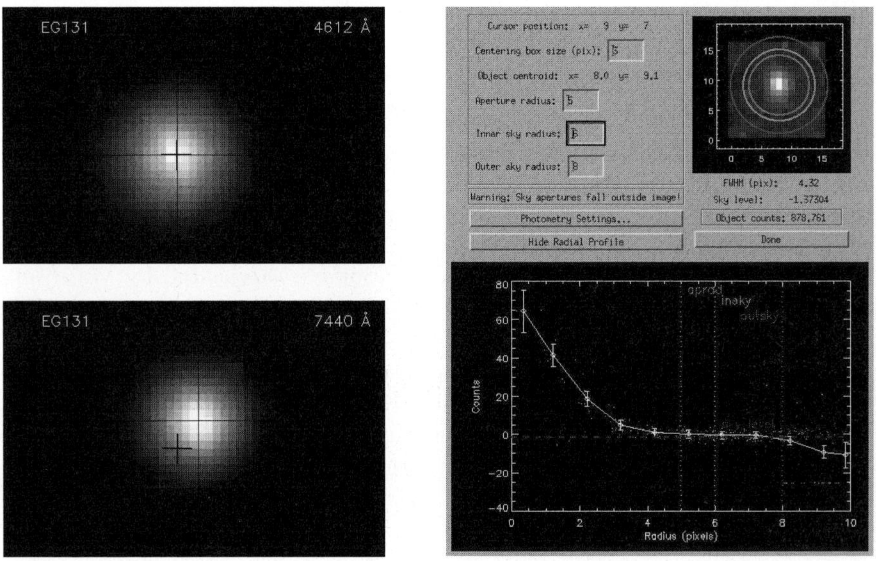

Fig. 1. 2D images, extracted as datacube slices from integral field (3D) spectroscopy. As in CCD photometry, the property of possessing a well-defined point-spread-function (PSF) is used for extraction techniques like aperture and PSF-fitting photometry. The frame to the right shows the appearance of a software tool, displaying a map at 4959Å and corresponding radial plot of an XPN in Leo A, observed with the PMAS instrument (see text). The two panels to the left show maps of the standard star EG131 at the wavelengths of 4612Å and 7440Å, respectively, obtained from a datacube taken with the GMOS-IFU at GEMINI (courtesy James Turner). The offset is due to differential atmospheric refraction.

combination with a fit to the XPN point-spread-function is indeed an extremely useful tool to overcome the limitations of classical slit spectroscopy (see Fig. 1). Our new results for XPN in common with data from the literature have shown, in particular, that conventional slit spectroscopy tends to overestimate low excitation emission line intensities in e.g. [N II], [S II], [O I], and [O II], whenever a contamination from the diffuse ISM background must be suspected. It was also shown that the coincidence of the background H_β absorption line with the XPN emission line, for example, gives rise to systematic errors when the background intensity cannot be assessed accurately enough because of a spatially variable superposition of unresolved stars of different spectral types.

3 Recent Results

Encouraged by our pilot study in M31, we have continued observations in other local group galaxies, e.g. Leo A, Sextans-A and Sextans-B. As an example, we present recent results of XPN candidates in Leo A (Fig. 2), obtained with PMAS at the Calar Alto Observatory 3.5m telescope. While we were able to confirm candidate No.1 as an XPN, candidate No.2 turns out to be a misidentification, i.e. a faint continuum source (Fig. 3).

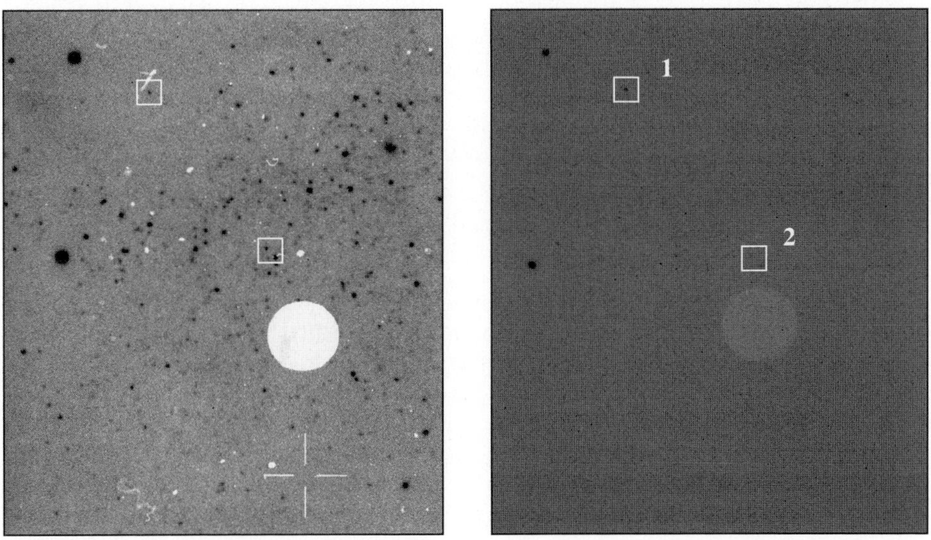

Fig. 2. XPN candidates in Leo A. Object no.1 is the XPN observed spectroscopically by Skillman et al. 1989. Object No.2 was identified by Jacoby & Lesser (1981) in a narrow-band imaging survey of local group galaxies. The two frames were obtained with the PMAS A&G CCD camera. Left: Johnson R-band, right: narrow-band [O III] λ5007. Note that the white circular aperture and the cross-hair are artificial features of the optical system.

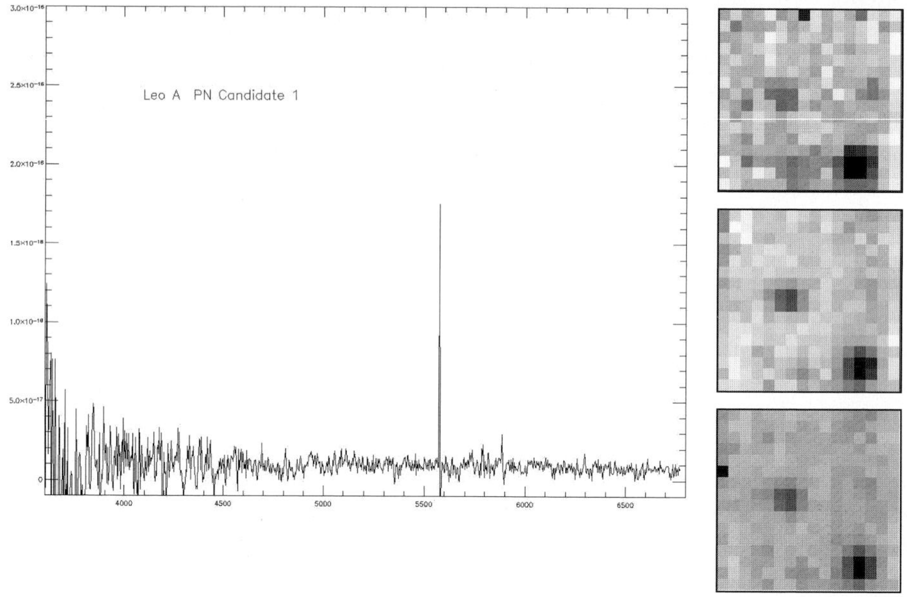

Fig. 3. XPN candidate No.2 in Leo A. The spectrum shows no trace of [O III] $\lambda\lambda$5007, 4959 above a flux level of a few 10^{-17}erg/cm^2/sec. Already inspection of the images in Fig. 2 indicates that candidate No.2 cannot be an XPN, further confirmed by the U, V, and R-band maps, which were synthesized from the datacube (top to bottom, right): No.2 is seen as a continuum point source left of center of the frame.

4 Outlook

Using PMAS at the Calar Alto 3.5m telescope, we will continue with a scheduled observing programme on background-limited M31 bulge XPN. Other local group galaxies with substantial background contamination problems, e.g. M33, are equally promising and challenging targets. The use of 3D instruments at 8m class telescopes, e.g. the VIMOS and FLAMES IFUs at the VLT, or the GMOS-IFUs at the GEMINI-N and -S telescopes, would appear to provide a significant advantage in terms of efficiency. Moreover, we think that IFUs can offer huge benefits for observations of very faint objects with uncertain positions, e.g. intracluster planetary nebulae in Virgo, since – contrary to slit spectroscopy – the pointing accuracy of the telescope is relatively unimportant. A completely new domain for the study of resolved stellar populations in nearby galaxies will be opened with novel instruments like the MUSE 3D spectrograph for the VLT, covering a 1 arcmin^2 field-of-view with seeing-limited spatial sampling (Bacon et al. 2002).

Acknowledgements

This work was supported by the German *Deutsche Forschungsgemeinschaft, DFG* under grant HA1850/10-3 , and by the German Verbundforschung des BMBF, grant 05AL9BA1. We would like to thank James Turner, GEMINI-S, for providing us with the EG131 standard star datacube obtained with the GMOS-IFU.

References

1. Arnaboldi, M. et al. 2002, AJ 123, 760
2. Bacon, R., et al. 2002, in *Scientific Drivers for ESO Future VLT/VLTI Instrumentation*, eds. J. Bergeron, G. Monnet, Springer, Berlin, p. 108
3. Becker, T. 2002, Thesis, University of Potsdam
4. Ciardullo, R., Rubin, V. C., Ford, W. K., Jacoby, G. H., Ford, H.C. 1988, AJ 95, 438
5. Ciardullo, R., Jacoby, G.H., Ford, H.C., Neill, J.D. 1989, ApJ 339, 53
6. Durrell, P.R., Mihos, J.C., Feldmeier, J.J., Jacoby, G.H., Ciardullo, R. 2003, ApJ 582, 170
7. Feldmeier, J. J., Ciardullo, R., Jacoby, G. H., Durrell, P. R. 2003, ApJS 145, 65
8. Ford, H., Jacoby, G.H. 1978, ApJS 38, 351
9. Gerhard, O., Arnaboldi, M., Freeman, K. C., & Okamura, S. 2002, ApJ 580, L121
10. Hook, R.N., Lucy, L.B. 1993, in *Proc. Science with the HST*, p. 245
11. Hui, X., Ford, H. C., Freeman, K. C., Dopita, M. A. 1995, ApJ 449, 592
12. Jacoby, G.H., Lesser, M.P. 1981, AJ 86, 185
13. Jacoby, G. H., Ford, H., Ciardullo, R. 1985, ApJ 290, 136
14. Jacoby, G. H., Kaler, J. B. 1993, ApJ 417, 209
15. Jacoby, G. H., Ciardullo, R. 1999, ApJ 515, 169
16. Kelz, A., Roth, M. M., & Becker, T. 2003, SPIE 4841, 1057
17. Lucy, L.B. 1994, in *The Restoration of HST Images and Spectra II*, eds. R. Hanisch & R. White, p. 79
18. Magrini, L., Perinotto, M., Corradi, R. L. M., Mampaso, A. 2003, A&A 400, 511
19. Richer, M. G., Stasińska, G., McCall, M. L. 1999, A&A Suppl. 135, 203
20. Roth, M. M., Seydack, M., Bauer, S., & Laux, U. 1997, SPIE 2871, 1235
21. Roth, M. M. et al. 2000a, SPIE 4008, 277
22. Roth, M. M., Becker, T., & Schmoll, J. 2000b, ASP Conf. Ser. 195: Imaging the Universe in Three Dimensions, 122
23. Roth, M.M., Becker, T., Böhm, P., Kelz, A. 2002a, Proc. Scientific Drivers for ESO Future VLT/VLTI Instrumentation, eds. J. Bergeron, G. Monnet, Springer, p. 136
24. Roth, M.M., Becker, T., Kelz, A., Schmoll, J. 2004, ApJ 603, 531
25. Skillman, E.D., Kennicut, R.C., Hodge, P.W. 19889, ApJ 347, 875
26. Si'lchenko, O., Afanasiev, V. 2000, A&A 364, 479
27. Walsh, J. R., Walton, N.A., Jacoby, G. H., Peletier, R.F. 1999, A&A 346, 753
28. Walsh, J.R., Jacoby, G.H., Peletier, R.F., Walton, N.A. 2000, SPIE 4005, p.131

Extragalactic PNe Observed with 3D Spectroscopy

Katrina Exter[1], Petra Böhm[2], Thomas Becker[2], and Martin Roth[2]

[1] Inst. de Astrofisica de Canarias, c/ Via Láctea, La Laguna, Tenerife, Spain
[2] Astrophys. Institut Potsdam, An der Sternwarte 16, D-14482 Potsdam, Germany

1 Introduction

The observations of extragalactic PNe (XPNe) can be very difficult when they are superimposed on the bright and crowded background of a galaxy. We explain here why spatially resolved, 3D spectroscopy can be very useful in disentangling the flux of the point-source XPNe from the background.

2 3D Spectroscopy in Brief

For more information on what 3D spectrographs do, we refer to the paper by M. Roth in these proceedings and to the web site of the Euro3D consortium, *www.aip.de/Euro3D*. One can also consult the special issue of *Astronomische Nachrichten*, 325, 2 (2004), dedicated to science with 3D spectroscopy.

3D instruments allow for spatially continuous spectral observations of a target. Slice the data cube along one axis (Fig. 1) to obtain the spectrum of a unique patch of the observed field, slice perpendicular for an image of the field in any one wavelength. The observational advantages over slit or multi-object spectroscopy are: (i) no slit losses; (ii) no fiddling with slit widths to compensate for wavelength dependent seeing or refraction; (iii) the background contamination *immediately around the source* can be removed. This last point in particular is important in the observations of XPNe in crowded fields.

3 Observations of Point Sources

We observed 4 XPNe in M31 [1], and Fig. 1 demonstrates the advantage of spatially continuous spectroscopy. We used point-spread-function fitting to extract the point source from the background (contaminating) flux, which includes the absorption line spectrum from the background stars, and emission lines from the interstellar medium. For our example PN in Fig. 1, the background emission is highly inhomogeneous, with emission very close to, but *not* coincident with the PN. A slit orientated in different directions would sample completely different parts of this background and, as we have shown [1], would not result in the correct de-contamination for this PN.

The PSF fitting technique [2] uses, among others, the *cplucy* algorithm [3], available within IRAF. The PSF itself is defined either from a clear point source

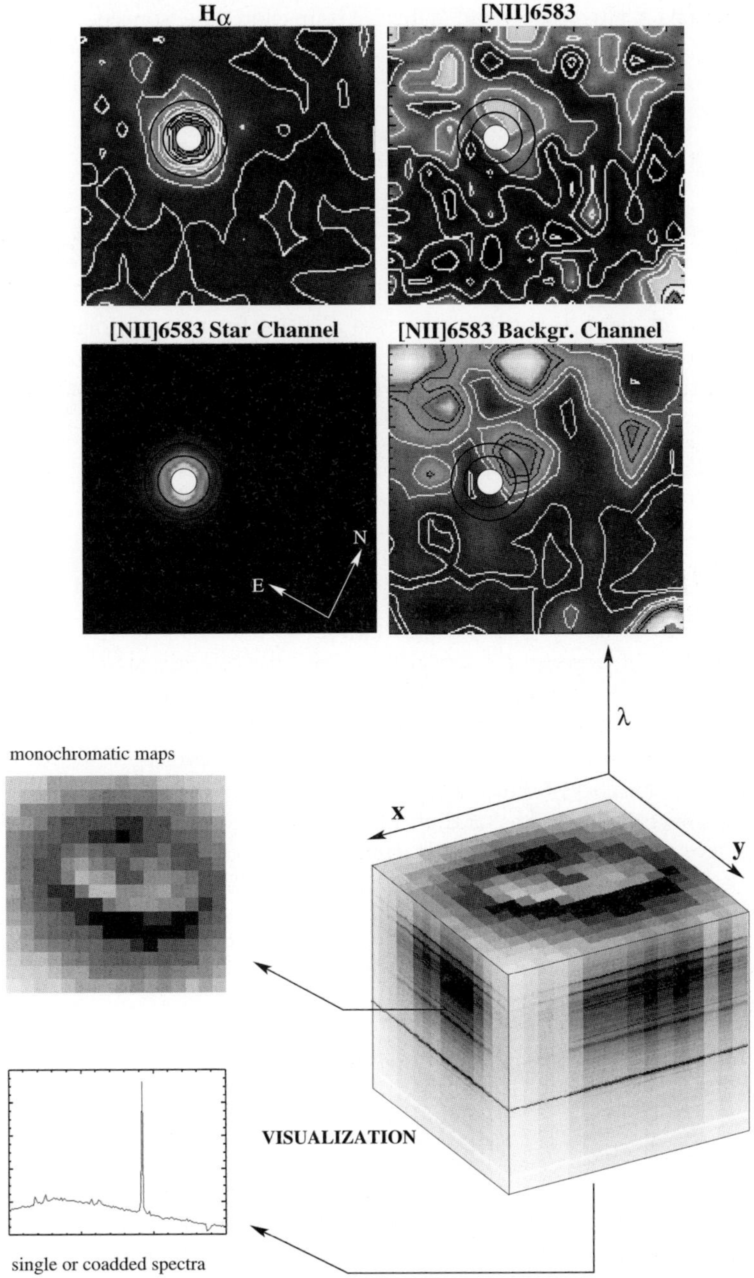

Fig. 1. **UPPER** PN 27 in M31, **top** showing reconstructed images at Hα and [N II], **bottom** *left* showing this point source in the line [N II] and *right* the remaining background 'contamination'. **LOWER** A 3D data cube is shown. RA and Dec are along the bottom axes, and wavelength along the third.

in the field, or in this case from a bright emission line source. Changes with wavelength are followed, changes with spatial coordinates could be also. Testing of the technique [2] has shown that it works very well even with faint sources.

4 Our M31 PNe

Observed were 4 PNe: PN 27, 29, 56 and 181 [6]. The details are given in [1], and here are our *preliminary* abundance calculations. T_e and N_e were calculated from the O and S line ratios, with default values when necessary. The ionisation correction factors were taken from [4]; see also [5] for more information on the abundance calculation method. We obtained or adopted the following for 3 of the PNe (abundances are given as log+12, linear for H, and N_e in log).

- **PN 27**: $T_e = 16000$ K, $N_e = 4.5$: $\mathbf{O^{2+}}$:8.2, $\mathbf{N^{+}}$:6.0, **He**:0.08.
- **PN 29**: $T_e = 12000$ K, $N_e = 3.6$: **O**:8.6, **N**:8.3, **Ne**:7.9, **Ar**:6.1, **S**< 5.8, **He**: 0.09.
- **PN 56** $T_e = 16000$ K, $N_e = 3.7$: $\mathbf{O^{2+}}$:8.1, $\mathbf{Ne^{2+}}$:6.8, $\mathbf{He^{2+}}$:0.016.

PN29 provides an instructive example: when observed before[6], the density from the [S II] ratio was too low for the [O III] line fluxes. We found that, in fact, *all* the [S II] emission is background emission.

References

1. M. Roth, T. Becker, A. Kelz, J. Schmoll: ApJ, 603, 531 (2004)
2. T. Becker, S. Fabrika, M. Roth: AN, 325, 155 (2004)
3. L. Lucy: 'Image Restorations of High Photometric Quality'. In: *The Restoration of HST Images and Spectra II*, ed. by R. Hanisch, R. White (Baltimore: STScI, 1994), p.79
4. R. Kingsburgh, M. Barlow: MNRAS, 27, 257 (1994)
5. K. Exter, M. Barlow, R. Clegg, N. Walton: MNRAS, 349, 129 (2004)
6. G. Jacoby, R. Ciardullo: ApJ, 515, 169 (1999)

Spectroscopy of Planetary Nebulae in Sextans A and Sextans B

Laura Magrini[1], Pierre Leisy[2,3], Romano L.M. Corradi[3], Mario Perinotto[1], Antonio Mampaso[2], and José Vílchez[4]

[1] Dipartimento di Astronomia e Scienza dello Spazio, Universitá di Firenze, Italy
[2] Instituto de Astrofísica de Canarias, Tenerife, Canarias, Spain
[3] Isaac Newton Group of Telescopes, La Palma, Canarias, Spain
[4] Instituto de Astrofísica de Andalucía, Granada, Spain

Abstract. Sextans A and Sextans B are two dIr galaxies situated in the outskirts of the Local Group (LG), in which both PNe and H II regions have been detected (Jacoby & Lesser [5], Magrini et al. [6], [7]). We present spectroscopic observations of PNe and HII regions in these two galaxies obtained with the VLT. Preliminary results on the PNe physico-chemical properties are presented.

1 The Galaxies: Sextans A and Sextans B

Sextans A and Sextans B are both dwarf Irregular galaxies (Ir V and Ir IV-V morphological types, respectively, cf. van den Bergh [11], hereafter vdB00) with approximately the same V luminosity, belonging to the outer fringes of the Local Group. Their distances from the barycenter of the LG are quite similar (1.60 and 1.72 Mpc, see vdB00) and their separation on the sky is relatively small ($\sim$ 10 degrees), which corresponds to about 280 kpc. Moreover their velocity difference is only 23 $\pm$6 km s^{-1}. These results support the hypothesis of a common formation of these two galaxies, probably together with NGC 3109 and the Antlia galaxy. Considering the mean distance of the four galaxies from the barycenter of the LG, 1.7 Mpc, this sub-group is located beyond the zero velocity surface of the LG (cf. vdB00) and thus it can be considered the nearest external group of galaxies. As said by Mateo ([9]) "No two Local Group dwarfs have the same star-formation history". This is true also when comparing very similar galaxies. In fact, in spite of a probable common formation, Sextans A and Sextans B show a different star formation history as indicated by the different amounts of stars in the various evolutionary phases, and also reflected in the different number of PNe and HII regions observed. Sextans A shows a large contribution from an old stellar population, whereas it has modest intermediate (4-10 Gyrs) and recent star formation (1-4 Gyrs ago). On the other hand, Sextans B exhibits very strong recent star formation (1-4 Gyrs), together with a very old population (c.f. the review by Mateo ([9] about the star formation history of the LG dwarfs).

In this context, PNe represent an useful tool to trace the chemical enrichment history of galaxies. Moreover, in contrast to chemical abundances derived from H II regions, they are tracers of stellar population over a wide range of ages, from old to intermediate age.

2 Observations and Data Reduction

PNe (1 in Sextans A and 5 in Sextans B) and H II regions (10 and 9, respectively) have been observed in December 2003 with VLT (ESO, Paranal) equipped with FORS2 spectrograph in multi-object spectroscopy (MOS) observing mode. The FORS2 spectrograph was used with two setups: the 300V and 300I gratings providing a dispersion of 3.0 Å/pixel. The total resulting spectral range, from ~3200Å to ~10000 Å, includes basic lines needed for the determination of chemical abundances. For each galaxy, a total of six exposures were taken, three with the 300V grating (total exposure time 5400 s) and three using 300I (3600 s). A spectrophotometric standard star, GD 108, was observed once during each night. The data were reduced using the IRAF LONGSLIT package and complemented with MIDAS (MOS and LONG packages). The spectra are presented in Figs. 1 and 2. Identified emission lines are marked with the name of the corresponding ion and/or wavelength.

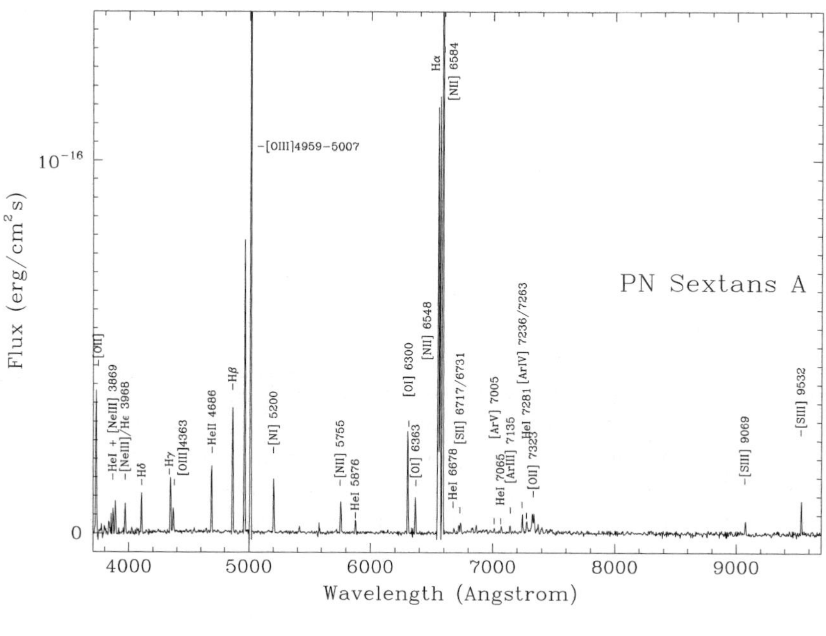

Fig. 1. VLT spectra of the planetary nebula in Sextans A

3 Chemical Abundances of Planetary Nebulae

Chemical abundances have been derived with the classic Ionization Correction Factors (ICFs) method, following Kingsburgh & Barlow ([4]). As the [O III]

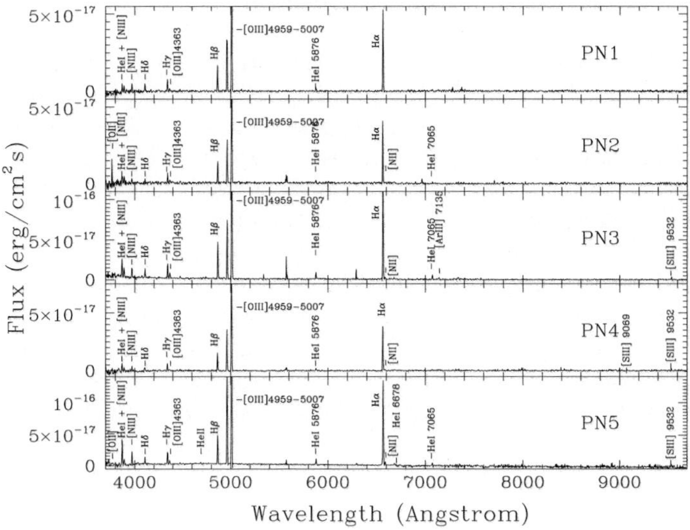

Fig. 2. VLT spectra of the five planetary nebulae in Sextans B. Identification numbers are from Magrini et al. ([6])

4363Å emission line was always measurable with a sufficiently high signal-to-noise ratio, the [O III] electron temperature was derived for all PNe. As the nitrogen emission lines are so intense in Sextans A, the [N II] 5755Å is observed and an electron temperature is derived. [S III] infrared emission lines were detected in 4 PNe. They allow a better accuracy in the measurement of sulphur abundance, especially in low metallicity environments where [S II] lines are extremely faint.

In addition, the PNe have been modelled with the photoionization code CLOUDY 94.00 (Ferland et al. [2]), assuming a blackbody central star with effective temperature derived using the Ambartsumian's ([1]) or Gurzadyan's ([3]) methods, and a spherical nebula with constant density, derived from [S II] 6717/6731 Å flux ratio when available or assumed equal to 3000 cm^{-3} in the other cases. For further details on the modelling procedure see Magrini et al. ([8]) and Perinotto et al. (these proceedings). The chemical abundances derived with the two methods are in good agreement with each other. The CLOUDY model also allowed some nebular and stellar parameters, such as the radius of the nebula, the stellar luminosity and temperature to be derived (see Table 2). The preliminary chemical abundances and their errors, derived with the ICF method, are shown in Table 1.

Although the definition of Type I PNe is a function of metallicity of the host galaxy (cf. Magrini et al. [8]), as a first approximation, we applied the Galactic Type I definition by Kingsburgh & Barlow ([4]), log(N/O)>-0.1. With this definition, the Sextans A PN is a Type I PN, while the Sextans B PNe

Table 1. Chemical abundances of Sextans A and Sextans B planetary nebulae derived with the ICF method. Metal abundances are expressed in 12+log(X/H).

Name	He/H	N/H	O/H	Ne/N	Ar/H	S/H
SexA-PN	0.085±0.01	8.45±0.1	8.0±0.1	6.7±0.2	5.2±0.3	5.8±0.1
SexB-PN1	0.082±0.02	-	8.0±0.2	7.0±0.3	-	-
SexB-PN2	0.05±0.03	-	7.65±0.3	6.7±0.3	-	-
SexB-PN3	0.083±0.01	5.4†	7.5±0.2	6.9±0.3	5.5±0.3	-
SexB-PN4	0.084±0.02	5.5†	8.0±0.3	7.0±0.3	-	5.4‡
SexB-PN5	0.118±0.02	8.0±0.3	8.2±0.1	7.65±0.2	-	-

Note: † this is the ionic N^+ abundance. ‡ this is the ionic S^{2+} abundance. In both † and ‡ cases, ICF(N) and ICF(S), respectively, were not computed because [O II] lines were not detected.

Table 2. Some nebular and stellar parameters of Sextans A and Sextans B planetary nebulae. † Derived with the CLOUDY model. $T_\star$ marked with : are upper limits to the stellar effective temperature obtained when the He II 4686 Å was not measurable.

Name	N_e	$T_{e[O\ III]}$	$T_{e[N\ II]}$	Radius(pc)†	$T_\star$†	$\log(L_\star/L_\odot)$†	$c\beta$
SexA-PN	2700	21000	13400	0.13	190000	3.8	0.22
SexB-PN1	-	12900	-	0.08	50000:	3.2	0.11
SexB-PN2	-	17300	-	0.02	73000:	3.3	-
SexB-PN3	-	17800	-	0.03	74000:	3.5	0.11
SexB-PN4	-	13300	-	0.03	70000:	3.1	-
SexB-PN5	-	12700	-	0.01	84000	3.4	0.14

are non-Type I. Using the stellar luminosity and temperature derived with the CLOUDY model, we compared the location of the six PNe in the $\log T_\star$-$\log L_\star$ space with the H-burning evolutionary tracks from Vassiliadis & Wood ([12]). We found a more massive progenitor for the Sextans A PN (~2.5$M_\odot$) than those of Sextans B PNe (~1-1.5$M_\odot$), in agreement with its Type I nature.

Another interesting aspect evident from Table 1 is the large range of O/H, ~0.7 dex, of Sextans B PNe in spite of their similar evolutionary age. Since oxygen is not processed by PN progenitors, this represents a sign of the inhomogeneity of the interstellar medium at the time of the formation of the PN progenitors. We plan to compare this inhomogeneity with the spatial distribution of H II region chemical abundances.

4 The Metallicity-Luminosity Relation using PNe

In Fig. 3, the relationship of V-luminosity vs. oxygen abundance from PNe in the LG is shown. Galaxies are identified with different symbols according to their morphological type. A different behaviour of star forming (irregulars and spirals) and non-star forming (spheroidals) galaxies was already noted by several

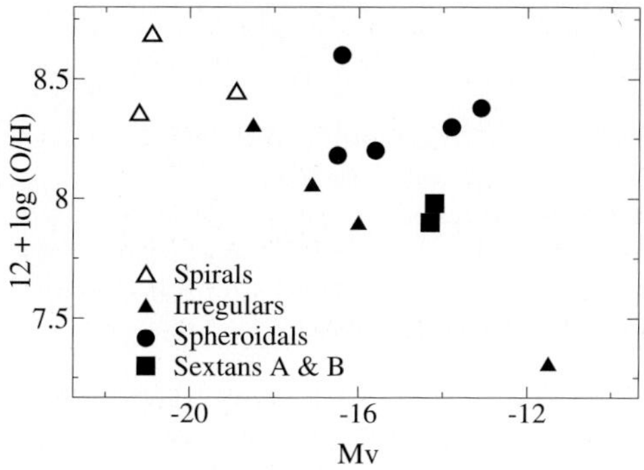

Fig. 3. Relationship of V-luminosity vs. oxygen abundance from PNe in the LG. PN abundances: Sextans A and Sextans B from this work; for other galaxies from data in the literature.

authors (cf. Mateo [9]): spheroidal galaxies are generally more metal rich than irregular galaxies with the same V luminosity. There are several hypotheses concerning this behaviour (cf. Pilyugin [10]), but a satisfactory understanding is still missing. The uniform way to derive chemical abundances with PNe, which are present in galaxies of every morphological type, will help in understanding the origin of this fundamental relation.

References

1. Ambartsumian, V. A., 1932, Poulkovo Obs. Circ 4, 8
2. Ferland, G. J., Korista, K. T., Verner, D. A., et al.: 1998, PASP, 110, 761
3. Gurzadyan, G. A., 1988, ApSS, 149, 343
4. Kingsburgh, R. L. & Barlow, M. J., 1994, MNRAS, 271, 257
5. Jacoby, G. H. & Lesser, M. P., 1981, AJ, 86, 185
6. Magrini, L., Corradi, R. L. M., Walton, N. A., et al., 2002, A&A, 386, 869
7. Magrini, L., Corradi, R. L. M., Greimel, R., et al., 2003, A&A, 407, 51
8. Magrini, L., Perinotto, M., Mampaso, A., Corradi, R. L. M., 2004, A&A in press, astro-ph/0406642
9. Mateo, M. L., 1998, ARA&A, 36, 435
10. Pilyugin, L. S., 2001, A&A, 374, 412
11. van den Bergh, S., 2000 in *The Galaxies of the Local Group*, Cambridge University Press
12. Vassiliadis, E. & Wood, P. R., 1994, ApJS, 92, 125

Local Group Census: The Dwarf Irregular Galaxy NGC 6822

P. Leisy[1,3], L. Magrini[2], R.L.M. Corradi[1], and A. Mampaso[3]

[1] Isaac Newton Group of Telescopes, La Palma, Canarias, Spain
[2] Dipartimento di Astronomia e Scienza dello Spazio, Universitá di Firenze, Italy
[3] Instituto de Astrofísica de Canarias, Tenerife, Canarias, Spain

Abstract. In the framework of the Local Group Census (LGC), an imaging programme to search for emission line objects in the Local Group, we present the detection of 13 new candidate planetary nebulae (PNe) and follow-up optical spectroscopic confirmation in the dwarf irregular galaxy NGC 6822. This substantially increases the PNe sample of NGC 6822 to 17. Our sample of PNe is complete within 3.5 mag below the brightest PN. Using the PN luminosity function (PNLF), we find a distance modulus of $m - M = 23.32 \pm 0.45$. The spectroscopic data help to discriminate between the different emission line objects (PNe, SNR, HII regions). The spectra with the highest S/N ratio allow us to determine abundances for 6 PNe and 4 HII regions. For the first time, we can compare the PN abundances of helium, nitrogen and oxygen in many LG galaxies. We find that the nitrogen is not enhanced in NGC 6822 (nor in M 33) as much as was expected from their respective metallicities. In contrast, the helium is largely enhanced in all the galaxies. The oxygen abundances in NGC 6822 are high (close to those of the H II regions), which supports the idea that most PNe originated from relatively massive and young progenitors.

1 NGC 6822

NGC 6822 is a barred dwarf galaxy (type Ir IV-V) with a total luminosity of $M_V = -15.96$ and a mass of $1.9\,10^9\,M_\odot$ (van den Bergh [19]). It is brighter, larger (2.9 kpc × 2.9 kpc) and closer (490 kpc) than any other dwarf irregular in the LG (Mateo [17]), allowing deeper studies of the properties of individual stars and emission-line objects. Some further extension (4 kpc × 6 kpc) is revealed by a study of Carbon stars (Letarte et al. [13]) or by the H I content (6 kpc × 13 kpc, de Blok & Walter [4]). NGC 6822 is gas rich, with a recent star formation whose rate has been more or less continuous but slow during the last Gyr. Gallart et al. [8] concluded that NGC 6822 started forming stars from the beginning (12-15 Gyr), with possible small enhancements 400 and 100-200 Myr ago.
Eight PNe were previously known (Killen & Dufour [11]), but we only confirm 4. Only limited spectroscopic information for two of them is available in the literature prior to this work.

2 Spatial Distribution

With our imaging program (LGC) we have discovered 13 new candidate PNe in NGC 6822. Most of the candidate PNe lie in the bar of NGC 6822. The bar

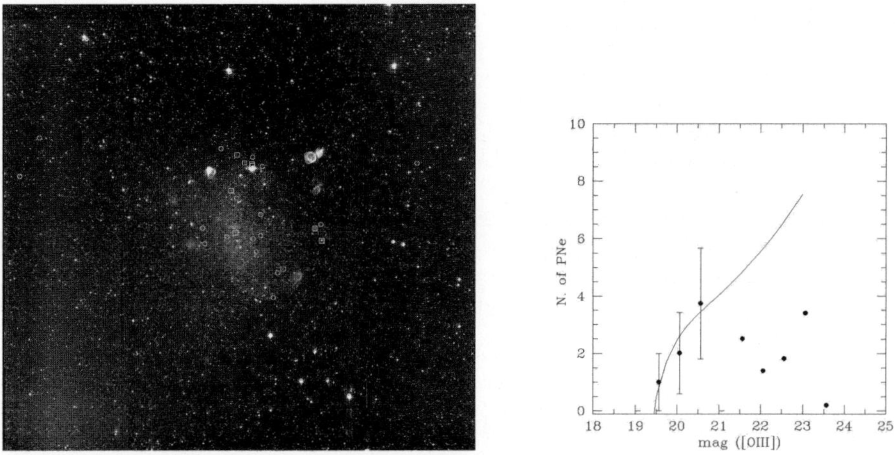

Fig. 1. (**a**): Image of NGC 6822 showing spatial distribution of the 17 PNe. (**b**): PNLF of the 17 PNe in NGC 6822.

(0.9 kpc $\times$ 0.9 kpc) is mostly composed of young blue stars (de Blok [5]) and its spatial distribution matches well that of H II regions (Battinelli et al. [1]). Therefore the PNe of the bar probably originate from massive and young progenitors. This is also supported by the abundance determination described in Sect. 4. For a sample of PNe in a galaxy which suffered a recent star burst ($<$ 1Gyr), the PNLF is not affected by metallicity effects (Marigo et al. [16]). This should be the case for NGC 6822.

Notably, three PNe have been discovered at large galactocentric distances (see Fig. 1a), about $\sim$2.2 kpc from the center of NGC 6822. NGC 6822 is surrounded by a large elliptical neutral hydrogen envelope (Hutchmeier [9]). These three PNe could be the first stellar representatives associated with this large H I envelope, but at least the two eastern PNe seem not to be spatially associated. They are probably inter-galactic PNe, or were produced by tidal streams extending faraway from the core of the galaxy.

In each galaxy of the LG the number of candidate PNe scales with the total luminosity of the galaxy (Corradi & Magrini *these proceedings*). The number of PNe found in NGC 6822 also falls close to this relation, making us confident of having found all the brightest objects. Assuming full completeness for $S/N > 10$, it is reached over the range of the brightest 3.5 magnitude.

3 Planetary Nebula Luminosity Function

Since 1989, the [O III] Planetary Nebula Luminosity Function (PNLF) has been widely used as a distance indicator (Ciardullo et al. [2]). The distance determination through the PNLF must be taken with caution when the number of PNe is quite small as the statistical significance is low (cf. Méndez et al. [18], Magrini et al. [14]). The sample of 17 PNe was fitted to the "universal" PNLF

(Jacoby [10]). Figure 1b presents the PNe histogram and the best fit for these points. An Eddington correction has been made and the magnitude error was also taken into account.
If one uses the brightest PN and assumes that its absolute magnitude is that of the cutoff of the PNLF ($M_* = -4.47$), the distance modulus derived would be 24.3. The fit to the PNLF sample gives a distance modulus value of 24.12±0.36. Correcting for foreground and mean internal reddening (Av∼0.8), and associating the error in quadrature, we derive a true distance modulus of NGC 6822 of m-M=23.32±0.45 or a distance of $D = 461\,^{+106}_{-86}\,kpc$. This is in agreement with the mean value of 23.48 (van den Bergh [19]) and also with the recent RR Lyrae determination of Clementini et al. [3] (23.36±0.17).

4 Spectroscopy – Abundances

Spectroscopic confirmation was obtained for most of the candidate PNe identified in the images. Some long-slit spectra were taken with EMMI at the NTT (ESO). Additional, albeit less deep, confirmation spectra were obtained with AutoFib2 at the 4.2m WHT (La Palma). Figure 2 presents part of the blue and red spectra which have been used to discriminate between SNR, HII regions and PNe. We used the criterion of an object's size to determine if it is an H II region or a PN: HII-018 for example is extend and so not a PN. We also checked if the observed continuum is part of the object (for an HII region/SNR) or is only associated with an underlying object (e.g. for PN-019). Figure 2 also shows the diagnostic diagram (revisited by Delgado [6]) that we also used to tentatively classify objects. The different zones are defined for our galaxy. We also superposed the zone for HII regions of NGC 6822 (dashed lines), derived from line measurements available in the literature.

Figure 3 shows samples of high quality spectra of 20 mn exposures taken with the NTT under very good seeing conditions (∼0.4″). In Fig. 4 we present the N/O abundance ratios versus Helium or Oxygen for PNe in NGC 6822 and for some PNe in M33 (Magrini et al. [15]). In NGC 6822 we have determined abundances for 6 PNe and 4 HII regions.

In the LMC and SMC, Leisy & Dennefeld [12] found a large enrichment of He, N, C for PNe. Given the metallicity of NGC 6822 we would expect to find more or less the same results. To our surprise the element enhancements are different with respect to the LMC, SMC or the Milky Way. As in M 33, in NGC 6822 the nitrogen seems not to be produced in large quantities. On the contrary, the LMC or SMC have produced huge amounts of nitrogen via the 3rd dredge-up. Nevertheless helium is largely enhanced in all galaxies (maybe more in M 33 than in NGC 6822).

The PNe oxygen abundances in NGC 6822 are very high (close to those of the H II regions). This also supports an origin from massive and young progenitors for the PNe in NGC 6822. Surprisingly the behavior of the oxygen abundance is different from all the well studied LG galaxies (LMC, SMC and M 33).

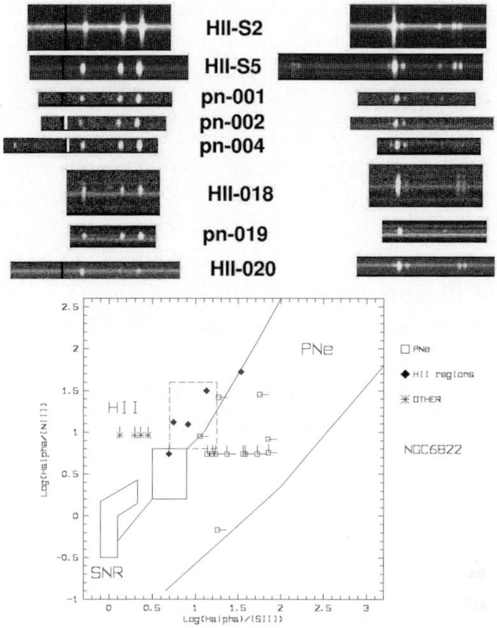

Fig. 2. 2-D NTT spectra: (**upper**): (*left*) blue spectra shortward of [OIII] 5007Å; and (*right*) red spectra in the H_α region. (**bottom**) [NII]/H_α-[SII]/H_α diagnostic diagram to disentangle the different emission line objects.

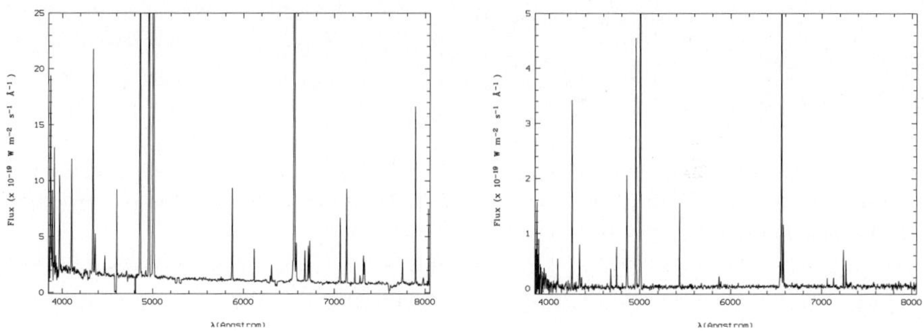

Fig. 3. Sample of 2 NTT spectra (**Left**): H II region KD_S2. (**Right**):PN-004

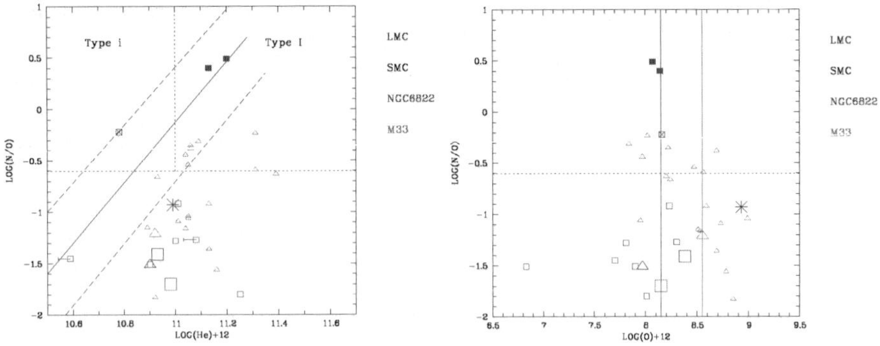

Fig. 4. Abundance diagram (squares for NGC 6822, triangles for M 33) (**a**) N/O versus He/H (**b**) N/O versus O/H. Mean LMC (square) and SMC (triangle) are also plotted

5 Perspectives

The new detections of the Local Group Census have substantially increased the number of known PNe in the LG. The new PNe are clearly an invaluable resource for high S/N spectroscopic follow-up aimed at determining their physical and chemical properties and also of their host galaxies. These detections will increase the number of galaxies with PNe in galaxies of different metallicities, mass and evolution, for which we can make a direct comparison. This task is appropriate to 8-10m class telescopes like the VLT, which can provide high S/N spectra in a reasonable exposure time (see Magrini et al., these proceedings).

References

1. Battinelli, P., Demers, S., Letarte, B. (2003) A&A *405*, 563
2. Ciardullo, R., Jacoby, G.H., Booth, J., Ford, H.C., (1989) ApJ *344*, 715
3. Clementini, G., Held, E.V., Baldacci, L., Rizzi, L., (2003) ApJ *588*, L85
4. de Blok, W.J.G., Walter, F., (2000) ApJ *537L*, 95
5. de Blok, W. J. G., Walter, F., (2003) MNRAS *341*, L39
6. Delgado, G., et al. (2004) Instituto Astrofisico de las Canarias *Master Thesis*
7. Freedman, W.L. & Madore, B. F., (1990) ApJ *365*, 186
8. Gallart, C., Aparicio, A., Vilchez, J.M., (1996) Astron. J *112*,1929
9. Hutchmeier, W.K., (1979) A&A *75*,170
10. Jacoby, G.H., (1989) ApJ *339*, 39
11. Killen, R.M., Dufour R.J., (1982) PASP *94*,444
12. Leisy, P., Dennefeld, M., (2004) A&A *submitted*
13. Letarte, B., Demers, S., Battinelli, P., Kunkel, W.E., (2002) AJ *123*, 832
14. Magrini, L., Corradi, R.L.M., Greimel, R., et al: (2003) A&A *407*, 51
15. Magrini, L., Perinotto, M., Mampaso, A., Corradi, R.L.M., (2004) A&A *in press*
16. Marigo, P., Girardi, L., Weiss, A., et al., (2004) A&A *423*, 995
17. Mateo, M., (1998) A&A Rev *36*, 435
18. Méndez, R. H., Kudritzki, R. P., Ciardullo, R., et al.: (1993) A&A *275*, 534
19. van den Bergh, S., (2000) The galaxies of the Local Group, Cambridge Astrophysics Series, vol no:35

Spectroscopy of PNe in Sextans A, Sextans B, NGC 3109 and Fornax

Alexei Y. Kniazev[1], Eva K. Grebel[2], Alexander G. Pramskij[3], and Simon A. Pustilnik[3]

[1] Max-Planck-Institut für Astronomie, Königstuhl 17, D-69117 Heidelberg, Germany
[2] Astronomisches Institut, Universität Basel, Venusstrasse 7, CH-4102 Binningen, Switzerland
[3] Special Astrophysical Observatory, Nizhnij Arkhyz, Karachai-Circassia, 369167, Russia

Abstract. Planetary nebulae (PNe) and HII regions provide a probe of the chemical enrichment and star formation history of a galaxy from intermediate ages to the present day. Furthermore, observations of HII regions and PNe permit us to measure abundances at different locations, testing the homogeneity with which heavy elements are/were distributed within a galaxy. We present the first results of NTT spectroscopy of HII regions and/or PNe in four nearby dwarf galaxies: Sextans A, Sextans B, NGC 3109, and Fornax. The first three form a small group of galaxies just beyond the Local Group and are gas-rich dwarf irregular galaxies, whereas Fornax is a gas-deficient Local Group dwarf spheroidal that stopped its star formation activity a few hundred million years ago. For all PNe and some of the HII regions in these galaxies we have obtained elemental abundances via the classic $T_{\rm e}$-method based on the detection of the [OIII] λ4363 line. The oxygen abundances in three HII regions of Sextans A are all consistent within the individual rms uncertainties. The oxygen abundance in the PN of Sextans A is however significantly higher. This PN is even more enriched in nitrogen and helium, implying its classification as a PN of Type I. The presumably unaffected PN abundances of S and Ar are well below those in the HII regions, indicating a lower metallicity at the epoch of the PN progenitor formation. For two HII regions in Sextans B, the oxygen abundances do not differ within the rms uncertainties. The third one is, however, twice as metal-rich, providing evidence for the inhomogeneity of the current metallicity distribution in Sextans B. For the PN in Sextans B we measured an O/H that is consistent with that of the low-metallicity HII regions. For NGC 3109 our preliminary results indicate that the oxygen abundances of PNe and HII regions are all within a small range of $\pm$0.15 dex. For the PN in Fornax, Ne, Ar and S abundances suggest that the ISM metallicity was $\sim$0.3 dex lower at the epoch of the PN progenitor's formation, compared to the O/H value derived for the PN.

1 Introduction

Understanding how the elemental abundances of galaxies have changed over time is an essential issue of galaxy evolution studies. Abundance measurements constrain theoretical models, providing important clues to how galaxies evolve. In particular, combining the data on PNe numbers and on PN and HII region elemental abundances can allow one to derive an approximate enrichment and star formation history of a galaxy from intermediate ages to the present day. While HII regions indicate the present-day gas-phase elemental abundances ($\sim$ 10 Myr), PNe, although they cannot be accurately age-dated, reveal the chemical composition of a galaxy at "intermediate" ages of a few 100 Myr to a few Gyr. Furthermore, if HII regions and PNe at different locations are studied, this permits us to test the homogeneity with which heavy elements are/were distributed within a galaxy. For the Local Group and other nearby galaxies these data can be combined with star formation histories derived from color-magnitude diagrams of resolved stars, thereby yielding deeper insights on galaxy evolution. The goal of our present work was to improve our understanding of metallicities both at the current epoch (from HII regions), and in previous periods of star formation (from PNe) on the basis of new high-quality NTT spectrophotometry for three members of Antlia-Sextans group of dwarf galaxies [27] and for the Fornax dwarf spheroidal galaxy.

2 Observations and Reduction

Spectrophotometric observations of HII regions and PNe in the targeted galaxies were conducted with the NTT at ESO, La Silla, in February 2004. The observations were performed with the Red Arm of the EMMI multipurpose instrument with a long slit of $8' \times 2''$. For these observations the CCD rows were binned by a factor of 2, yielding a final spatial sampling of $0.33''$ pixel^{-1}. The seeing during the observations was very stable and varied from night to night in the range of $0.4''$ to $0.6''$. The whole spectral range covered by the two grisms was of 3800 – 8700 Å with the sampling of 1.6 Å pixel^{-1} for the blue (3800 – 7000 Å) and 1.4 Å pixel^{-1} for the red part (5750 – 8670 Å) of the spectra. Each galaxy was observed with slit positions which covered several HII regions and also one PN in Sextans A, one PN in Sextans B, four PNe in NGC 3109 and one PN in Fornax. The PNe positions have been taken from [21] for Sextans A, from [20] for Sextans B, from [23] for NGC 3109 and from [4] for Fornax

All primary reductions were done using the IRAF and MIDAS environments. The method of emission line measurements and the calculation of element abundances was described in detail in [10,11,14–16]. For the majority of PNe and HII regions in these galaxies O, N, S, Ar and He abundances were obtained with the classic $T_{\rm e}$-method after the detection of the [OIII] λ4363 line.

3 PN Abundances vs. HII Region Abundances

HII region metallicities mainly provide information about abundances of α-process elements, produced predominantly in short-lived massive stars. In contrast to HII regions, some elemental abundances in PNe are affected by nucleosynthesis in PN progenitors. It is well known that newly synthesized material can be dredged up by convection in the envelope, significantly altering abundances of He, C and N in the surface layers during the evolution of 1–8 $M_\odot$ stars on the giant branch and asymptotic giant branch (AGB) [18,2,9]. In addition, if during the thermally pulsing phase of AGB evolution, convection "overshoots"

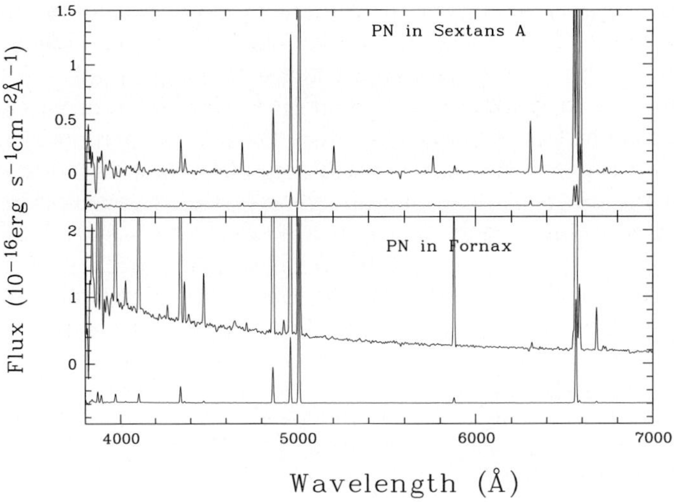

Fig. 1. Planetary nebulae in Sextans A and Fornax. The emission-line EMMI/NTT spectra were obtained with grism #5 and cover a wavelength range of 3800 and 7000 Å. The spectra at the bottom of each panel is scaled by 1/10 (top) and 1/30 (bottom) and shifted to show the relative intensities of strong lines.

into the core, significant amounts of ^{16}O can be mixed into the inter-shell region and may be convected to the surface [8,1]. In combination, these factors mean that surely only Ne, S and Ar abundances, observed in both HII regions and PNe, can be used to probe the enrichment history of galaxies.

4 Sextans A

Sextans A is a dIrr at a distance of 1.32 Mpc for which deep HST images have permitted the measurements of stars to a limiting absolute magnitude of $M_V \sim +1.9$ [5]. The gas abundance of heavy elements for Sextans A was first determined by [24]. This study presented only an empirical estimate of O/H, which has a characteristic rms uncertainty of ∼50%. While this estimate was successfully used in various statistical studies to combine the data on O/H with the CMD-derived SF history and metallicities, such accuracy is not sufficient.

We have derived an average value of 12+log(O/H)=7.57 from three HII regions in Sextans A. The values of N/O, S/O and Ar/O are consistent in all three HII regions to within their rms observational uncertainties and are close to those derived for the subgroup of the most metal-poor HII galaxies [11]. New, high accuracy chemical abundances for HII regions in Sextans A allow us to probe metallicity homogeneity across the body of the galaxy. For the three HII regions observed in Sextans A, the measured abundances show no differences exceeding 0.1 dex. Moreover, these metallicities are in good agreement with those in three A-supergiants in Sextans A [13].

Chemical abundances derived for the PN in Sextans A show that this is a Type I object (see Figure 1), with a highly elevated nitrogen abundance: N/O $\sim$2. Its 12+log(O/H) = 8.02 is about a factor of 3 higher than in the HII regions, which implies significant self-pollution by the PN progenitor. Since the abundances of S and Ar should not be altered in a PN progenitor (see discussion above), their values indicate that the ISM metallicity was $\sim$0.5 dex lower at the epoch of the PN progenitor's formation, compared to the current metallicity in HII regions and A-supergiants.

5 Sextans B

Sextans B is a dIrr at a distance of 1.36 Mpc [12], in which red giants, intermediate-age stars, and young stars are found. As far as its global properties are concerned, Sextans B is considered to be a "twin" to Sextans A, but Sextans B is of particular interest because of a discrepancy in oxygen abundances between the HII measurements of [25,24,22].

We confirmed with good accuracy the low ISM metallicity of Sextans B with 12+log(O/H) = 7.53±0.04 in two HII regions. The element abundance ratios of O, N, S and Ar are well consistent with the respective patterns of very metal-poor HII galaxies. We have found that one HII region is significantly enriched, with an excess of O, N, S and Ar abundances, relative to the mean value of two other HII regions, by a factor of 2.5±0.5. The elemental abundances of the observed PN in Sextans B are consistent with those of the two HII regions with the low metallicity.

6 NGC 3109

NGC 3109, at a distance of 1.33 Mpc, is the most massive dIrr galaxy in Antlia-Sextans group. Compared to galaxies of similar total luminosity (e.g., the SMC), the optical extent is approximately two times larger. The H I gas, stellar content and star formation history of this galaxy have been studied, e.g., by [3,7]. The underlying old stars are metal-poor ([Fe/H] = −1.7 dex), as measured by [6]. [17] reported an oxygen abundance for one HII region with the T_e-method, although the S/N in the [OIII] λ4363 line was low ($\sim$ 2.5). Our preliminary measurement of O/H for this HII region is quite consistent with their value, but with higher S/N=6 for the [OIII] λ4363 line. We found that oxygen abundances of PNe and HII regions are all in a narrow range near 12+log(O/H) $\sim$ 7.65±0.15.

7 PN in Fornax

The PN in the Fornax dwarf spheroidal galaxy was the only one found by [4], who were the first to observe it spectroscopically. Our data are considerably better (see Figure 1), yielding S/N=25 for the [OIII] λ4363 line. We have calculated 12+log(O/H) = 8.28 for this PN, which is slightly lower than the O/H value 8.38 from [19] that was calculated on the basis of line intensities from [4]. With our new data we have for first time detected weak [SII] $\lambda\lambda$6717,6731 lines (I(6717+6731) $<$ 0.015 I(Hβ)) and have determined an electron number density, N_e(SII) = 750 cm^{-3}. With the newly determined O, N, Ne, Ar and S abundances and knowledge of mean heavy-element abundance ratios for these elements from [11] we have found that the PN is enriched with oxygen and the ISM metallicity was ~0.3 dex lower at the epoch of the PN progenitor formation compared to the value O/H derived for the PN. This result is consistent with the chemical evolution scenario for the Fornax dwarf derived by [26]. The Fornax PN shows a Wolf-Rayet blue bump in the spectrum that provides important constraints on the central star evolutionary status. It is worth noting that both PNe from the Sagittarius dwarf spheroidal galaxy from [28] also show Wolf-Rayet features in their spectra.

References

1. Blöcker, T. 2003, in IAU Symp. 209, Planetary Nebulae: their evolution and role in the Universe, eds. S.Kwok, M.Dopita, & R.Sutherland, 101
2. Boothroyd, A.I., & Sackmann, I.-J. 1999, ApJ, 510, 232
3. Carignan, C. 1985, ApJ, 299, 59
4. Danziger, I.J., Dopita, M.A., Hawarden, T.G., & Webster, B.L. 1978, ApJ, 220, 458
5. Dolphin, A., Saha, A., Skillman, E. et al. 2003, AJ, 126, 187
6. Grebel, Gallagher, & Harbeck 2003, AJ, 125, 1926
7. Greggio, L., Marconi, G., Tosi, M., & Focardi, P. 1993, AJ, 105, 894
8. Herwig, F. 2000, A&A, 360, 952
9. Henry, R.B.C., Kwitter, K.B., & Bates, J.A. 2000, ApJ, 531, 928
10. Izotov, Y.I., Thuan, T.X., & Lipovetsky, V.A. 1994, ApJ, 435, 647
11. Izotov, Y.I., & Thuan, T.X. 1999, ApJ, 511, 639
12. Karachentsev, I.D., Sharina, M.E., Makarov, D.I., et al. 2002, A&A, 389, 812
13. Kaufer, A., Venn, K.A., Tolstoy, E., Pinte, C., & Kudritzki, R.-P. 2004, AJ, 127, 2723
14. Kniazev, A.Y., Pustilnik, S.A., Masegosa, J., et al. 2000, A&A, 357, 101
15. Kniazev, A.Y., Grebel, E.K., Hao, L., Strauss, M.A., Brinkmann, J., & Masataka Fukugita 2003, ApJ, 593, L73
16. Kniazev, A.Y., Pustilnik, S.A., Grebel, E.K., Lee, H., & Pramskij, A.G. 2004, ApJS, 153, 429
17. Lee, H., McCall, M.L., Kingsburgh, R.L., Ross, R., & Stevenson, C.C. 2003, AJ, 125, 146
18. Leisy, P., & Dennefeld, M. 1996, A&AS, 116, 95
19. Maran, S.P., Gull, T.R., Stecher, T.P., Aller, L.H., & Keyes C.D. 1984, 280, 615
20. Magrini, L., Corradi, R.L.M., Walton, N.A., et al. 2002, A&A, 386, 869
21. Magrini, L., Corradi, R.L.M., Greimel, R., et al. 2003, A&A, 407, 51
22. Moles, M., Aparicio, A., & Masegosa, J. 1990, A&A, 228, 310
23. Richer, M.G., & McCall, M.L. 1992, AJ, 103, 54
24. Skillman E.D., Kennicutt, R.C., & Hodge, P.W. 1989, ApJ, 347, 875
25. Stasinska, G., Comte, G., & Vigroux, L. 1986, A&A, 154, 352
26. Tolstoy, E., Venn, K.A., Shetrone, M., et al. 2003, AJ, 125, 707
27. van den Bergh 1999, AJ, 517, L97
28. Walsh, J.R., Dudziak, G., Minniti, D. & Zijlstra, A.A. 1997, ApJ, 487,651

Planetary Nebula Abundances in NGC 5128 with VLT FORS

J.R. Walsh[1], G. Jacoby[2], R. Peletier[3], and N.A. Walton[4]

[1] European Southern Observatory, D-85748 Garching, Germany
[2] WIYN, Tucson, USA
[3] Kapteyn Astronomical Institute, University of Groningen, Groningen, The Netherlands
[4] Institute of Astronomy, University of Cambridge, Cambridge, England

Abstract. Previously-catalogued planetary nebulae in the nearest large elliptical galaxy NGC 5128 (Centaurus-A) have been observed with the VLT FORS1 spectrograph in multi-slit mode. Three fields with offsets from the galaxy centre of 4.0, 7.5 and 14.5 arcmin were selected and 21, 20 and 9 planetary nebulae (PN) respectively were detected. Low dispersion blue (3500-5700Å) and red (4500-8000Å) spectra were measured covering the strong abundance diagnostic emission lines. The electron temperature diagnostic line [O III]4363Å was clearly detected in two PN and marginally detected in a further six. Combined PN spectra in the three fields are analysed and the abundance patterns are compared with the stellar abundances.

1 Introduction

NGC 5128 (=Centaurus A) provides an ideal target for planetary nebulae studies. It is well situated for southern hemisphere telescopes, it is nearby (3.5Mpc) and is the nearest large elliptical galaxy. Hui et al. (1993) first catalogued 785 planetary nebulae (PN) in NGC 5128 from the [O III] on/off band technique and then proceeded to measure velocities of 433 of these PN and fit the potential to dark matter halo models (Hui et al. 1995). Subsequently Peng et al. (2004 and these proceedings) detected more PN bringing the total to over 1000 and showing that PN as far as a projected distance of 30kpc could be located belonging to the halo of the galaxy. The direct connection of the PN with the resolved stellar population can also be drawn since for NGC 5128 there is photometry of long period variables (Rejkuba et al. 2003) and red giants (Harris et al. 2002).

Some of the PN in NGC 5128 are bright enough that spectra could be obtained with 4m class telescopes. Walsh et al. (1999) used the ESO 3.6m and EFOSC for low dispersion optical spectra of five PN, including the brightest so far detected #5601 (Hui et al. 1993). Photoionization modelling of the strong emission lines allowed abundances to be obtained. That initial study has now been extended using the FORS1 spectrograph on the ESO VLT. The spatial density of PN in NGC 5128 matches very well the multi-slit capability of FORS (19 slitlets) and three fields at increasing projected radius were chosen. The aims of the spectroscopy are threefold:
a) determine the spectral properties of the PN in a galaxy with a very different

star formation history from the Milky Way or Andromeda spirals or the lower luminosity Local Group members, for example;
b) measure PN abundances of the most common elements (He, O, N, Ne, S, Ar, etc.) and chart trends in element abundances with radius;
c) compare the abundances of the PN with those of the stars and contribute to determining the star formation history of a large elliptical galaxy.

2 Observations

Three fields were observed with the FORS1 multi-slits configured to maximize the number of PN per field. The fields chosen were centred at 4.0, 7.5 and 14.5 arcmin offset from the galaxy nucleus, to provide a range in galactocentric distance. The fields are labelled with the region number from Hui et al. (1994), F42, F56 and F34 respectively. Figure 1 shows the assignment of slitlets for the field closest to the centre (F42) as designed using the FIMS tool. The inner two fields could be almost fully packed with PN whilst the outer had 9 PN allocated to slitlets. The gratings used were 600V in the blue giving spectra at a dispersion of 1.1Å/pix from about 3500 to 6000Å, depending on the position of the multi-slit in the field. Red spectra at identical positions were obtained with the 300V grism providing a wavelength coverage from 4500 to beyond 7500Å at a dispersion of 2.8Å/pix. All data were taken with $0.8''$ slit width and in service mode with typical integration times of 2 hours per field per grism.

The PN, as distinguished by 5007Å line emission, were located on the MOS spectra and extracted. In the two inner fields, six PNe were detected which were not in the catalogue of Hui et al. (1993). These PN were faint, typically $m_{5007} > 27.0$. For the field at $4.0'$ offset, 21 PN were detected with 18 slitlets; for the field at $7.5'$ offset (which included the brightest known PN in NGC 5128 - #5601 from Hui et al. (1993)), 20 PN were detected. The outer field had nine PNe measured, making a total of 50 PN detected spectroscopically.

3 Spectra and Abundances

Emission lines were fitted to the detectable lines and the blue and red spectra were merged using the line fluxes in the overlap region (Hβ and [O III]). The reddening was determined from the Hα and Hβ line ratio. Lines of [O II], [Ne III] He II, [O III], He I, [N II], [S II] and [Ar III] were observed in the spectra. In two PN the [O III]4363Å line was detected at signal-to-noise > 3. It was further marginally detected in six other PN down to a 5007Å magnitude of 25.2 in the outer field, where the underlying galaxy continuum is lowest. Thus given ideal conditions, and in particular low stellar continuum - which translates to the outer regions of galaxies - measurement of [O III]4363Å from the brightest PN to $\sim$10Mpc is feasible with 8-10m class telescopes. The individual PN spectra, excluding the faintest sources, were summed for each region, after correction for the radial velocity of each PN, and are shown in Figure 2. In these higher signal-to-noise spectra, weaker lines can be measured, for example He I 4471Å

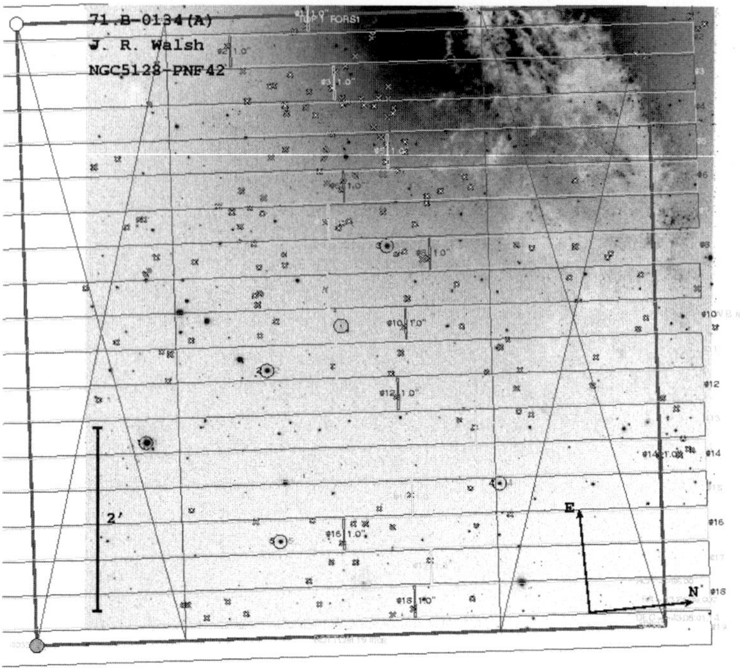

Fig. 1. A FORS [O III]5007Å pre-imaging view of a region adjacent to the dust lane in NGC 5128 showing the assignment of FORS1 slitlets for the field 4.0′ NW from the galaxy centre.

and [S II]6716Å. Analysis of the individual PN spectra will be presented in a future paper; here the three coadded spectra will be discussed.

Table 1 presents a summary of the summed spectra. The total Hβ flux, the observed [O III]5007Å magnitude and c, the extinction coefficient at Hβ, are listed. Empirical abundances of He, O, N, Ne, Ar and S were determined from the dereddened line ratios using directly measured [O III] electron temperatures and an assumed electron density of 10^4 cm^{-3}. Ionization correction factors from Kingsburgh & Barlow (1994) were employed.

4 PN and Stellar Abundances

Since the oxygen abundance remains almost unchanged through the red giant and asymptotic red giant phases for the more common lower mass stars, the PN oxygen abundance can be used to study the stellar abundance in the progenitors at the time of their formation (Richer 1993). In NGC 5128 only photometry of individual stars or absorption line indexes of the integrated starlight are available. Figure 3 compares the PN O abundance from the three regions v. the metallicity distribution values from Harris et al. (2002) in three fields with offsets of 7, 18 and 27 kpc (for a distance of 3.5Mpc, Hui et al. 1993).

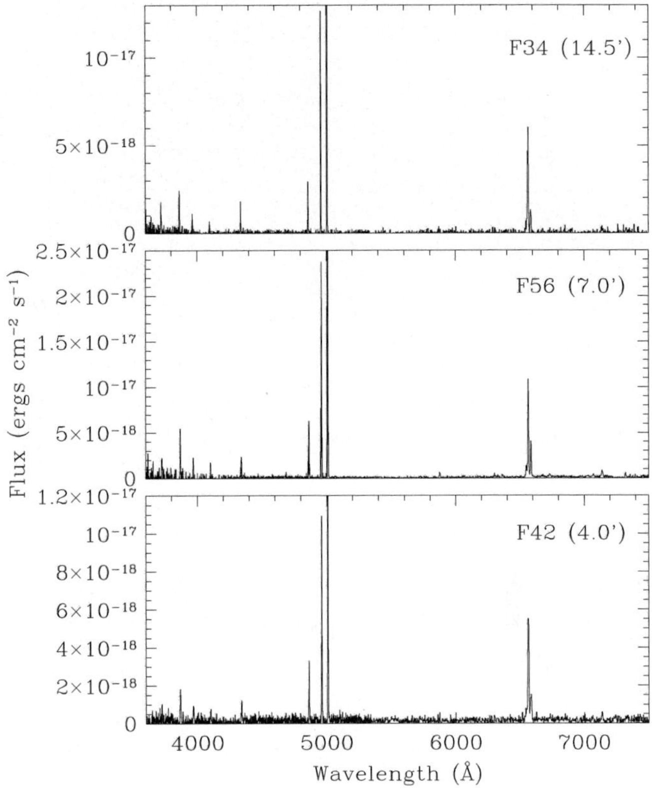

Fig. 2. Summed PN spectra from the three FORS fields. The field name and distance of the centre of the multislit from the nucleus of NGC 5128 is indicated.

Diagnostic	F42	F56	F34
Log F(Hβ)	-16.78	-16.44	-16.90
m_{5007A}	22.55	21.89	23.35
c	0.23	0.05	0.51
T_e	13900	12100	12300
12 + He/H	10.90	11.03	10.89
12 + O/H	8.2	8.4	8.6
12 + N/H	8.0	8.0	7.5
12 + Ne/H	7.5	7.8	8.0
12 + Ar/H	6.3	6.5	6.3
12 + S/H	6.7	6.7	6.6
Offset (kpc)	4.1	7.6	14.3
No. PNe summed	17	17	9

Table 1. Diagnostics and abundances for the summed PNe spectra in the three regions.

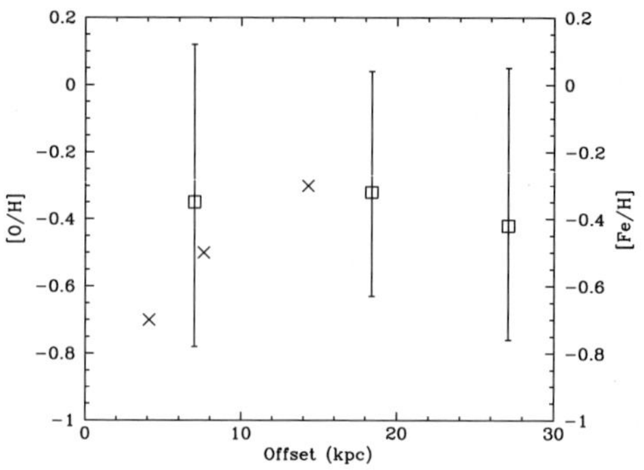

Fig. 3. Radial variation of the PN [O/H] abundances (crosses) compared with the peak of the red giant star metal-rich [Fe/H] metallicity distribution (squares, from Harris et al. 2002). The error bars refer to the width of the peak of the metal-rich stellar distribution, but a tail to much lower metallicities is present at all three positions.

The behaviour of the O gradient is quite unexpected in the sense that the O/H ratio *increases* outward. This result of course depends on the sample of PN spectra summed in the three regions and some clues to the origin of the PN is available from their radial velocities (Hui et al. 1995; Peng et al. 2004; Rejkuba & Walsh, these proceedings). In the outermost field, the nine PN all have velocities close to the systemic and thus belong to the 'bulge' population. In the innermost region, where the mean O abundance is lowest (Fig. 3), six of the 21 PN have radial velocities differing from the systemic velocity by more than 250kms^{-1}. These high velocity PN may thus be part of a distinct 'halo' population, with lower O abundance. Harris et al (2002) found a tail to lower red giant abundances ([Fe/H] < -1) in all three observed fields. The individual correlation of abundances with position and radial velocity of the PN in the galaxy may provide clues to their formation history.

References

1. Harris, W. E., Harris, G. L., AJ, **123**, 3108 (2002)
2. Hui, X., Ford, H. C., Ciardullo, R., Jacoby, G. H., ApJS, **88**, 423 (1993)
3. Hui, X., Ford, H. C., Freeman, K. C., Dopita, M. A., ApJ, **449**, 592 (1995)
4. Kingsburgh, R., Barlow, M.J., 1994, MNRAS, **271**, 257
5. Peng, E. W., Ford, H. C., Freeman, K. C., ApJS, **150**, 367 (2004)
6. Rejkuba, M., Minniti, D., Silva, D. R., A&A, **406**, 75 (2003)
7. Richer, M., ApJ, **415**, 240 (1993)
8. Walsh, J. R., Walton, N. A., Jacoby, G. H., Peletier, R. F., A&A, **346**, 753 (1999)

Part VII

PNe as Kinematic Probes of Galaxies

Galaxy Dynamics and the PNe Population

Nigel G. Douglas

Kapteyn Astronomical Institute, Postbus 800, 9700 AV Groningen, The Netherlands

Abstract. This review attempts to place the observations of extragalactic planetary nebulae in the context of galactic dynamics. From this point of view only the radial velocities of the PNe is important. We have built a specialised instrument to detect PNe in distant galaxies and measure their radial velocities in one step. This is explained in some detail, along with classical techniques for obtaining kinematic information. The review includes a vision of possible future developments in the field.

1 Spiral Galaxies

Many of the dynamical properties of galaxies are most easily seen in the case of spirals. Their defining property, which is a bright stellar population in coherent rotation, makes it relatively easy to determine the size and (apart from a sign ambiguity) the orientation of the galaxy and its rotation speed as a function of radius (the "rotation curve"). It has been found that a large fraction of the neutral hydrogen rotates along with the stars, allowing sensitive radio telescopes to map the disk to fainter limits (see Fig. 1 for a state of the art H I image).

The non-coherent motions of the stars, usually expressed as the standard deviation of the three velocity vectors in cylindrical coordinates, could be determined, at least for bright galaxies, and this gave clues as to the total mass density of the disk. The rotation curve itself provides a measure of the enclosed mass, and these diagnostics lead to the conclusion that in general considerable dark matter must be present. Observations of the neutral gas out to large radius from the centre, where the stellar contribution to the mass density is very small, suggest that a dark matter halo becomes dominant in most cases. Note that much of the progress in understanding spiral galaxies is possible because the geometry and orientation of the galaxy is evident, and to a lesser extent because radio as well as optical observations can be employed.

2 Elliptical Galaxies

The situation is quite different in the case of elliptical galaxies (frequently, and unfortunately, referred to as 'early-type' galaxies in the literature). Most of the following remarks apply to the intermediate type S0 galaxies as well. With their characteristic paucity of neutral hydrogen, the kinematics of these galaxies have traditionally been obtained from stellar spectroscopy alone, measuring line-of-sight integrals which are then deconvolved into velocity distributions by

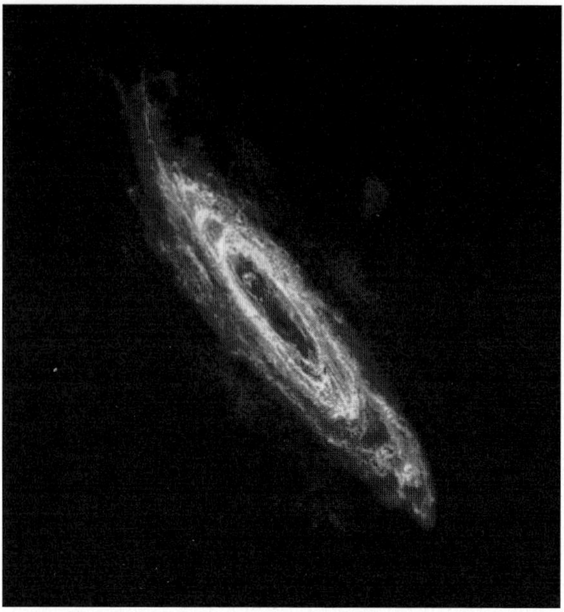

Fig. 1. This radio image of M31 is the most detailed view of an H I disk ever made, and spans 6° on the sky. It can be found in 'colour' at [3]

template-fitting. The effort involved in doing this is considerable. Moreover, it turns out that the velocity dispersion is usually comparable to, or larger than, any systematic rotation which may be present (see for example Fig. 2). This means that, contrary to intuition, rotation alone is not in general responsible for the observed ellipticity of the galaxy. Much of the cause must be sought in incoherent motions, expressed as anisotropy in the velocity distribution of the stars. The anisotropy is difficult to determine from the projected velocity information and the orientation and axis ratio cannot be determined unambiguously from the projected surface brightness. It is difficult even to determine the shape and orientation of the galaxy - an apparently round, spherically symmetric galaxy may be elongated along the line-of-sight and even then this axis can be shorter (oblate galaxy) or longer (prolate) than the tangential axes. Note, in passing, that it can be shown by simple order-of-magnitude arguments, that the optical depth of stars along a random line of sight through a galaxy is small: galaxies are transparent. Thus, the integrated spectrum of the starlight contains roughly equal contributions from the "back" and the "front" of the galaxy.

Despite these difficulties, integrated light stellar spectroscopy is responsible for virtually all of the basic information that we have on the kinematics of elliptical galaxies. The crucial limitation has turned out to be that, beyond about 2 R_e (the effective radius R_e being the mean radius of a contour containing half of the galaxian light) the surface brightness becomes so low that accurate sky subtraction becomes impossible. The problem is exacerbated for relatively nearby

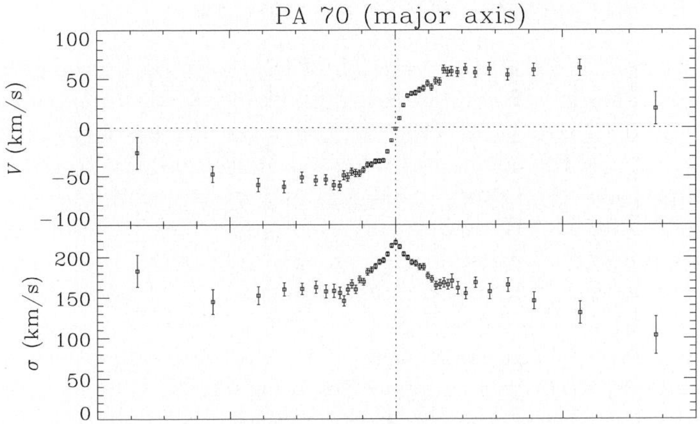

Fig. 2. Major axis rotation in NGC 3379 (from Fig. 3a of [19])

galaxies such as we are talking about here, since the galaxy is so extended that the sky spectrum generally cannot be determined in the same exposure, so that that both spatial and temporal variations degrade the resulting sky subtraction. This means that the outer parts of the galaxies are in effect inaccessible to integrated light spectroscopy. This is unfortunate as it turns out that the circular velocity curves[1] of nearly all elliptical galaxies are flat, like the rotation curves of spiral galaxies, out to the last measured point, indicating that they too have extended halos. Moreover, some theories of galaxy formation predict that the outer parts of elliptical galaxies possess considerable angular momentum. For this and other reasons we need to be able to explore the kinematics of elliptical galaxies at large radius.

3 Kinematic Tracers

One powerful way to circumvent the limitations of integrated light spectroscopy is to look for bright objects in closed orbits in the vicinity of the galaxy - their motion gives clues to the gravitational potential of the galaxy and hence to the total enclosed mass. Globular clusters (GC) are one such probe, recent examples being studies in NGC 4472 and Cen A (NGC 5128) [6,15] respectively. In these galaxies it turns out that the metal-rich and metal-poor GC are kinematically distinct. In the first, the metal-rich GC show net rotation and the metal-poor ones do not, while in the second the converse is true.

Other possible kinematic tracers are satellite galaxies and, in nearby galaxies such as M31 the kinematics of individual stars can be determined (see the article by Ferguson et al. elsewhere in these proceedings).

[1] the CVC is derived from the sum of the velocity and dispersion in quadrature.

4 Planetary Nebulae

During the workshop, spectacular images of PNe have been presented. Even in the Magellanic Clouds, the complicated morphology and exhilarating colours of PNe are easily detected. At greater distances, the PN envelope reduces to an unresolved point and the colours combine to a murky brown. To the extragalactic community, then, the beauty of a PN is only apparent in its spectrum, with its characteristic lines of Hydrogen and 'Nebulium.' The value of PNe as kinematic tracers in external galaxies was demonstrated by Nolthenius and Ford in 1986 [13], combining radial velocity measurements with dynamical modelling of M32. Each PN is seen as a discrete object and it is easy to measure the radial velocity of that object by taking the spectrum. At the specific wavelength of the [O III] emission line (5007Å) a single PN can far outshine the local background of integrated starlight (the emission lines are inherently narrow and usually have a measured linewidth of around 20 km/s, or 0.3Å, as a result of the expansion velocity of the envelope). PNe are, in principle, located throughout the body and halo of the galaxy, wherever progenitor stars are to be found. Unlike 'external' probes such as globular clusters, PNe can be expected to share the orbital properties of the underlying stellar population, which means that kinematical information from PNe can be directly combined with stellar kinematics. By 1993 several projects had been completed and the principal theoretical tools were in place, as summarised in a seminal paper by Hui et al. [10]. Of special note is the dramatic demonstration, in the case of NGC 3379, of how just a few tens of PNe can constrain the dynamical models based on integrated light spectroscopy [4].

Comparisons have been made between PNe and other discrete kinematic tracers. It is interesting that in the cases mentioned in §3 the PNe in Cen A are found to exhibit significant rotation [14], as do the metal-rich GCs in that galaxy, while data acquired on NGC 4472 show that the PNe population has little rotation, a different result but again consistent with the metal-rich GCs. The total gravitating mass of Cen A derived from the GCs has also found to be in excellent agreement with the value derived from stellar (PN) kinematics [14].

It is often stated, not least in our own papers [9], that PNe are good tracers of the "old" stellar population. I believe that it is worthwhile keeping in mind that there are more assumptions in this statement than is usually realised. The main-sequence lifetime of progenitor stars is *not* relevant, at least in the case where the galaxy is quiescent with only very low-level continuous star formation. In the steady state the creation rate of PNe is then *independent* of main-sequence lifetime, depends only on the initial mass function, and for a typical IMF is as likely to involve a star in the range 2-5$M_\odot$ as one in the range 1.5-2$M_\odot$. With mean ages of 0.6 Gyr and 2 Gyr respectively, these populations may well have different kinematics. What presumably tips the scale in favour of the older stars is the *lifetime* of the PNe phase, which is a strong function of progenitor mass [21]. On the other hand, PNe formed from high-mass progenitors may be brighter, and hence preferentially detected in surveys. The point is that, in special circumstances, one needs to at least question the veracity of the assumption.

The search for possible "young" PNe in Cen A (see Peng's paper elsewhere in this volume) is therefore commendable.

5 Obtaining Extragalactic PN Velocities

The traditional technique, pioneered by Hui, Arnaboldi, Ciardullo and others, has been to identify the PNe by means of a two-filter survey. One is a narrow-band [O III] filter centred on the systemic velocity of a target galaxy - both PNe and foreground stars show up as bright point-like objects - and the other is a continuum filter in which the stars show up and the PNe do not, or are very faint. This enables PN candidates to be identified, which are then re-observed using fibre or multi-object spectroscopy (MOS) to confirm the identification and to measure the radial velocity from the [O III] line. For sheer volume of effort and persistence over a number of years, the compilation of 780 confirmed PNe in Cen A must be viewed as one of the exemplary projects using this technique [14]. Amongst other results, the existence of PNe out to $15R_e$ showed that the stellar halo extends that far (see also Peng, these proceedings).

Another noteworthy approach is the use of a Fabry-Perot interferometer, which requires no candidate selection and which identifies the PNe in the entire field (typically a few arcmin) and returns their velocity (typically with an accuracy of ~ 30 km s^{-1}) in one observing run [20].

6 Slitless Spectroscopy and CDI

Counter-dispersed Imaging (CDI) is a type of slitless spectroscopy in which anti-symmetric images are used to determine velocities. The historical antecedents go back to Fehrenbach (see [9] and references therein) who realised that this procedure not only doubles the effective dispersion, but also diminishes systematic errors. Once again, Cen A plays a role in the story, as it was the first extragalactic object studied using CDI, by modifying an existing instrument at the Anglo-Australian Telescope. The idea was to compare PNe velocities and positions found using CDI with those determined earlier by Hui et al. [11] using the survey/fibre technique. The results [7] of the tests were very encouraging: for the 24 PNe found in a certain field the astrometry agreed well with the published data except for three cases, for which there was a discrepancy of more than 1 arcsec, and for precisely these three objects Hui et al. had been unable to measure a velocity. In retrospect, it is clear that the fibres missed the object. CDI, on the other hand, is insensitive to astrometric error and *always* returns a velocity if the PN is detected. The first 'new' results with CDI on an extragalactic source, using the ISIS spectrograph at the WHT, were published shortly afterwards [8]. The results of slitless spectroscopy at an 8-m telescope are of course dramatic, as exemplified by the detection of 535 PNe in NGC 4697 (see [12] and Méndez' contribution elsewhere in this volume). Amongst other results was a distance determination of around 10.5 Mpc. The relative merits of CDI and other techniques are discussed in [9].

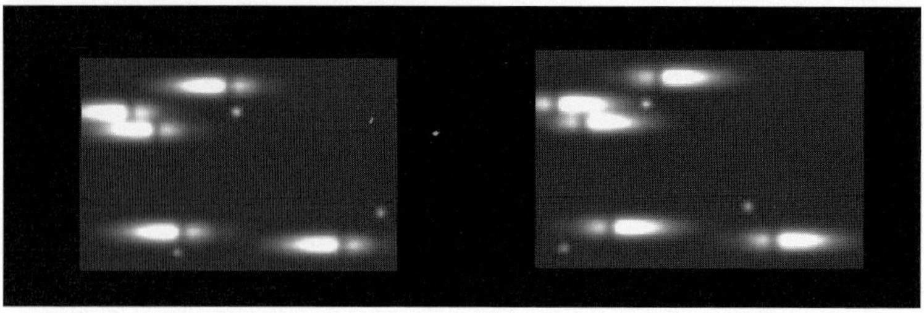

Fig. 3. A simulated pair of CDI images: the extended trails are stellar spectra, with an absorption line, convolved by the [O III] filter profile, while the PNe appear as dots. The illustration is intended to show how, in principle, the relative displacement of the two images of a given PN, in combination with suitably calibrated filter profiles, can return the radial velocity of the PN. In practice calibration is done with a special mask.

7 The PN.Spectrograph

The PN.Spectrograph (PN.S), which was developed between 1995 and 2001, is intended to optimally exploit the advantages of CDI described in the previous section. This required (a) that the counter-dispersed images should be obtained simultaneously, not sequentially, (b) that we achieved the highest possible efficiency in the vicinity of the λ5007 [O III] line, and (c) that we would make use of a 4-m class telescope, for which relatively large amounts of time are becoming available. Simultaneity of the CDI images is important because any systematic effects (such as seeing) which might affect the centroiding of the PNe images, and hence the velocity measurement, are to first order cancelled out. Moreover one *always* has a matched set of data, so that one clear night guarantees that velocities can be measured, which was not the case with our earlier work [8] and similar programs. Figure 4 shows how we decided to implement simultaneous CDI: a single collimator illuminates an optical pupil in which two gratings are located, edge to edge, each diffracting light into its own camera.

The PN.S was commissioned 2001 after a nail-biting delay caused by the degeneracy of place-names within the Canary Islands - specifically, the instrument was sent to the wrong island. There were however no further problems: the spectrograph worked "straight from the box" with a system efficiency of 33%, including telescope, detector and [O III] filter. For further technical details see [9]. Figure 5 shows the PN.S on the telescope. The first major result with the PN.S was the confirmation of the paucity of dark matter in the outer parts of NGC 3379, as suggested by [4] on the basis of 29 PNe. We have now obtained a total of 400 PNe in three similar galaxies [17]. This is part of our core program, namely the study of a sample of apparently round, intermediate-luminosity ellipticals. A side-benefit the PN.S is that effectively *all* of the observing time is spent in photometric mode, so that the data should yield good luminosity functions and thus distance estimates from the PNLF. On the other hand, the restriction to just the [O III] line makes it difficult to recognise interlopers such as H II regions, which makes the study of later-type galaxies more problematic.

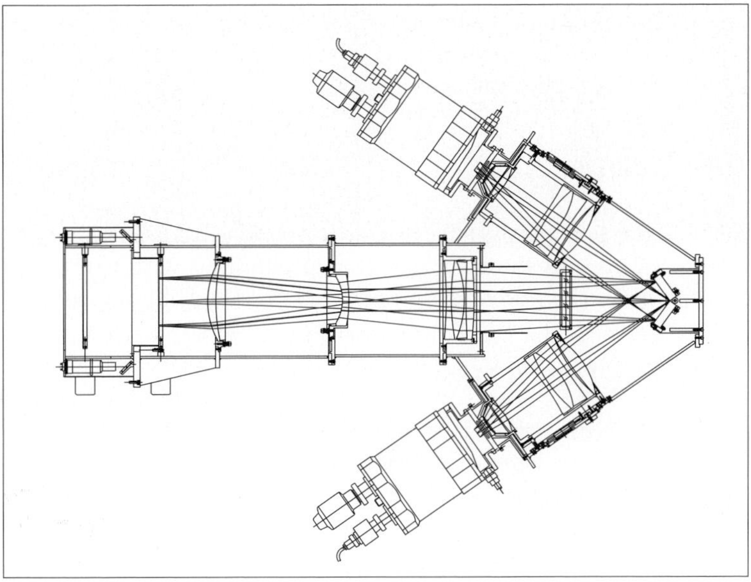

Fig. 4. Schematic of the PN.S - the telescope (not shown) is to the left.

To redress this we are in the process of adding a "piggy-back" H-α camera with about the same sensitivity as the main instrument. As has been shown [5], the relative flux of these two lines alone provides a useful discriminant against H II regions .

8 The Accuracy of PNe Radial Velocities

The limitation on the accuracy with which PNe velocities can be measured is, as usual, set by a combination of the intrinsic linewidth, instrumental accuracy, and the signal-to-noise which is obtained. For example, in the pioneering work of [11], in which Cen A (distance 3 Mpc) was observed with fibre spectroscopy at the ATT, typical 1σ errors were 3–13 km s^{-1}, depending on the number of counts obtained (~ 50). In the case of the PN.S the velocity determination of a given PN depends on the accurate centroiding of each of the two images obtained, as well as, as usual, on the accuracy of the wavelength calibration. We have established the centroiding accuracy which can theoretically be obtained ([9], Appendix) and it turns out that for a double detection at the S/N= 5 level the separation of the CDI pairs should be measurable with an accuracy of about 0.6 pixels, corresponding to 20 km s^{-1} with our modest dispersion of 1.0 pixel Å^{-1}. Higher dispersion would help, but CDI does not lend itself to such a choice for reasons discussed elsewhere [9]. Initial results, in which we compared measurements of NGC 3379 PNe radial velocities produced by [4] with PN.S commissioning data on the same field, roughly confirm this - apart from a small systematic trend with velocity, which has yet to be investigated, the measurements agreed in absolute heliocentric velocity to an accuracy of 28 km s^{-1}, including a contribution of $\sim$ 7 km s^{-1} from the earlier measurements.

9 Future Trends

Encouraged by the speculative discussion initiated earlier in the workshop by Quentin Parker, I present some adventurous predictions with respect to technological advances with the potential to push forward the field of galaxy dynamics.

9.1 Wavelength-Tagged Detectors

It is surprising how few people are aware of the fact that detectors are potentially capable of registering not only the flux at a given position but also the time of arrival and wavelength of each photon! Some of these properties were indeed realised by the historic Image Photon Counting System (IPCS [2]), but the last parameter, wavelength, remains a challenge. If the wavelength of each photon can be tagged, then we can do away with filters. Any filter bandpass one wishes to specify can be synthesised *after the event.* The state-of-the-art technology in this area is currently the Superconducting Tunnel Junction (STJ, [16]) which yields a wavelength resolution of about 600Å in the visible region. This is somewhat better than a V-band filter. Hopefully, with this or perhaps some other technology, we may be able to get down to 60Å resolution, at which point one could measure the Hα, [O III], and "off-band" flux from every point in the field, all in one observation, and without having to purchase a single filter

Fig. 5. PN.S at the telescope

(the PN.S filters cost $US 10,000 each). This technology would rather radically alter our data-collecting strategy.

9.2 Whole-Body IFU

The integral field unit (IFU) was mentioned during this meeting as an alternative to slit and multi-object spectroscopy in the investigation of extragalactic PNe (see articles by Exter, and by Roth, and their colleagues, elsewhere in this volume). As such, the IFU must be of potential interest to those of us measuring radial velocities. An IFU consists of more-or-less contiguous cluster of fibres, or else a lens array, covering an area of the field and providing a spectrum of each position. The field size is in principle limited only by the 'real estate', in the form of spectrographs and detectors, that one is prepared to provide. However, this is a non-trivial issue. For galaxy kinematics with PNe one needs to obtain spectra at large galaxian radius and, even though the angular size of each aperture could be made larger to compensate, the number of spectrographs required will still grow quickly. At large galaxian radius, where the background is low, it may be remain cost-efficient to survey for the PNe with two-band images or with slitless spectroscopy. Still, the idea remains exciting since the concept of '3D spectroscopy', the registration of a complete spectrum at each position on the sky, is hard to beat. Fortunately, until such monsters are actually built, there will still be a place for the PN.S.

9.3 Tangential Velocities[2]

However well we measure PNe radial velocities, we are still measuring only one of the three components of the velocity vector. From an analytic point of view this is *much* worse than measuring the radial velocities for only a third of a sample, say. To illustrate this consider the extreme but not contrived case of an S0 galaxy viewed pole-on. The coherent rotation of the stellar population around the short axis is entirely in the plane of the sky and thus undetectable. In such cases, radial-velocity information alone makes it difficult to unambiguously determine anisotropy *however many PNe you detect.* Tangential velocities (i.e. proper motions) would lift this ambiguity completely. In the example just given it does not much matter whether the proper motions are determined from the *same* objects as those whose radial velocities are known, but in general it would be best if all three components could be determined for a set of test particles. Such a test particle then has only one undetermined parameter, its distance along the line-of-sight, whereas if the tangential and radial velocity information is spread over different particles then there are two unknowns.

At first glance, the prospect seems totally beyond the realm of the feasible. Proper motions at extragalactic distances are tiny: at the distance of Cen A

[2] This is work in progress with Jean-Pierre Macquart, until recently working at the Kapteyn Institute and now at the NRAO, Socorro.

(about 3 Mpc) one could wait approximately 100 years to see 1 milliarcsecond of proper motion. However, the next generation of centimetre-wavelength radio telescopes, of which the Square-kilometre Array (SKA) is the design prototype, together with the existing technology of Very Long Baseline Interferometry (VLBI) may put this within reach.

The performance goals of the SKA translate to a continuum sensitivity over the wavelength range of operation, and for a one hour integration with 512 MHz bandwidth (roughly equal to the internal linewidth of a PN), of 72 nJy beam^{-1}. We will see that this is easily sufficient to be able to detect PNe at extragalactic distances, but we ideally also want to be able to detect the *same* PNe optically to obtain radial velocities. I therefore took the 962 Galactic PNe from the Strasbourg-ESO catalogue [1] for which a reliable [O III] flux measurement was available, and correlated this with the radio catalogue of Siodmiak and Tylenda [18]. This catalogue has 264 entries, of which 232 have measured [O III] (λ5007) fluxes. Next, I asked which of the PNe would potentially be visible at extragalactic distances. In optical work, a PN magnitude[3] of $m \sim 28$ is usually the practical limit. Assuming for simplicity that the Galactic PNe are at a distance of 10 kpc, I shifted them to the distance, indicated by D28 in table 9.3, at which the [O III] flux would be reduced to $m = 28$ and calculated the corresponding 5 GHz flux. To be conservative I rejected those PNe for which the flux at D28 was less than 500 nJy. This left 47 objects, as listed in the table. No statistical conclusions can be drawn from this inhomogeneous procedure, but the point to note is that there exist at least *some* PNe whose radio and optical fluxes would be detectable out to several Mpc. The question of whether the observation is feasible in practice is still to be investigated - it may be that, at this flux level, confusion rather than sensitivity may be the main problem.

How does the astrometric resolution offered by the SKA compare to that required to observe the transverse motions of planetary nebulae? We would like to measure the transverse velocity of a PN to within 10%, which implies that we measure the angular displacement to the same accuracy. Assume that the typical transverse speed of a planetary nebula is $v \sim 200\,\mathrm{km/s}$. For a galaxy at a distance D a typical PN then moves an angular displacement:

$$\Delta\theta = 42 \left(\frac{T}{1\,\mathrm{yr}}\right) \left(\frac{v}{200\,\mathrm{km/s}}\right) \left(\frac{D}{1\,\mathrm{Mpc}}\right)^{-1} \quad \mu\mathrm{as} \tag{1}$$

over a time interval T. If SKA is part of a VLBI network with a 5000 km baseline we estimate that at 5 GHz and with an attainable S/N of 100 we would obtain a positional accuracy of 7 μas. Note that there is no incentive to improve much on that, since the error in D places a limit of about 10% on the accuracy to which v can be measured. SKA would reach this accuracy for nearby ($\sim$ 1Mpc)

[3] PN magnitudes are defined by eqn(2) in the article by Robin Ciardullo elsewhere in this volume; the apparently arcane zero-point is chosen so that a PN of a given magnitude appears equally bright as a star of that magnitude seen through a standard V-band filter.

galaxies in just a couple of years. A more serious concern is the stability of local emission. The appearance or disappearance of local hot spots on the ring could offset the emission centroid significantly. It would be ironic if the beautiful internal structure of PNe, ignored by the galaxy kinematicists, were to return in this way to haunt them.

Table 1. Optical and Radio fluxes of some PNe - the columns give name, measured flux at 5007 and 5 GHz, D28 (the distance at which the apparent magnitude at λ5007 would be $m = 28$) and the correspondingly recalculated 5 GHz flux.

Name	Flux 5007Å	5GHz (mJy)	D28 (Mpc)	5GHz (μJy)	Name	Flux 5007Å	5GHz (mJy)	D28 (Mpc)	5GHz (μJy)
000.7+03.2	3.90e-13	15.6	1.4	0.80	060.5+01.8	5.85e-13	26.9	1.7	0.92
000.7+04.7	5.87e-14	12.8	0.5	4.36	062.4-00.2	6.60e-13	17.0	1.8	0.52
001.3-01.2	1.80e-14	9.6	0.3	10.68	064.7+05.0	8.40e-12	245.0	6.5	0.58
001.7+05.7	9.12e-13	24.4	2.1	0.53	067.9-00.2	8.52e-14	17.3	0.7	4.06
002.8+01.7	9.93e-15	13.8	0.2	27.78	088.7+04.6	4.04e-14	15.2	0.4	7.52
003.6+03.1	1.36e-13	23.7	0.8	3.49	093.3-00.9	9.38e-13	37.3	2.2	0.80
005.5+06.1	8.98e-14	11.8	0.7	2.63	093.5+01.4	3.93e-13	372.0	1.4	18.94
006.2-03.7	2.06e-13	11.4	1.0	1.11	095.2+00.7	1.07e-12	59.9	2.3	1.12
008.3-01.1	3.66e-12	163.6	4.3	0.89	096.3+02.3	5.24e-13	16.9	1.6	0.65
011.9+04.2	2.51e-12	70.5	3.5	0.56	098.2+04.9	1.07e-12	28.1	2.3	0.52
015.9+03.3	6.29e-14	58.6	0.6	18.62	232.8-04.7	1.45e-13	26.1	0.9	3.61
016.4-01.9	1.77e-12	78.4	3.0	0.89	235.3-03.9	4.02e-13	22.2	1.4	1.10
020.9-01.1	5.33e-13	249.0	1.6	9.34	253.9+05.7	1.99e-13	6.0	1.0	0.60
024.1+03.8	7.75e-13	49.1	2.0	1.27	355.9+02.7	4.44e-14	21.6	0.5	9.74
024.8-02.7	4.93e-14	13.6	0.5	5.51	356.5-03.9	4.47e-14	11.1	0.5	4.96
027.6+04.2	5.00e-13	20.9	1.6	0.84	358.3+03.0	1.25e-14	6.9	0.2	11.06
034.0+02.2	1.21e-13	31.0	0.8	5.10	358.5+05.4	2.31e-12	280.0	3.4	2.43
043.1+03.8	1.20e-13	21.3	0.8	3.54	358.6+01.8	1.35e-13	6.0	0.8	0.89
045.9-01.9	1.37e-14	9.1	0.3	13.29	358.9+03.2	4.67e-15	27.3	0.2	117.01
048.1+01.1	3.29e-13	14.2	1.3	0.86	358.9+03.4	2.41e-13	12.4	1.1	1.03
052.9+02.7	3.00e-13	17.3	1.2	1.15	359.3-03.1	8.20e-14	11.4	0.6	2.78
055.2+02.8	4.84e-13	23.2	1.6	0.96	359.7-01.8	6.83e-13	21.5	1.8	0.63
056.0+02.0	1.14e-13	14.6	0.8	2.57	359.8+03.7	3.51e-14	15.6	0.4	8.89
058.9+01.3	6.68e-13	17.2	1.8	0.51					

10 Acknowledgements

Thanks to R. Braun of ASTRON for making available Fig. 1, to T.S. Statler of Ohio University for helpful comments and for making available Fig. 2, and to N. Napolitano for useful discussions. We are grateful for the continuing support of the Isaac Newton Group of telescopes at La Palma.

References

1. A. Acker, F. Ochsenbein, B. Stenholm, R. Tylenda, J. Marcout, and C. Schohn : Strasbourg - ESO catalogue of galactic planetary nebulae, VizieR Online Data Catalog, (1994)
2. A. Boksenberg, and W. L. W. Sargent : ApJ **198**, 31
3. R. Braun, E. Corbelli, R.A.M. Walterbos and D. Thilker : ASTRON Newsletter **17**, 3 (October 2002)
4. R. Ciardullo, G. H. Jacoby, and H. B.Dejonghe : ApJ **414**, 454 (1993)
5. R. Ciardullo, J. J. Feldmeier, G. H. Jacoby, R. Kuzio de Naray, M. B. Laychak, and P. R. Durrell : ApJ **577**, 31 (2002)
6. P. Côté, D. E. McLaughlin, J. G Cohen, and J. P. Blakeslee : ApJ **591**, 850 (2003)
7. N.G. Douglas, K. Taylor: MNRAS **307**, 190 (1999)
8. N.G. Douglas, J. Gerssen K. Kuijken and M.R. Merrifield : MNRAS **316**, 795 (2000), **328**, 1200 (2002)
9. N.G. Douglas, M. Arnaboldi, K.C. Freeman, K. Kuijken, M.R. Merrifield, A.J. Romanowsky, K. Taylor, M. Capaccioli, T. Axelrod, R. Gilmozzi, J. Hart, G. Bloxham and D .Jones : PASP **114**, 1234 (2002)
10. X. Hui : PASP **105**, 1011, (1993)
11. X. Hui, H. C. Ford, K. C. Freeman, and M.A. Dopita, : ApJ **449**, 592, (1995)
12. R. H. Méndez, A. Riffeser, R.-P. Kudritzki, M. Matthias, K. C. Freeman, M. Arnaboldi, M. Capaccioli, and O. E. Gerhard : ApJ **563**, 135, (2001)
13. R. Nolthenius and H. Ford : ApJ **305**, 600 (1986)
14. E. W. Peng, H. C. Ford, and K. C. Freeman : ApJ **602**, 685, (2004)
15. E. W. Peng, H. C. Ford, and K. C. Freeman : ApJ **602**, 705, (2004)
16. M. A. C. Perryman, F. Favata, A. Peacock, N. Rando, and B. G. Taylor : A& A **346**, L30 (1999)
17. A.J. Romanowsky, N.G. Douglas, M. Arnaboldi, K. Kuijken, M.R. Merrifield, N.R. Napolitano, M. Capaccioli, and K.C. Freeman : Science **301**, 1698 (2003)
18. N. Siodmiak, and R. Tylenda : Radio emission from planetary nebulae, VizieR Online Data Catalog (2001)
19. T. S. Statler and T. Smecker-Hane : AJ **117**, 839 (1999)
20. B. Tremblay, D. Merritt, and T. B. Williams : ApJL **443**, L5 (1995)
21. E. Vassiliadis and P.R. Wood : ApJS **92**, 125 (1994) and priv. comm, P. Wood (1999)

Mapping the Stellar Dynamics of M31

Helen Merrett[1], M. Merrifield[1], K. Kuijken[2,3], A. Romanowsky[1], N. Douglas[3], N. Napolitano[3], M. Arnaboldi[4], M. Capaccioli[5], K. Freeman[6], O. Gerhard[7], D. Carter[8], N.W. Evans[9], M. Wilkinson[9], C. Halliday[10], and T. Bridges[11]

[1] School of Physics & Astronomy, University of Nottingham, University Park, Nottingham, NG7 2RD, UK
[2] Leiden Observatory, Leiden, The Netherlands
[3] Kapteyn Institute, Groningen, The Netherlands
[4] Osservatorio Astronomico di Torino, Pino Torinese, Italy
[5] Osservatorio di Capodimonte, Naples, Italy
[6] Research School of Astronomy and Astrophysics, Australian National University, Canberra, Australia
[7] Astronomisches Institut, Universität Basel, Switzerland
[8] Astrophysics Research Institute, Liverpool John Moores University, UK
[9] Institute of Astronomy, Cambridge, UK
[10] Osservatorio Astronomico di Padova, Padova, Italy
[11] Physics Department, Queen's University, Kingston, Canada

Abstract. Using the Planetary Nebula Spectrograph, we have observed and measured the velocities for 2764 PNe in the disk and halo of the Andromeda galaxy. Preliminary analysis using a basic ring model shows a rotation curve in good agreement with that obtained from HI data out to ~ 20 kpc. Some substructure has also been detected within the velocity field, which can be modeled as the continuation of the tidal–remnant known as the Southern Stream, as it passes through Andromeda's disk.

1 Introduction

The Andromeda galaxy, the nearest large spiral galaxy ($D \sim 770$ kpc), provides a unique observational target. M31's size and proximity allow surface structures to be resolved and studied in unprecedented detail. However, due to its large angular size, until now it has not been practical to perform a large survey of M31's stellar disk dynamics. Previous dynamical studies have either measured gas in the disk (e.g. HI [2] and HII [11]) or tracer populations in the halo (e.g. globular clusters [9]; and RGB stars [10] and [5]).

With the commissioning of the Planetary Nebula Spectrograph (PN.S) [3] at the William Herschel Telescope in La Palma, it has become possible to perform such a survey. Nine nights of observations (13 allocated) with the PN.S and 6 nights surveying the halo with the Wide Field Camera on the Isaac Newton Telescope have led to a preliminary catalogue of 2764 PNe, delving ~ 4.5 magnitudes into the PN luminosity function, extending $\sim 2.5°$ along the major axis, $\sim 1.5°$ along the minor axis, and covering various elements of substructure in the halo's stellar population density, such as the Southern Stream and Northern Spur [4].

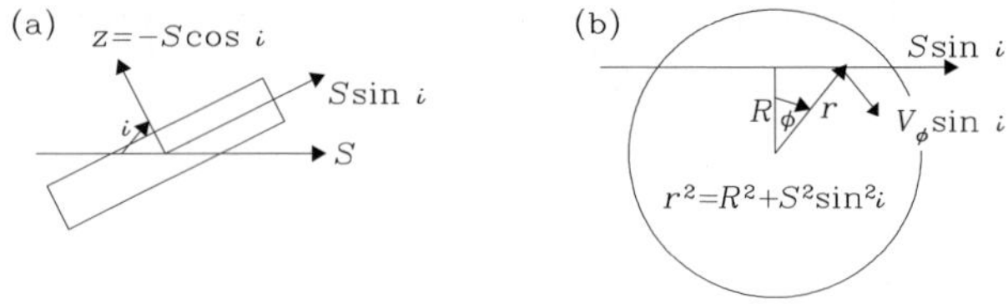

Fig. 1. Model for line–of–sight integration through the disk, shown with the disk **(a)** edge on, **(b)** face on

2 Rotation

A raw rotation curve for the PN.S data has been extracted by averaging PN velocities using a flat ring model (10 concentric rings of width 3.36 kpc), where PNe are assumed to lie in a thin disk and move on circular orbits. The PNe within a ring are weighted according to their angular distance from the major axis (ϕ) and those that are closer to the minor axis than the major axis are omitted, as are PNe close to identified external objects. A 3σ–clipping routine has been used to eliminate PNe with discrepant velocities, such as halo, foreground or background contaminants.

As the Andromeda galaxy is highly inclined ($i = 77°$) a correction to the raw rotation velocities must be made to account for the line–of–sight integration through the disk. This tends to reduce the observed velocities, $V_{\phi_{\rm obs}}$, with respect to the rotation velocity, V_ϕ, by a factor $f(R', z', i)$, where the parameters are as shown in Fig.1 , and

$$V_\phi = \frac{V_{\phi_{\rm obs}}}{f\left(R', z', i\right)} \tag{1}$$

where

$$f\left(R', z', i\right) = \frac{\int_0^\infty \nu(r', z') \cos\phi {\rm d}S'}{\int_0^\infty \nu(r', z') {\rm d}S'}. \tag{2}$$

Here, ν is the number density of PNe, and the primed quantities have been defined in units of the photometric disk scale length, r_0, i.e. $R' = \frac{R}{r_0}$, $r' = \frac{r}{r_0}$, $z' = \frac{z}{r_0}$ and $S' = \frac{S}{r_0}$. An expression for $f(R', z', i)$ can be found by combining (2), the geometrical relations in Fig.1 and the expression

$$\nu(r', z') = \nu_0 \exp\left(-r' - \frac{z' r_0}{z_0}\right) \tag{3}$$

which assumes that the disk is exponentially declining both radially and perpendicularly to the plane, and z_0 is the disk scale height.

A somewhat more significant correction is that of the asymmetric drift, which accounts for the fact that not all PNe are moving on circular orbits. The difference between the true circular rotation velocity, V_c, and V_ϕ, or asymmetric drift, is related to the velocity dispersion. A correction for this has been performed using the method in [8], whereby it is assumed that

$$V_c^2 = V_\phi^2 + \sigma_\phi^2 \left(2R' - 1\right) \tag{4}$$

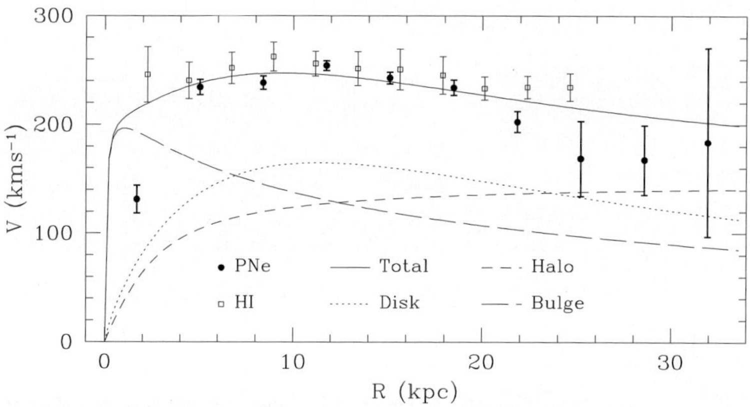

Fig. 2. PNe rotation curve. The filled circles are the PN.S data (binned into 10 concentric rings with the PNe nearest the minor axis eliminated), corrected for the line–of–sight integration through a disk of scale height z_0 and the asymmetric drift. The open squares show the averaged HI data presented in [12]. The lines are model fits for a three component model, fit to all the corrected PNe beyond 3.36 kpc

In order to perform this correction the velocity dispersion has been fitted as an exponential function of the form $\sigma_\phi = \sigma_0 e^{-r/2r_0}$. The dynamical scale length has been set to twice the photometric value to maintain a constant disk scale height with radius [1].

The fully corrected rotation curve is shown in Fig.2, alongside the averaged HI rotation curve given in [12]. With the exception of the first data point (where a number of bulge PNe are likely to be present and hence the asymmetric drift correction breaks down) it is clear that the PN.S data are in very good agreement with the HI rotation curve and are of very high quality out to ∼20 kpc. Beyond this point there is marginal evidence that the PN rotation curve drops to a lower velocity than is seen in the HI data. However, there are very few PNe per ring beyond 20 kpc and we may simply be seeing halo contamination.

Figure 2 also shows a three component model fit to the PN data from all but the first ring. This shows that the PN rotation curve just reaches a radius where the dark matter halo becomes dominant.

3 PNe in Halo Structures

A number of PNe from this survey lie in the region of the Northern Spur (*circles*, Fig.3a), an over-density of stars in Andromeda's halo [4]. This over-density is also seen in the PNe counts (similarly positioned halo fields in the opposite quadrant have 1 or 2 PNe, while Northern Spur fields contain up to ∼6). These planetaries have velocities similar to the maximum observed rotation velocities in the disk (*circles*, Fig.3b). This would suggest that the Northern Spur is not a kinematically distinct substructure, but in fact forms part of the disk, implying that the stellar disk must be severely warped.

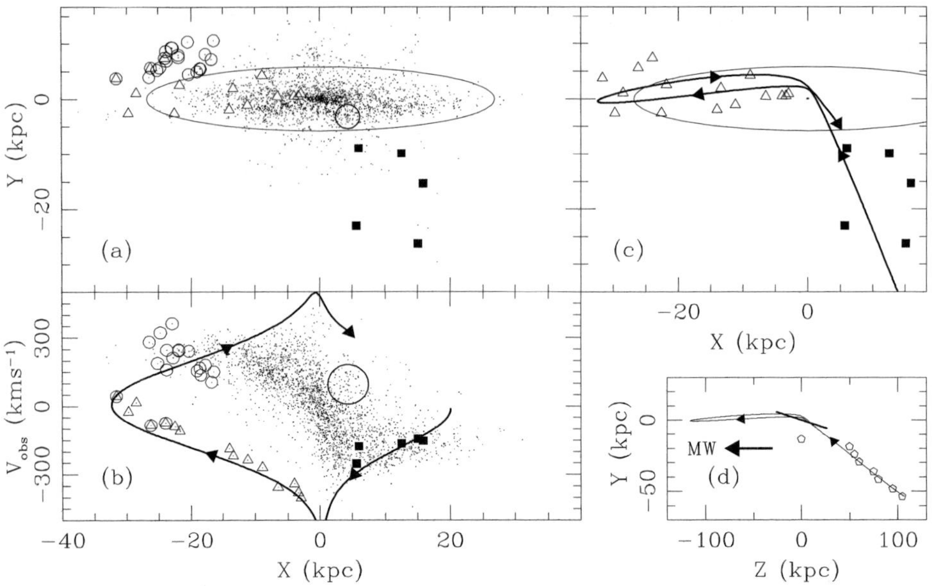

Fig. 3. Velocity Structures. **(a)** PNe positions. To the lower right (*squares*) are PNe in the region of the Southern Stream, with velocities below -130 kms^{-1}; to the upper left (*circles*) are PNe in the Northern Spur region; and scattered through the left side of the disk (*triangles*) are PNe selected for their unusual velocities. The ellipse marks the 2° radius of the disk and the large circle marks the position of M32. **(b)** Velocity versus distance along the major axis and the orbit location; PNe are marked as above. **(c)** Orbit location in the $X-Y$ plane. **(d)** Orbit location in the $Z-Y$ plane. The arrow indicates the direction of the Milky Way; the short heavy line is the location of M31; and the points (*pentagons*) show the measured depth to the stream from [6].

PNe in the region of the Southern Stream [4] have also been targeted by this survey (*squares*, Figs.3a). Those PNe with velocities below -130 kms^{-1} seem to form a coherent kinematic structure (*filled squares*, Fig.3b).

4 Substructure in the Velocity Field

Clearly a number of PNe in the sample have velocities which imply that they are not just normal disk PNe. One such population has been identified whereby the PNe lie below a line for which there is no equivalent population in the opposite side of the disk (*triangles*, Fig.3b).

These PNe have a low dispersion ($\sim$23 kms^{-1}, including an instrumental dispersion of $\sim$15-20 kms^{-1}) about a straight line fit. Such low dispersions are indicative of tidal streams. We therefore propose that these PNe are associated with a continuation of the Southern Stream as it passes into the disk of M31, where it can no longer be tracked via photometric methods (see also [7]).

The locations of the unusual PNe suggest a tidal stream lying almost straight along the major axis. We have therefore generated a simple orbit model (shown in Figs.3b, c and d) using a flattened singular isothermal potential, $\Phi(R,z) = \frac{1}{2}v_c^2 \ln\left(R^2 + \frac{z^2}{q^2}\right)$ where R and z are polar coordinates aligned with M31's disk plane; v_c is set to the upper envelope of PNe velocities (250 kms^{-1}); and a value of $q = 0.9$ is adopted for the flattening. The orbit has been set to fall in along the path defined in [6] (Fig.3d), and turn into the disk. This orbit agrees reasonably well with the velocities of RGB stars in the Southern Stream region [5] and clearly picks up the unusual PN velocities we have detected.

The change in projection angle seen in Fig.3d between the orbit's entry into the disk and as it moves across the disk, can explain the significantly higher number of Stream PNe seen in the disk portion of the orbit (15 PNe) compared to the southern region (probably ~ 5).

5 Conclusion

We have mapped the dynamics of the Andromeda galaxy's planetary nebula system for PNe some 4.5 mags into the luminosity function. From this we have constructed a stellar rotation curve which agrees well with the HI data. A simple three component model has been fit to the PN data out to the realm where the dark halo becomes dominant. It is important to measure this using the stellar population, as there can be no ambiguity as to the nature of the force producing the rotation—unlike gas, stars are subject only to gravitational forces.

Velocities have been measured for the halo structure known as the Northern Spur, supporting the hypothesis that it represents a severe warp in the stellar disk.

A number of PNe have been identified as forming a substructure within the velocity field. These have a very low velocity dispersion, suggesting they are part of a tidal stream. We have shown they can be modeled as a continuation of the Southern Stream as it turns into M31's disk where it would no longer be detectable via photometric methods.

References

1. R. Bottema: A&A **275**, 16 (1993)
2. R. Braun: Ap.J **372**, 54-66 (1991)
3. N. Douglas, M. Arnaboldi, K. C. Freeman, et al.: PASP, **114**, 1234 (2002)
4. A. Ferguson et al.: A.J. **124**, 1452-1463 (2002)
5. R. Ibata et al.: astro-ph/0403068 (2004)
6. A. McConnachie et al.: MNRAS **343**, 1335-1340 (2003)
7. H. R. Merrett et al.: MNRAS **346**, L62-66 (2003)
8. E. Neistein et al.: AJ **117**, 2666-2675 (1999)
9. K. Perrett et al.: AJ, **123**, 2490-2510 (2002)
10. D. Reitzel and P. Guhathakurta: AJ, **124**, 234-265 (2002)
11. V. Rubin and W. Ford: ApJ **159**, 379 (1970)
12. L. Widrow et al.: ApJ **588**, 311-325 (2003)

A Keck/Deimos Survey of Red Giant Branch Stars in the Outskirts of M31

Annette M. N. Ferguson[1], Scott Chapman[2], Rodrigo Ibata[3], Mike Irwin[4], Geraint Lewis[5], and Alan McConnachie[4]

[1] Max-Planck-Institut für Astrophysik, Postfach 1317, 87541 Garching, Germany
[2] California Institute of Technology, Pasadena, CA 91125, USA
[3] Observatoire de Strasbourg, 11, rue de l'Université, Strasbourg F-67000, France
[4] Institute of Astronomy, Madingley Road, Cambridge CB3 0HA, UK
[5] School of Physics, University of Sydney, NSW 2006, Australia

Abstract. We are using the DEIMOS multi-object spectrograph on the Keck II 10 m telescope to conduct a spectroscopic survey of red giant branch stars in the outskirts of M31. To date, velocities have been obtained for most of the major substructures in the halo as well as at several positions in the far outer disk and inner halo. First results concerning the giant stellar stream and major axis substructures are presented here.

1 Introduction

Within the popular ΛCDM model for structure formation, massive galaxies are built up through the merger and accretion of smaller subsystems and through the smooth accretion of gas. Under the assumption that at least some of the accreted satellites contain significant stellar components, one expects observable signatures of this process in the form of tidal streams and other stellar inhomogeneities. While such features have been previously found in our Milky Way, little has been known, until recently, about their frequency in other galaxies. One of the main goals of the Isaac Newton Telescope Wide-Field Camera (INT WFC) survey of M31 [9,4,11] has been to search for signatures of satellite disruption in the outer halo of our nearest giant neighbour.

During the past four years, we have used the INT WFC to map $\approx$ 40 sq. deg. (163 contiguous pointings) around M31 with 800–1200 sec exposures in the V and i passbands [9,4,11]. This depth is sufficient to reach V$\sim$ 24.5, $i\sim$ 23.5, and enables us to probe the top three magnitudes of the red giant branch (RGB). Our current coverage extends to 4 degrees ($\approx$55 kpc) and 2.5 degrees ($\approx$30 kpc) along the major and minor axes respectively. Figure 1 shows the distribution of "blue" RGB stars across our survey area. As already apparent from our interim results [9,4], the distribution of giant branch stars at large radius is very far from uniform. Many of the clumpy features visible in the stellar distribution have effective surface brightnesses $\Sigma_V \gtrsim 28$ mag sq. arcsec – a value typically unobservable by traditional techniques but possible here due to the fact we are resolving individual stars at the bright end of the luminosity function.

The mere existence of stellar overdensities in the outskirts of M31 indicates an active accretion history, however many important questions remain. Are the

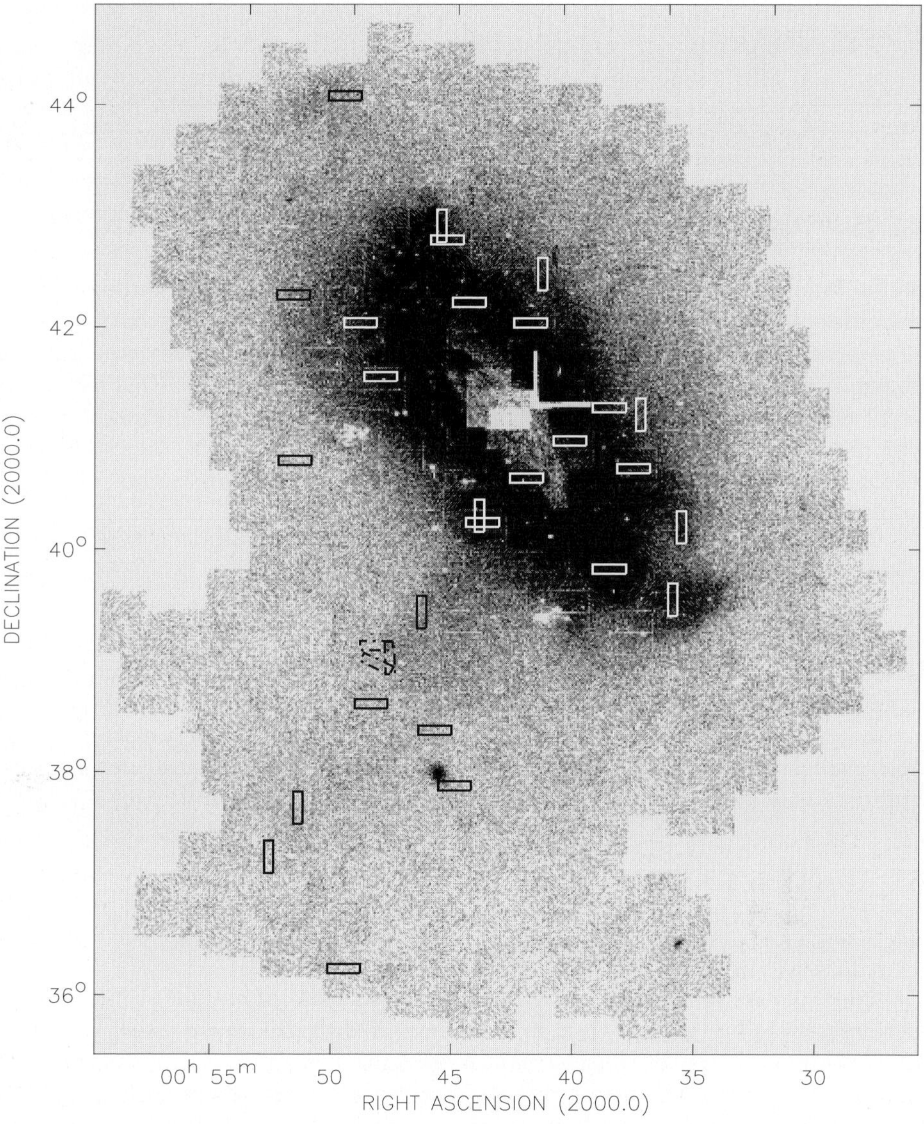

Fig. 1. The distribution of "blue" red giant branch stars in a ≈ 40 square degree (125 x 95 kpc) area centered on M31, as mapped with the INT WFC. Overlaid are the 30 Keck/DEIMOS pointings observed as of September 2003; these mainly target major regions of stellar substructure and a few locations in the far outer disk/inner halo. The three DEIMOS fields of [8] are shown as dashed rectangles.

overdensities the result of many small accretions, or one large one? How much of the structure is simply the result of a warped/perturbed outer disk? What is the relationship between various stellar overdensities, such as a the giant tidal stream, and M31's innermost satellites? To address these issues, we are pursuing a comprehensive follow-up program consisting of a large radial velocity survey with Keck II/DEIMOS and a detailed stellar populations study with HST/ACS.

2 Surveying the M31 Outer Halo with Keck/DEIMOS

Stellar kinematics at the distance of M31 can be probed directly through radial velocities of individual giant stars, or indirectly through tracer populations, such as planetary nebulae (PNe). While PNe have been used successfully to study the kinematics of the disk and inner halo (e.g. [14,7]), the efficiency of this technique declines rapidly in the low surface brightness outer regions. If one assumes the canonical $\alpha_{2.5} \sim 50 \times 10^{-9}$ PNe per unit V-band luminosity [3], then ≈ 50 PNe would be expected per square degree at $\Sigma_V \sim 24$ mag sq. arcsec but only ≈ 1 would be expected within the same area at $\Sigma_V \sim 28$ mag sq. arcsec.

We have been conducting our radial velocity survey using the DEIMOS multi-object spectrograph on the KeckII 10m telescope (see Fig. 1 for the pointings observed to date). Our strategy involves both standard multi-slit masks, resulting in ≈ 100 targets per $16.9' \times 5.0'$ DEIMOS pointing, as well as the use of a custom-built narrow-band filter to limit wavelength coverage and increase multiplexing in high density regions, typically allowing ~ 400 targets per mask. We exploit the near-IR Calcium II triplet lines (~ 8500Å) to provide information about both radial velocities (typical uncertainty here 5–10 km/s) and metallicities. Full details of our survey strategy and results so far are given in [10,13,2].

3 Results To Date

3.1 The Giant Stellar Stream

Radial velocities for stars lying at four locations along the giant stellar stream were presented by [10] and used in combination with distance data [12] to place constraints on the orbit of the stream progenitor and on the total mass of M31. Figure 2 shows velocity histograms at these locations for stars with good S/N and good quality cross-correlation measurements. Gaussians distributions have been used to describe the components attributed to the stream (the dominant component in all fields significantly displaced from the inner halo (i.e. fields S01, S02, S06)) and the centroids and widths are indicated. Note that considerably more stars are now present in S08 than in [10] due to a reanalysis of the original data. A smooth gradient in both radial (heliocentric) velocity and line-of-sight velocity dispersion (uncorrected for the instrumental error) is apparent along the stream. The measurements of [8] for fields lying between S02 and S06 are fully consistent with these trends. The southern extent of the stream is essentially at rest with respect to M31 ($\Delta V_{\rm M31} \sim -25$ km/s) and has the largest observed

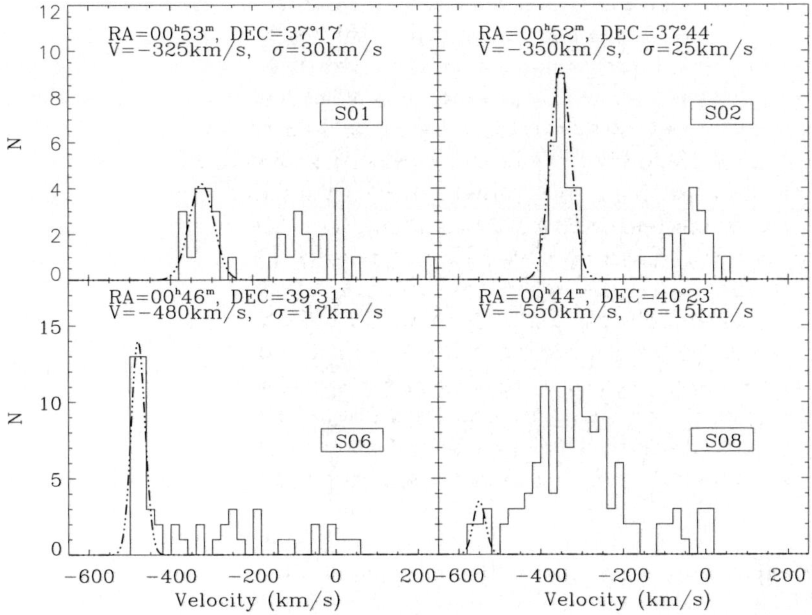

Fig. 2. Stellar radial (heliocentric) velocity distributions at four locations lying along the giant stellar stream (see [10]). The dashed-dotted lines represent gaussian fits to the components attributed to the stream stars. The measured velocity dispersion of the stream declines with distance to M31. The systemic heliocentric velocity of M31 is −300 km/s.

velocity dispersion ($\sigma_v \sim 30$ km/s) – these observations suggest this location is close to apocentre [10,6].

The best-fitting progenitor orbit (in the best-fitting potential, see [10]) is highly radial and viewed close to edge-on. It passes near the centre of M31 before looping around to a position north-east of the galaxy centre (also supported by detailed studies of the stellar populations in these parts [4,5]). Before velocity data were available, a prime candidate for the progenitor was the compact dE, M32, which projects directly onto the stream, has a comparable line-of-sight distance and shares a similarly high metallicity. This association no longer appears likely however, since the radial velocity of M32 ($\Delta V_{\rm M31} = +100$ km/s) is inconsistent with the expected velocity of the stream at this position in the current orbital phase ($\Delta V_{\rm M31} \sim -280$ km/s). Similar arguments make an association with NGC 205, M31's second closest luminous satellite, equally unlikely.

3.2 Substructures Along the Major Axis

Significant substructure has been identified along both the north-eastern and south-western major axes (Fig. 1), termed the "northern spur" and the "G1 clump" respectively. These stellar overdensities are unlikely to be related to

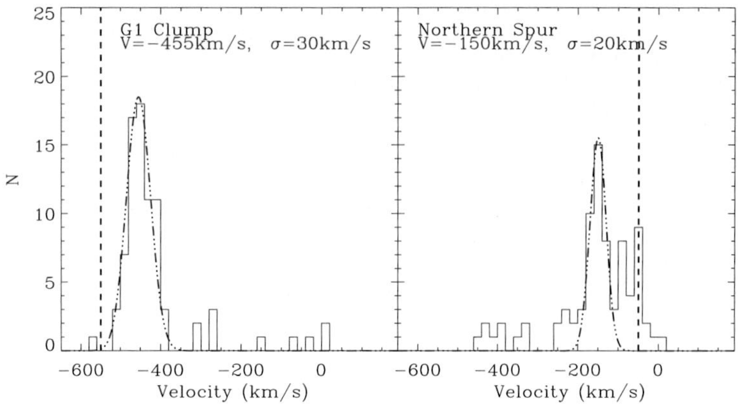

Fig. 3. Radial velocity distributions for regions lying close to the north-eastern and south-western major axes. The dashed-dotted lines represent gaussian fits to the dominant component in these fields, while the vertical dashed lines indicate simple expectations for the disk rotation velocity at these locations along the major axis.

each other – for example, representing debris from a satellite orbiting within the disk plane – given the different colours exhibited by their constituent stellar populations [4,5]. Figure 3 presents radial velocity histograms for these regions constructed from stars with good S/N measurements. The G1 clump, located at a radius of $\sim$ 35 kpc, exhibits a clear velocity peak at -455 km/s (see also [15]). On the other hand, the G1 globular cluster, which is projected near the edge of the overdensity, has a sufficiently different radial velocity (-331 km/s), making it unlikely that the two are related. The northern spur region, located at a radius of $\sim$ 25 kpc, exhibits a clear velocity peak at -150 km/s. While it is tempting to associate the major axis substructures with the outer disk, we note that simple expectations for the disk velocity (assuming $V_{rot} = 250$ km/s, based on HI observations [1]) are -550 km/s and -50 km/s respectively for the clump and the spur – velocities which differ from those observed by $\sim$ 100 km/s (Fig. 3).

4 Summary

Our spectroscopic survey of RGB stars in the outskirts of M31 aims to constrain the nature and origin of the stellar substructure observed in these parts, and to quantify the kinematic and metallicity structure of the far outer disk and inner halo. Our results so far have enabled M32 and NGC 205 to be ruled out as the progenitors of the giant stellar stream; instead, they suggest an orbit which connects the stream to the diffuse overdensity located north-east of the galaxy centre. The kinematics of the substructure observed along the major axes defy easy interpretation at present. Future observations of radial velocities in unperturbed regions of the M31 outer disk will be crucial.

References

1. R. Braun: ApJ **372**, 54 (1991)
2. S. C. Chapman et al. : in preparation (2004)
3. R. Ciardullo, G. H. Jacoby, H. C. Ford & J. D. Neill: ApJ **339** 53 (1989)
4. A. M. N. Ferguson et al. : Astron. Journal **124**, 1452 (2002)
5. A. M. N. Ferguson et al. : in preparation (2004)
6. A. Font et al. : ArXiv Astrophysics e-prints, astro-ph/0406146
7. D. Hurley-Keller et al. : ArXiv Astrophysics e-prints, astro-ph/0408137
8. P. Guhathakurta et al. : ArXiv Astrophysics e-prints, astro-ph/0406145
9. R. Ibata et al. : Nature **412**, 49 (2001)
10. R. Ibata et al. : MNRAS **351**, 117 (2004)
11. M. J. Irwin et al. : in preparation (2004)
12. A. W. McConnachie et al. : MNRAS **343**, 1335 (2003)
13. A. W. McConnachie et al. : MNRAS **351**, L94 (2004)
14. H. R. Merrett et al. : MNRAS **346**, L62 (2003)
15. D. B. Reitzel, P. Guhathakurta, & R. M. Rich: Astron. Journal **127**, 2133 (2004)

Planetary Nebulae in NGC 5128 with VLT FLAMES

Marina Rejkuba and Jeremy R. Walsh

European Southern Observatory, Karl-Schwarzschild-Str. 2, D-85748 Garching, Germany

1 Introduction and Observations

FLAMES, the Fibre Large Array Multi-Element Spectrograph is the multi-object, intermediate and high resolution spectrograph of the VLT with a 25' field of view. In the MEDUSA mode it can observe up to 130 objects with $1''.2$ fibres at a resolution R=6000 to 30000 using the GIRAFFE spectrometer. Simultaneously up to 8 additional targets can be observed with the red arm of UVES (resolution 47000 for 476-683 nm). It is thus ideally suited for kinematical studies of large samples of planetary nebulae (PN) in nearby galaxies.

NGC 5128 (=Centaurus A) is the nearest giant elliptical galaxy, and is a unique testing ground for stellar population models. During the FLAMES Science Verification run in 2003 February we used the combined GIRAFFE+UVES mode to observe known PN and globular clusters (GCs) in NGC 5128. The GIRAFFE spectrograph was used at low resolution L3 setting (R=7500, spectral range 450-508 nm). This interval contains the strongest PN emission lines of [O III]4959 and 5007Å, providing radial velocities, but also includes the Hβ and He II lines which provide line ratio and excitation information in addition.

PN were selected from the Hui et al. (1993) catalogue and the GCs are from various published sources. In total 529 PN and 48 globular clusters (GCs) were observed through MEDUSA fibres in the central region and the SW part of the galaxy. Six bright GCs were observed with UVES in all the exposures of the central field. The central field was observed with 4 different fibre setups and the SW field once with exposures of 45-90min for a total time of 7.5hr. Only the GIRAFFE observations will be presented here. These results will be published together with an NTT survey for more PN in NGC 5128 and accompanying FLAMES spectra with the same GIRAFFE setup.

2 Reduction and Results

The GIRAFFE spectra were reduced using the automatic pipeline (`http://www.eso.org/instruments/flames/drs.html`) which performs wavelength calibration and box extraction of the 1D spectra from each fibre. The emission lines were interactively fitted by Gaussians. The absolute velocity precision was found to be better than 3 kms^{-1} from repeat observations.

Figure 1 shows a summary of the results for the PN. The histogram of the PN (heliocentric) radial velocities (upper left) indicates a systematic velocity

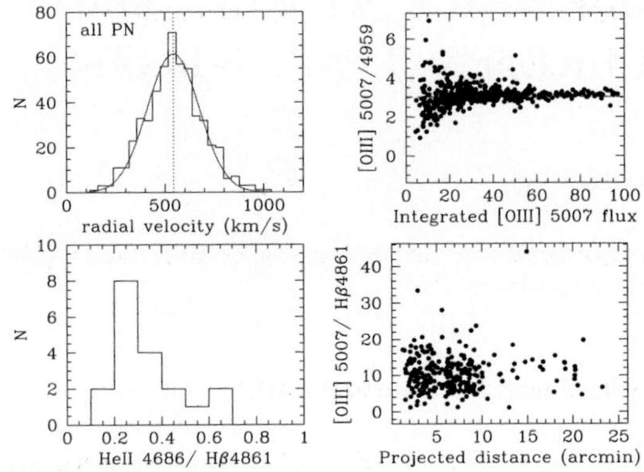

Fig. 1. Top left: Radial velocity distribution of observed PN is shown. The vertical line indicates the systemic velocity. **Top right:** The [O III]5007 and 4959Å doublet ratio is shown as a function of the integrated [O III] 5007Å line flux (in ADUs). **Bottom left:** The He II 4686Å/Hβ ratio could be measured for 19 PN. **Bottom right:** The [O III]5007Å/Hβ ratio is plotted as a function of projected distance.

of 543 kms^{-1} from a Gaussian fit, to be compared with the value of 541$\pm$ 7 kms^{-1} determined by Hui et al. (1995) from 433 PN. The quality of the spectra can be assessed by comparing the observed ratio of the [O III] doublet with the theoretical value of 3.03; the doublet ratio is plotted against 5007Å line strength (upper right). In 19 of the PN, the He II 4686Å line was detectable in the GIRAFFE spectra and the ratio to Hβ is plotted (lower left). In order to investigate any effects of ionization or metallicity, the [O III]5007Å/Hβ is plotted versus radial distance (lower right). No clear trends are apparent although there are fewer large ratio values in the outer regions.

The modest spectral resolution of the GIRAFFE observations enable emission lines with a separation of 80 kms^{-1} to be resolved and hence allows line-of-sight blends to be detected. Two PN, 525 and 4417 (Hui et al. 1993), show double components; neither of these PN are in the brightest 1 mag of the PN Luminosity Function and, with one component weaker, they they will have little effect on the PNLF. Two multiple PN were found towards the globular clusters H6 and G169 (Harris et al. 1992); one already detected by Minniti & Rejkuba (2002) and Peng et al. (2004). For the two PN in G169, the redder one has a velocity similar to the globular cluster and may indeed be associated with it.

References

1. Harris, G. L., Geisler, D., Harris, H. C., Hesser, J. E., 1992, AJ, **104**, 613
2. Hui, X., Ford, H. C., Ciardullo, R., Jacoby, G. H., 1993, ApJS, **88**, 423
3. Hui, X., Ford, H. C., Freeman, K. C., Dopita, M. A., 1995, ApJS, **88**, 423
4. Minniti, D., Rejkuba, M., 2002, ApJ, **575**, L59
5. Peng, E. W., Ford, H. C., Freeman, K. C., 2004, ApJS, **150**, 367

Probing Halos with PNe: Mass and Angular Momentum in Early-Type Galaxies

Aaron J. Romanowsky

School of Physics and Astronomy, University of Nottingham, University Park, Nottingham NG7 2RD, England

Abstract. We present an observational survey program using planetary nebulae, globular clusters, and X-ray emission to probe the halos of early-type galaxies. We review evidence for scanty dark matter halos around ordinary elliptical galaxies, and discuss the possible implications. We also present measurements of rotation in the halos.

1 Introduction

Studying early-type galaxies (ellipticals and lenticulars) in their outer parts (much outside $\sim R_{\rm eff}$, the half-light radius) can provide powerful clues about their structure and formation – but the very low stellar surface densities and the general lack of cold gas have precluded a convenient observational approach, and progress in this area has been slow. However, the situation is now changing with recent advances in instrumentation (described below) and the use of planetary nebulae (PNe), globular clusters (GCs), and X-ray emission as halo probes.

As discussed in Sect. 3, some initial results are puzzling: PN and GC dynamics in ordinary L^* ellipticals imply much less dark matter (DM) than expected in theories of galaxy formation. However, other lines of observational evidence – using satellite dynamics and weak lensing [1] – imply more dominant DM halos in L^* ellipticals. Also, spiral galaxies and very bright ellipticals show massive DM halos [2]. Thus, not only is it necessary to study a larger sample of galaxies for better statistics, but it is also important to cross-check the different mass probes for consistency. After identifying which probes are reliable, we can combine them for much stronger constraints on the mass and orbit distributions in galaxy halos. To explore these issues, we are studying a representative sample of near-L^* ellipticals using integrated stellar kinematics, PNe, GCs, and X-rays.

2 Halo Probes: PNe, GCs, and X-Rays

PN kinematics are a very promising probe, tracing the main stellar population in galaxy halos which is otherwise too diffuse to observe. The *PN.Spectrograph* is the breakthrough instrument for this technique, making it now possible to obtain hundreds of PN velocities in L^* galaxies out to distances of ~ 25 Mpc [3]. Four L^* ellipticals have been studied using PNe, and their projected velocity dispersions have been found to decline markedly with galactocentric radius (see Fig. 1, *left*), suggesting very low DM content.

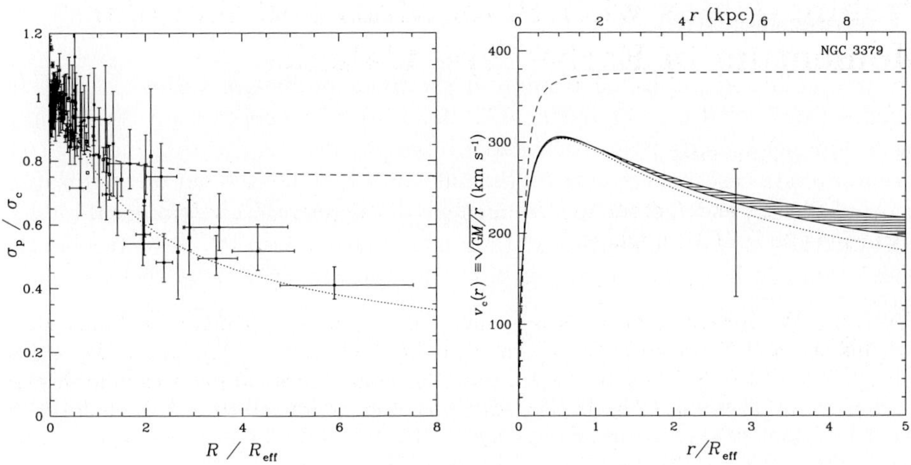

Fig. 1. *Left:* Projected velocity dispersion profiles with radius for four elliptical galaxies [4,5], scaled and stacked. Simple model predictions show an isothermal halo (dashed line) and a constant mass-to-light ratio (Υ) galaxy (dotted line). *Right:* Modeled circular velocity profile of NGC 3379. Shown are a constant Υ solution (dotted line); results from PN dynamics (shaded region [5]), from an X-ray analysis (dashed line [6]), and from GC dynamics assuming orbital isotropy (error bar [7]).

Like PNe, **GCs** are useful as bright point-like sources in galaxy halos. GC systems are typically more extended than the main galaxy light, allowing them to probe to even larger radii than PNe. Furthermore, GCs are interesting in their own right as simple stellar systems formed at early times: their properties can reveal events in their host galaxy's history which are difficult to discern in the jumble of the main galaxy light. One generic prediction of most GC formation scenarios is that the more metal-rich GCs are associated with the main stellar population – which can be verified by comparisons of metallicities and kinematics between the halo stars (represented by the PNe) and the GCs.

While it has long been possible to study GC kinematics in very GC-rich galaxies like M87, it is only with the latest spectrographs that this is viable for a large sample of typical galaxies. These instruments (e.g., VLT+FLAMES,VIMOS; Magellan+IMACS; Gemini+GMOS; Keck+DEIMOS) have the advantages of a wide field of view, multi-object capability, and 8-meter-class collecting area.

In principle, thermal **X-ray emission** from hot gas trapped in the potential well of a galaxy offers an excellent probe of its halo mass, as it is not subject to the systematic anisotropy uncertainty plaguing stellar systems. In practice, it is difficult and crucial to remove contaminating point sources, to check the dynamical equilibrium of the gas, and to measure the radial temperature variation. With their large collecting areas and high angular resolution, the *Chandra* and *XMM-Newton* observatories have now made it possible to rigorously pursue mass studies based on X-rays. We note the importance of not selecting the galaxy sample based on L_X as this will introduce a bias in the halo mass results.

3 Mass Results

We have collected published results on the mass profiles of various early-type galaxies (M87, NGCs 1399, 3379, 4472, 4636, 5128) to check their consistency; the techniques include PNe, GCs, and X-rays [8]. The results are generally consistent, even in cases where parts of the halo appear to be well out of equilibrium.

A case study is NGC 3379, for which we have constructed dynamical models to fit long-slit stellar kinematical data in combination with PN velocities [5]; the models fully include the variations in orbital anisotropy which otherwise make the mass profile very uncertain. We find a total B-band mass-to-light ratio at 5 $R_{\rm eff}$ of $\Upsilon_B = 7.1 \pm 0.6$ (solar units); for comparison, standard models of stellar populations imply $\Upsilon_{*,B} = 6$–11, leaving very little room for DM inside 5 $R_{\rm eff}$.

A major remaining systematic uncertainty in this analysis is the assumption that NGC 3379 is approximately spherical. But if it in fact contained a significant disk component (as in an S0 galaxy) viewed face-on, the large-radius stellar kinematics could appear "cold" even with a massive DM halo. Although the galaxy's inner parts do not show the disk-like kinematics expected in such a scenario [9], a more convincing test is to use independent mass probes (Fig. 1, *right*). The GCs are a good candidate since GC systems are rarely flattened; GC kinematics in NGC 3379 reveal a low halo velocity dispersion consistent with the PN result [7]. A ring of HI gas at large radius also shows kinematics consistent with the low DM content [10]. On the other hand, a preliminary analysis of the galaxy's X-ray emission using *Chandra* data indicates a higher DM content [6]. We will see if these various analyses, when complete, converge to similar results.

NGC 3379 seems a convincing case for a low-DM system, but how common is this situation? Another candidate is the S0 galaxy NGC 5128, where PN and GC dynamics imply a total $\Upsilon_B \sim 20$ inside the virial radius [11]. But it and the other three L^* ellipticals still need to be modeled in as much detail as NGC 3379 to verify the apparent low DM content. Note also that one of these, NGC 4494, has very faint gaseous X-ray emission, consistent with the low-DM scenario [12].

Firm conclusions await a larger sample size along with rigorous analyses, but if the apparent low DM content of these galaxies turns out to be a widespread phenomenon, what could be the explanation? The idea that these are all face-on S0s is unlikely from a statistical point of view and because of the cross-checks in NGC 3379. The galaxies may have lost much of their DM halos through interactions with other galaxies, but most of them are not in the high-density environments thought necessary for this. Modified Newtonian Dynamics has been advanced as an explanation [13], but this alternative gravitational theory has difficulty explaining the large mass discrepancies found in brighter ellipticals.

A higher DM content in these galaxies may be possible if baryonic processes have driven more DM into the galaxy centers than we have assumed using standard ΛCDM halos [14]. But this would contradict many other lines of evidence that the DM mass fraction in the inner parts of ellipticals is fairly low, and would involve lower values for Υ_* than are plausible using current stellar population synthesis models. An additional possibility is that a decline with radius of Υ_* could mask the increase from a DM halo; however, this would require either a metallicity log gradient of ~ -1.4 – which is much larger than what is implied by typical studies of integrated stellar light and by the abundance studies of halo PNe presented at this workshop – or a young halo population, which is excluded in NGC 3379 by study of the color-magnitude diagram of resolved stars [15].

A more arguable scenario is that these galaxies do in fact have large amounts of DM residing even farther out in the halo than we have been able to probe. Such halos would have much lower concentrations (or equivalently, lower central DM mass fractions) than predicted by ΛCDM. This conclusion has also (controversially) been reached with many dynamical studies of late-type galaxies and with strong lensing studies of early-types, and could be explained by new baryonic processes or by yet more exotic DM theories [16]. We are now quantifying constraints on DM halo models vis-à-vis empirical results on Υ gradients [17].

4 Angular Momentum Results

It is well known that elliptical galaxies exhibit much smaller specific angular momenta λ than spirals. But studies in ellipticals have been confined to their central parts, and it is surmised that the "missing" λ may be found in their outskirts. Indeed, most simulations of elliptical formation through galaxy major mergers predict that their outer parts should rotate rapidly.

With PN data in five ellipticals, we can now examine the rotation of their stellar halos (Fig. 2, *left*). In none of these galaxies do we find substantial rotation: typically, $v/\sigma \sim 0.2$. On the other hand, NGC 5128 has a rapidly rotating halo

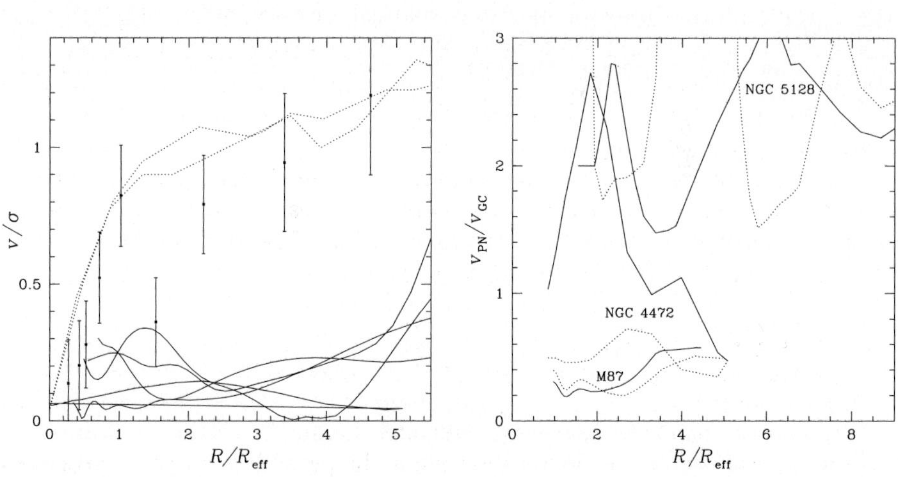

Fig. 2. *Left:* Rotational parameter with radius for five elliptical galaxies (M87, NGC 821, NGC 3379, NGC 4472, and NGC 4494) observed with PNe (solid lines [5,18]), for NGC 5128 (error bars [11]), and for a simulated merger remnant (dotted lines [19]). *Right:* Ratio between PN and GC rotational velocities at a given radius, in three galaxies. Dotted lines are for metal-poor GCs, and solid lines for metal-rich GCs.

($v/\sigma \sim 1$), consistent with a dissipationless merger model [19]. A caveat is that these ellipticals have been partially selected for their roundness, biasing against observably high rotation; more secure results will come with corrections for projection effects, and with a wider ellipticity range in the galaxy survey. Also, the theoretical expectations are not yet entirely clear; inclusion of baryonic physics seems to produce galaxies with lower λ than in dissipationless simulations [20].

We can also compare the rotational behavior of the stars (via the PNe) with the GCs (Fig. 2, *right*). However, with the few cases so far available for study, we do not yet see a consistent pattern (such as $v_{\rm PN}/v_{\rm GC} \sim 1$ for metal-rich GCs).

5 Summary

Comparisons of early-type galaxy halo mass results from independent probes (PNe, GCs, and X-rays) yield generally consistent results. A surprising trend is seen for L^* ellipticals to show low dark matter content, which may be indicative of low-concentration halos. We find so far little evidence for the strong rotation that may be expected in ellipticals' stellar halos.

References

1. F. Prada et al.: ApJ **598**, 260 (2003); J. Guzik, U. Seljak: MNRAS **335**, 311 (2002)
2. M. Persic, P. Salucci, F. Stel: MNRAS **281**, 27 (1996); M. Loewenstein, R. White: ApJ **518**, 50 (1999); T. Treu, L. Koopmans: ApJ **611**, 739 (2004)
3. N. Douglas: these proceedings; N. Douglas et al.: PASP **114**, 1234 (2002)
4. R. Méndez, A. Riffeser, R.-P. Kudritzki, M. Matthias et al.: ApJ **563**, 135 (2001)
5. A. Romanowsky, N. Douglas, M. Arnaboldi et al.: Science **301**, 1696 (2003)
6. E. O'Sullivan, A. Romanowsky, T. Ponman et al.: in preparation
7. M. Beasley et al.: in prep; G. Bergond, A. Romanowsky, S. Zepf et al.: in prep
8. A. Romanowsky, C. Kochanek: ApJ: 553, 722 (2001); K. Matsushita et al.: A&A **386**, 77 (2002); Y. Ikebe et al.: Nature **379**, 427 (1996); C. Jones et al.: ApJ **482**, 143 (1997); M. Paolillo et al.: ApJ **565**, 883 (2002); N. Napolitano et al.: A&A **383**, 791 (2002); T. Richtler et al.: AJ **127**, 2094 (2004); S. Schindler et al.: A&A **343**, 420 (1999); A. Kronawitter et al.: A&AS **144**, 53 (2000); K. Matsushita et al.: ApJ **499**, L13 (1998); M. Loewenstein, R. Mushotzky: `astro-ph/0208090`; Y. Schuberth et al.: these proceedings
9. T. Statler: AJ **121**, 244 (2001)
10. S. Schneider: ApJ **288**, L33 (1985)
11. E. Peng, H. Ford, K. Freeman: ApJ **602**, 685 (2004); ApJ **602**, 705 (2004)
12. E. O'Sullivan, T. Ponman: MNRAS **349**, 535 (2004)
13. M. Milgrom, R. Sanders: ApJ **599**, L25 (2003)
14. J. Navarro, C. Frenk, S. White: ApJ **490**, 493 (1997)
15. M. Gregg, H. Ferguson, D. Minniti, N. Tanvir, R. Catchpole: AJ **127**, 1441 (2004)
16. J. Ostriker, P. Steinhardt: Science **300**, 1909 (2003); D. Rusin et al.: ApJ **595**, 29 (2003); H. Mo, S. Mao: MNRAS **353**, 829 (2004)
17. N. Napolitano: these proceedings; MNRAS, in press (`astro-ph/0411639`)
18. A. Romanowsky, N. Douglas, K. Kuijken, M. Arnaboldi: in preparation
19. M. Weil, L. Hernquist: ApJ **460**, 101 (1996)
20. J. Sommer-Larsen et al.: ApJ **596**, 47 (2003); L. Wright et al.: `astro-ph/0310513`

Modelling Kinematics and Dark Matter: The Halos of Elliptical Galaxies

Ortwin Gerhard

Astronomisches Institut, Universität Basel, Venusstrasse 7, CH-4102 Switzerland

Abstract. This review is focussed on the outer halos of elliptical galaxies. Its emphasis is on: (i) planetary nebulae as test particles to trace the stellar kinematics at large radii; (ii) the observed angular momentum in elliptical galaxy halos and its theoretical relevance; (iii) dynamical modelling of stellar-kinematic data; and (iv) a discussion of the evidence for dark matter halos in ellipticals from a wide range of measurements.

1 Introduction

In current hierarchical theories, elliptical galaxies represent an advanced stage of the galaxy formation process: they are massive galaxies found in dense environments. They reflect 'early' conditions in galaxy formation as well as subsequent merging processes, making the study of their properties particularly interesting.

However, their old stellar populations and complicated orbital structure make many important questions harder to answer for ellipticals than for spirals. This includes cosmological questions about their ages and precise formation mechanism, about the concentration of their dark matter halos, and the segregation of baryonic matter, as well as dynamical questions about the orbit distribution and total angular momentum, to name only a few.

The subject of this review is the halos of elliptical galaxies and how planetary nebulae observations and dynamical modelling can help answer some of these questions. Several other papers in these proceedings elaborate on the topics discussed here.

2 Planetary Nebulae as Kinematic Tracers

Planetary Nebulae (PNe) are excellent test particles to trace the kinematics in the outer halos of elliptical galaxies. They occur as a brief stage in the late evolution of stars with masses between $\sim 0.8 - 8\,\mathrm{M}_\odot$, when these stars evolve from the Asymptotic Giant Branch (AGB) phase to their final white dwarf stage. Their nebular gas envelope converts up to 15% of the central star's radiation energy to photons in the characteristic λ5007 [O III] emission line [1]. From this emission line, PNe can be identified and their radial velocities measured spectroscopically at distances of up to 20–30 Mpc. With 8m class telescopes [2] or the special-purpose Planetary Nebula Spectrograph (PN.S) [3], several hundred PNe have been found in giant elliptical galaxies with distance up to the Virgo cluster. The PN number density approximately follows the surface density of stars; thus a few tens of PNe can be used to measure the stellar kinematics at several effective radii R_e, where the surface brightness is too faint for absorption line spectroscopy (ALS).

From these PN velocities, constraints on the angular momentum, dynamics, and mass distribution in the outer halos of ellipticals can be derived; these are important to compare with predictions from theories of galaxy formation.

In this analysis, it is always advantageous if additional information can be included. Especially, stellar kinematics from ALS are highly desirable: (i) to determine the mass-to-light ratio in the inner, presumably baryon-dominated parts of the galaxy; and (ii) to narrow down the range of permitted dynamical anisotropies at 1-2 effective radii [4]. Also, globular cluster (GC) velocities are now becoming available in substantial numbers around luminous ellipticals. While the spatial distribution and dynamics of the globular cluster system may be different from those of the stars, analysis of these velocities gives additional constraints on the common gravitational potential [5]. Lastly, an ideal case is when the galaxy potential is narrowly constrained from Chandra or XMM X-ray observations [6]. Then the information obtained with the PN kinematics can be used entirely for inferring the orbital structure of the outer stellar halo in the known potential. These issues are further discussed later; the present section is concerned with PNe as kinematic tracers.

2.1 Does the PN Number Density Follow Light?

Although PNe are representative for the bulk of all stars, there are some population effects that can bias their number density profile with respect to the galaxy's luminosity profile. The PN luminosity function (PNLF) is observed to have a near-universal form and bright cut-off magnitude [7]. Then the quantity $\alpha_{X,n}$ is defined as the ratio of the number of PN in a stellar population, n magnitudes down the PNLF from the cut-off magnitude, to its total stellar luminosity, in a specified wavelength band X. While the bolometric $\alpha_{\mathrm{bol},\infty}$ is expected to be nearly independent of stellar population [8], Hui *et al.* show that the blue $\alpha_{B,2.5}$ depends on colour B-V, decreasing by a factor of 4 from the M31 bulge to the reddest elliptical galaxies [9]. Age, or metallicity, or both, must influence $\alpha_{B,2.5}$. Theoretical models predict that the brightest PNe in B have central stars with progenitor masses $\sim 2.5\,\mathrm{M}_\odot$ [10], with luminosity also dependent on uncertain post-AGB evolution parameters and metallicity. The presence of a younger population in elliptical galaxies, whether made in situ or accreted, could thus change the number of bright PNe significantly. Real ellipticals can have colour and metallicity gradients. Therefore it is prudent to check carefully whether the observed (usually magnitude-limited) PNe samples do indeed trace light.

Observational tests of this question have been carried out in a number of galaxies, including M31, Cen A, the Leo group galaxies NGC 3377, 3379 and 3384, several Virgo ellipticals, the Fornax galaxies NGC 1399 and NGC 1404, and NGC 4697 [2,7,9,11]. Generally, the number of PNe as a function of isophote follows the luminosity distribution for radii outside $\sim 1-2'$, in some cases very well. At smaller radii some PNe are lost against the bright background of the galaxy, and the PNe no longer track the luminosity profile. A colour gradient such that the inner parts of the galaxy are redder with fewer PN per blue light may further decrease the number of PNe per unit light there; NGC 4697, for example, has a slight gradient of this kind [12].

2.2 Do the PN Kinematics Follow the Absorption Line Stellar Kinematics?

If the distribution of PNe follows light, the PNe can be considered as faithful tracers also for the stellar kinematics. A test of this is to compare the rotation velocities v and velocity dispersions σ obtained from PN radial velocities with those from ALS, taking into account the greater smoothing of the PN velocity field due to spatial averaging. Mendez used his sample of 535 PN velocities in NGC 4697 for a simple comparison [2]. His sample, restricted to a stripe around the major axis, indeed followed the rotation seen in the ALS data; however, the number of PNe per mean velocity point is ~ 10 and so the errors are large, $\sim 50\,\mathrm{km\,s^{-1}}$. Similarly, the PN velocity dispersions in major axis sectors were consistent with the ALS dispersions within $\sim 20\,\mathrm{km\,s^{-1}}$.

In such a case the PN velocities and the ALS velocity measurements can be simply combined in modelling the stellar kinematics. Even the incompleteness in the center does not present a problem, because the probability of losing a PN against the bright galaxy background is uncorrelated with its radial velocity. Then only the PN velocities, but not their radial distribution must be used in the modelling. If, on the other hand, because of an outward colour gradient and the observed dependence of α_B on colour, the distribution of PNe is more extended than the stellar light, say, then the PNe velocity dispersions would overestimate the stellar velocity dispersions, as can be seen most easily from the spherical Jeans equation. In this case, the PNe would have to be included in the modelling as a separate test particle distribution.

Another kinematic bias can occur for elliptical galaxies in galaxy clusters, in that the PN samples, particularly in the outer galactic halo, can be contaminated by intracluster PNe [13]. ICPNe would be difficult to disentangle from galactic PNe when the velocity dispersions of the galaxy and its host cluster are comparable, such as for NGC 1399 in the Fornax cluster.

2.3 How Many PNe Velocities Are Needed for Orbital and Potential Analysis?

With a few tens of PN velocities it is possible to detect a clear trend of rotation in the outer halo; see, e.g., the work of Arnaboldi *et al.* for NGC 4406 and NGC 1316 [13, 14]. To determine the detailed shape of the rotation velocity field requires substantially larger samples. Fig. 1, due to N. Sambhus, shows the results of fitting the velocities of approximately 200 PNe from PN.S observations [15] in the E0/1 galaxy NGC 3379 with a non-parametric spline method. The velocity field shows a peak-to-peak rotation of $\sim 100\,\mathrm{km\,s^{-1}}$ at $\sim 100''$, with a subsequent decline out to $\sim 300''$, confirmed by the rotation along the line of maximum gradient. However, both the detailed shape of the rotation curve and the reality of the apparent asymmetry in the velocity field remain uncertain. For comparison, SAURON data show a clear rotation gradient in the central 20", which is unresolved in the PN velocities [16]. The conclusion from this is that PN velocities should be used as an outward extension of the ALS velocity measurements whenever possible [14].

The second important application of PN velocities is to help determine the masses of ellipticals in their outer regions, where the dark matter halos should begin to dominate. Because of the large freedom in the orbital structure and velocity anisotropy in elliptical galaxies, this is not an easy task. If no other kinematic constraints are available, Merritt has estimated that at least many hundreds of PN velocities would be required [17]. Thus, it is again advantageous to use the PN data in conjunction with ALS velocities and other data. From ALS, data out to $\sim 2R_e$, including line profile shape parameters h_3 and h_4, a range of dynamical models and gravitational potentials can be determined that are consistent with these data. Within these models, the implied orbital anisotropy and halo mass are correlated. Then only a few tens of PN velocities beyond $2R_e$ are needed to further narrow down the permitted dynamics and mass distribution [4].

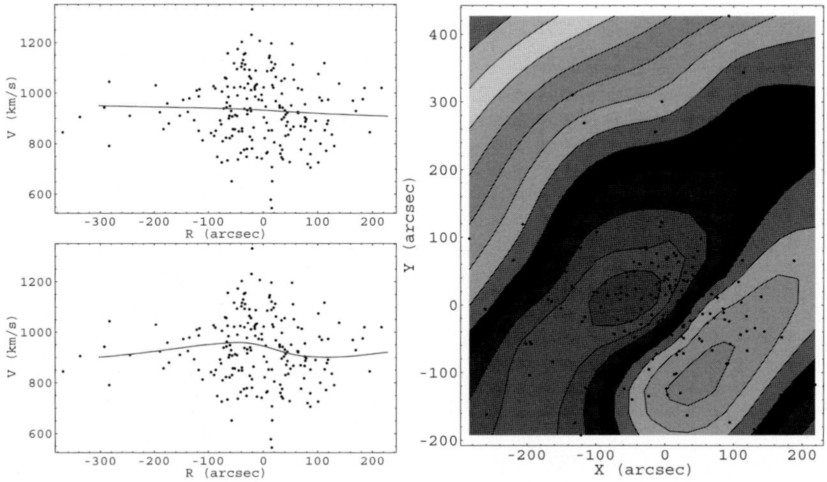

Fig. 1. Velocity field and streaming velocities from $\sim$ 200 PNe in NGC 3379. The right-hand figure shows a smoothed spline model for the projected mean velocity, derived from the PN velocities; contour spacing is $10\,\mathrm{km\,s^{-1}}$ and the galaxy center is at (0,0). The two plots on the left show fits with different smoothing to the distribution of PN velocities versus distance along the direction of maximum gradient.

3 Angular Momentum in the Outer Parts of Ellipticals

One of the most interesting early results obtained with PNe was the discovery of rapid rotation in the halos of the giant elliptical galaxies Centaurus A, NGC 1399, and NGC 1316 [18, 14]. The PNe radial velocity fields in these early-type galaxies show that their outer halos are rapidly rotating, and that these galaxies contain comparable angular momentum to spiral galaxies of similar luminosity. Recent work on several further ellipticals has shown that most of these do not rotate rapidly at large radii; see diagram in [19]. There appears to be a range of outer rotation properties observed.

The amount of rotation in the halos of elliptical galaxies is a valuable probe for how these systems formed. Elliptical galaxies are now believed in general terms to form by merging processes; what is less clear is the kind of progenitors that dominated in the formation of the present population of ellipticals. Spiral-spiral galaxy mergers, observed in the local Universe and following Toomre's original merger hypothesis [20], give remnants that morphologically and kinematically resemble elliptical galaxies in many ways [21]. Equal-mass mergers have low $v/\sigma \sim 0 - 0.2$ within $R \lesssim R_e$, as observed for giant ellipticals, due to angular momentum transfer from inner regions to the extended outer halos by dynamical friction in the merger. These remnants can, however, contain significant angular momentum at large radii, reaching $v/\sigma \sim 0.2 - 0.5$, even though a lot of spin angular momentum is carried away by material in the tidal tails. Unequal-mass mergers rotate faster than equal-mass mergers. As argued in [21], binary mergers of disk galaxies may be the main formation mechanism of low- and intermediate mass ellipticals.

An alternative merging channel to form an elliptical is through multiple major and minor mergers in a compact group of galaxies [22]. In this case the tidal forces are more effective in disrupting the progenitors before coalescence, so dynamical friction is less effective. As a result, the remnants have more angular momentum in their inner parts than spiral-spiral mergers, placing them not far from the oblate-isotropic line in the $v/\sigma - \epsilon$ diagram, and their outer parts may reach $v/\sigma \sim 1$.

Computations of the angular momentum of dark matter halos growing by merging and accretion in hierarchical universes result in low spin parameter λ and low v/σ [23]. Interestingly, the values of these parameters in the evolution of individual halos are most likely to increase in major mergers, and generally decrease in multiple accretion of satellites. If this is indicative for the luminous components also, then ellipticals that were last shaped by a major merger should contain the highest angular momenta. The rotation velocities in the remnant halos are fairly constant with radius, however, confirming that the dissipation and dynamical friction processes acting specifically on the baryonic component are crucial for shaping the angular momentum distributions in elliptical galaxies.

These results show that there is no simple, one-to-one correspondence between angular momentum at large radii and formation mechanism. For example, of the ellipticals with outer PN kinematics, Centaurus A is believed to have formed from the merger of two disk galaxies; in this galaxy v/σ rises to ~ 1

beyond $R = 15\,\mathrm{kpc}$ [18]. However, how much angular momentum resides in the outer halos of elliptical galaxies is clearly a key issue which, when understood for a representative sample of elliptical galaxies, will be crucial for determining the merging channel that dominated their formation. This is a research program that can ideally be tackled by PNe radial velocity measurements with the PN.S.

4 Dynamical Analysis of Kinematic Data

This Section gives a brief overview of how the gravitational potential of an elliptical galaxy can be inferred from stellar-kinematic data. These data may include PN velocities, which individually sample the stellar line-of-sight velocity distributions $L(v)$ (hereafter LOSVD) at their positions in the galaxy image, as well as ALS spectroscopy measurements of the first few moments of the LOSVD, at a set of 1D or 2D binned positions.

As is well known, velocity dispersion profile measurements (and streaming velocities, if the galaxy rotates) do not suffice to determine the distribution of mass with radius, due to the degeneracy with orbital anisotropy. With very extended measurements a constant M/L model can be ruled out (e.g., [18]), but the detailed $M(r)$ still remains undetermined. Thus mass determination in elliptical galaxies always involves determining the orbital structure at the same time, and requires a lot of data. LOSVDs from absorption line profile shapes provide additional data with which the degeneracy between orbit structure and mass can largely be broken. Simple spherical models are useful to illustrate this [24]: at large radii, radial orbits are seen side-on, resulting in a peaked LOSVD (positive Gauss-Hermite parameter h_4), while tangential orbits lead to a flat-topped or double-humped LOSVD ($h_4 < 0$). Similar considerations can be made for edge-on or face-on disks [25] and spheroidal systems [26].

One may think of the LOSVDs constraining the anisotropy, after which the Jeans equations can be used to determine the mass distribution. However, the gravitational potential influences not only the widths, but also the shapes of the LOSVDs (see illustrations in [24]). Furthermore, eccentric orbits visit a range of galactic radii and may therefore broaden a LOSVD near their pericentres as well as leading to outer peaked profiles. Thus, in practice, the dynamical modelling to determine the orbital anisotropy and gravitational potential must be done globally, and is typically done in the following steps:

(0) choose geometry (spherical, axisymmetric, triaxial);

(1) choose dark halo model parameters, and set total luminous plus dark matter potential Φ;

(2) write down a composite distribution function (DF) $f = \Sigma_k a_k f_k$, where the f_k can be orbits, or DF components such as $f_k(E, L^2)$, with free a_k;

(3) project the f_k to observed space, $p_{jk} = \int K_j f_k d\tau$, where K_j is the projection operator for observable P_j, and τ denotes the line-of-sight coordinate and the velocities;

(4) fit the data $P_j = \Sigma_k a_k p_{jk}$ for all observables P_j simultaneously, minimizing a χ^2 or negative likelihood, and including regularization to avoid spurious

large fluctuations in the solution. This determines the a_k, i.e, the best DF f, given Φ, which must be $f > 0$ everywhere;

(5) vary Φ, go back to (1), and determine confidence limits on the parameters of Φ.

If the mass distribution and gravitational potential are known from analysis of, e.g., X-ray data, step (5) can be omitted. Because this eliminates the degeneracy with orbital anisotropy, considerably fewer data are then needed.

Such a scheme has been employed regularly to model spherical and axisymmetric galaxies [e.g., 4,27], using mostly orbits ("Schwarzschild's method") and, more rarely, distribution function components. Discrete velocity data were modelled in [e.g., 5,17]. The modelling techniques used to constrain black hole masses from nuclear kinematics and dark halo parameters from extended kinematics are very similar.

Line-profile shape parameter measurements are now available for many nearby ellipticals, but those reaching to $\sim 2R_e$ are still scarce [e.g., 28, 4]. Modelling of the outer mass profiles of ellipticals from such data has been done for some two dozen round galaxies in the spherical approximation, and for a few cases using axisymmetric three-integral models.

From recent SAURON integral field measurements it has become clear that the kinematics of many elliptical galaxies show features incompatible with axisymmetry [16]. This has motivated Verolme *et al.* to extend Schwarzschild's method to triaxial models [29], a development that has only recently become possible with the increase of available computing power. In parallel, adaptive N-body codes are being developed that use the algorithm of Syer & Tremaine for training N-body systems to adapt to a specified set of data constraints [30]. These latter models have the additional advantage of simultaneously providing a check for the model's dynamical stability. One application has been to the rotating Galactic bar; work currently in progress is on axisymmetric and rotating triaxial galaxies.

5 The Dynamics and Dark Matter Halos of Elliptical Galaxies

5.1 Results from Dynamical Analysis of Absorption Line Profiles

Giant ellipticals with $L \gtrsim L_*$ are typically fit best by dynamical models that have near-isotropic to modestly radially anisotropic velocity dispersions at intermediate radii, $0.5 - 1R_e$ [4,27,31]. This is based on spherical and axisymmetric three-integral models of about two dozen ellipticals. In ellipticals less luminous than $M_B = -19.5$ rotation is important and, moreover, these galaxies become rotation-dominated ($v/\sigma \sim 2$) at $1 - 2R_e$, similar to disk-dominated S0 galaxies [32].

Most of the results on dark matter halos in ellipticals obtained from ALS data to $2R_e$ are based on spherical models for round ellipticals. [4,31] analyzed the line-profile shapes of a sample of 21 mostly luminous, slowly rotating, and nearly round elliptical galaxies in a uniform way, using spherical DF models. Intrinsic deviations from sphericity and embedded, near-face-on disks can play a role in only a small number of these bright galaxies. Nevertheless, a similar study using three-integral axisymmetric models of near-edge-on galaxies avoiding these issues will be very worthwhile. For these reasons the three-integral models of Matthias & Gerhard for the boxy E4 elliptical galaxy NGC 1600 provide some of the strongest evidence for radial anisotropy, because this galaxy must must be viewed nearly edge-on [27].

The sample of Kronawitter *et al.* [4] includes a subsample with mostly new extended kinematic data, reaching to $\sim 2R_e$, and a subsample based on the less extended older data of [28]. Based on these data and on photometry, non-parametric spherical models were constructed from which circular velocity curves, anisotropy profiles, and radial profiles of M/L were derived, including confidence ranges. The circular velocity curve (CVC) for test particles on circular orbits of varying radius is a convenient measure of the potential, even though luminous elliptical galaxies do not rotate rapidly. The CVCs of the elliptical galaxies analysed by [4,31] are flat to within $\simeq 10\%$ for $R \gtrsim 0.2R_e$ to $R \gtrsim 2R_e$, independent of luminosity (Fig. 2). This argues against strong luminosity segregation in the dark halo potential.

The dynamical models imply small to modest amounts of dark matter within $2R_e$ [4,31]. However, constant M/L models can be ruled out only for 7/21 ellipticals in this sample, at the 2σ level. There are ellipticals in this sample which are very well represented by constant M/L models, and no indication for dark matter within $2R_e$, and others where the best dynamical models result in local M/L_B's of 15-30 at $2R_e$. Likewise, Magorrian & Ballantyne [25] find evidence for additional dark matter only in a subset of their galaxies, using constant-anisotropy spherical modelling. Thus, despite the uniformly flat CVCs, there is a spread in the ratio of the CVCs from luminous and dark matter. As in spiral galaxies, the combined rotation curve of the luminous and dark matter is flatter than those for the individual components ("conspiracy").

In the models with maximum stellar mass, the dark matter contributes $\sim 10-40\%$ of the mass within R_e. The flat CVC models, when extrapolated beyond the range of kinematic data, predict equal interior mass of dark and luminous matter at $\sim 2-4R_e$, consistent with results from X-ray analyses. Even in maximum stellar mass models, the implied halo core densities and phase-space densities are at least ~ 25 times larger, and the halo core radii ~ 4 times smaller, than in maximum disk spiral galaxies with the same circular velocity [33]. This could imply that some elliptical galaxy halos collapsed at high redshifts or perhaps even that some of their dark matter might be baryonic.

Dust and Scattered Light? Baes & Dejonghe have argued that scattered light from a high-velocity nucleus might give rise to broad wings in the LOSVD at large radii, in such a way as to mimic a dark matter halo in dynamical analysis [34]. This requires an optical depth $\tau \simeq 1$, i.e., that elliptical galaxies are on the verge of opaqueness. No such effect on the surface density of background

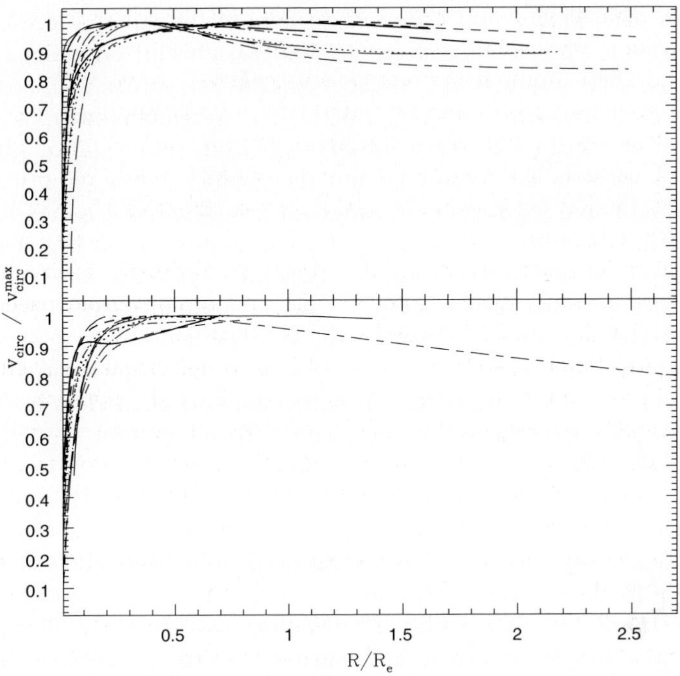

Fig. 2. "Best model" circular velocity curves of all galaxies from the sample of [4], plotted as a function of radius scaled by the effective radius R_e, and normalized by the maximum circular velocity. The upper panel shows the galaxies from the extended kinematics subsample, the lower panel the galaxies from the subsample with older data from [28]. (Figure from [31].)

galaxies is known to this author. A direct test is moreover possible when velocity measurements are available simultaneously from ALS data and from PNe, as the PN velocities are unaffected (e.g., NGC 4697 above), or when mass determinations are possible with independent techniques (e.g., NGC 1399 below). Also, the effect should be absent when the dispersion profile is constant; one such case in the sample of [4] is NGC 7626.

5.2 Results from Analyses Including PN or GC Velocities

Since the dark matter fraction inside $2R_e$ is still modest, and the orbit structure in the outer main bodies of ellipticals is not well-constrained by data that end at $2R_e$, it is important to include discrete velocity data from planetary nebulae (PN) or globular clusters (GC) that reach to larger radii.

The results so far are mixed. PNe velocity dispersions in several intermediate luminosity ellipticals were found to decline with radius, and to require little if any dark matter at $2 - 5R_e$ [2,15]. Two of these (NGC 3379, 4494) were also part of the sample of [4,31], but were consistent with constant M/L also in that study. NGC 3379 has been argued to be a face-on S0 and a weakly triaxial system viewed face-on [35]. This would not seem to be a likely explanation for several galaxies at the same time, unless selecting round galaxies introduces a significant bias in this sense. It is comforting that the near-edge-on, similar luminosity NGC 4697 also has little evidence for dark matter [2]. See Romanowsky's paper for further discussion [19]. In Cen A, the observed PN kinematics do require a dark halo, consistent with the GC velocities, but the implied M/L_B is noticeably low [18, 36] when compared with values determined from the hot gas in X-ray bright ellipticals. A similar, mild gradient in M/L_B was inferred from PN velocities in NGC 1316 [14]. In several further elliptical galaxies, globular cluster velocities were used to estimate host galaxy masses, and were found to increase outwards to several effective radii, but no M/L_B values were given [37].

Two interesting cases are the central galaxies of the Fornax and Virgo clusters, NGC 1399 and M87. In NGC 1399, the dynamical models for the ALS kinematics, predicting significantly increased M/L_B already at $1-2R_e$ [4], the PN and GC velocities [5,18,38], and the ROSAT and ASCA X-ray data [39] imply a steady outward rise in M/L_B. The PN and GC velocities are in the right radial range to allow a study of the transition between the potential of the central NGC 1399 galaxy and the potential of the Fornax cluster. Similarly, from a study of the stellar kinematics and the GC velocities around M87, and a comparison to the X-ray mass profile, Romanowsky & Kochanek [5, see also 40] find a rising circular velocity curve, and suggest that the potential of the Virgo cluster may already dominate at $r \sim 20\,\mathrm{kpc}$ from the center of M87. The dark matter distribution inferred by them is in agreement with that inferred from the X-ray data [6].

5.3 Results from Gas Rotation Measurements

Generally, elliptical galaxies do not contain cold gas disks or rings that can be used to measure their halo circular velocities, but there are some notable exceptions. One such example is the early-type galaxy IC 2006 in the Fornax cluster. This galaxy has a remarkably axisymmetric HI ring extending from $0.5-6R_e$, with circular velocity nearly constant in this range of radii. In this galaxy, M/L rises strongly with radius, signifying a nearly axisymmetric dark halo [41].

Other ellipticals with settled HI disks are described in [42]. These galaxies probably acquired the HI gas by accretion. The dynamical interpretation requires understanding the shape of the potential. In several of these ellipticals, M/L_B shows a substantial increase with radius, e.g., NGC 2974, NGC 3108, NGC 4278. In NGC 3108, again the rotation curve is flat, to $6R_e$, in agreement with the results from analysing stellar kinematics discussed above.

5.4 Results from Modelling X-Ray Data

The hot X-ray emitting gas in luminous ellipticals is another means to estimate the gravitational potential. This requires measuring the X-ray emissivity and temperature. Typical temperature profiles are nearly constant. Some problems are that point sources must be removed from the emission maps, requiring high spatial resolution, and that the assumed hydrostatic equilibrium may not be strictly fulfilled. However, any gas motions will have velocities of order the sound speed, so the error made with this assumption is not likely to be very large.

X-ray mass measurements have been made regularly as new X-ray missions were flown, but have tended to concentrate on a small number of galaxies. With the new high-spatial and spectral resolution and high-sensitivity data from Chandra and XMM this situation may soon change. In the detailed analyses so far, M/L ratios of ~ 50 at 10's of kpc have been found in, e.g., NGC 1399, M87, NGC 4472, NGC 4636, NGC 507, NGC 720 [6,39,43]; these galaxies are often the centers of galaxy clusters, however. Ubiquitous halos have furthermore been found in studies of the family properties of ellipticals from X-ray data [44].

5.5 Conclusion

In addition to the body of evidence reviewed here, studies of satellite dynamics from SDSS data indicate large $M/L \sim 100 - 200$ at radii $260h^{-1}$ kpc in $0.5 - 3L_*$ galaxies, and gravitational lensing studies indicate nearly isothermal mass profiles (constant CVCs) with a modest dark matter fraction inside $2R_e$ [e.g., 45]. Space does not permit a detailed discussion of these issues here.

Altogether the case is strong that, like all galaxies, elliptical galaxies have dark matter halos surrounding them, with the innermost circular velocity curves nearly flat. But there may be some intermediate luminosity objects in which the halos are more diffuse. Future work to understand this is clearly worthwhile.

References

1. M. Dopita, G.H. Jacoby, E. Vassiliadis: ApJ, **389**, 27 (1992)
2. R. Mendez *et al.*: ApJ, **563**, 135 (2001)
3. N. Douglas *et al.*: PASP, **114**, 1234 (2002) and this proceedings
4. R.P. Saglia *et al.*: AJ, **119**, 153 (2000); A. Kronawitter *et al.*: A&AS, **144**, 53 (2000)
5. M. Kissler-Patig *et al.*: AJ, **115**, 105 (1998); A.J. Romanowsky, C.S. Kochanek: ApJ, 553, 722 (2001)
6. D.A. Buote *et al.*: ApJ, **577**, 183 (2002); K. Matsushita *et al.*: A&A, **386**, 77 (2002)
7. G. Jacoby: ApJ, **339**, 39 (1989); R. Ciardullo *et al.*: ApJ, **339**, 53 (1989); R. Ciardullo: in *Stellar Candles for the Extragalactic Distance Scale*, ed. by D. Alloin, W. Gieren, Lect. Notes Phys., **635**, 243 (2003)
8. A. Renzini, A. Buzzoni: in *Spectral Evolution of Galaxies*, ed. by C. Chiosi, A. Renzini (Reidel, Dordrecht 1986), p. 195; A. Buzzoni, M. Arnaboldi: this proceedings
9. X. Hui *et al.*: ApJ, **414**, 463 (1993)
10. P. Marigo *et al.*: A&A, **423**, 995 (2004)
11. R. Ciardullo, G.H. Jacoby, H.C. Ford: ApJ, **344**, 715 (1989); G.H. Jacoby, R. Ciardullo, H.C. Ford: ApJ, **356**, 332 (1990); McMillan, R. Ciardullo, G.H. Jacoby: ApJ, **416**, 62 (1993)
12. P. Goudfrooij: A&AS, **104**, 179 (1994)
13. M. Arnaboldi *et al.*: ApJ, **472**, 145 (1996)

14. M. Arnaboldi *et al.*: ApJ, **507**, 759 (1998)
15. A. Romanowsky *et al.*: Science, **301**, 1696 (2003)
16. E. Emsellem *et al.*: MNRAS, **352**, 721 (2004)
17. D. Merritt: ApJ, **413**, 79 (1993)
18. X. Hui *et al.*: ApJ, **449**, 592 (1995); E.W. Peng, H.C. Ford, K.C. Freeman: ApJ, **602**, 685 (2004); M. Arnaboldi *et al.*: ESO Messenger, **76**, 40 (1994)
19. A. Romanowsky: this proceedings
20. A. Toomre: in *The Structure and Evolution of Normal Galaxies*, ed. by B. Tinsley, R. Larson, (Yale Univ. Obs., Newhaven, 1977), p. 401
21. O. Gerhard: MNRAS, **197**, 179 (1981); J. Negroponte, S.D.M. White: MNRAS, **205**, 1009 (1983); J. Barnes: ApJ, **331**, 699 (1988); ApJ, **393**, 484 (1992); L. Hernquist: ApJ, **400**, 460 (1992); J.S. Heyl, L. Hernquist, D.N. Spergel: ApJ, **463**, 69 (1996); T. Naab, A. Burkert: ApJ, **597**, 893 (2004)
22. J. Barnes: Nature, **338**, 132 (1989); M. Weil, L. Hernquist: ApJ, **460**, 101 (1996)
23. J. Barnes, G.P. Efstathiou: ApJ, **393**, 484 (1987); P.J. Quinn, W.H. Zurek: ApJ, **331**, 1 (1988); J.S. Bullock *et al.*: ApJ, **555**, 240 (2001); M. Vivitska *et al.*: ApJ, **581**, 799 (2002)
24. O.E. Gerhard: MNRAS, **265**, 213 (1993)
25. R. Bender: A&A, **229**, 441 (1990); J. Magorrian, D. Ballantyne: MNRAS, **322**, 702 (2001)
26. W. Dehnen, O.E. Gerhard: MNRAS, **261**, 311 (1993)
27. H.-W. Rix *et al.*: ApJ, **488**, 702 (1997); M. Matthias, O.E. Gerhard: MNRAS, **310**, 879 (1999); T.S: Statler, H. Dejonghe, T. Smecker-Hane: ApJ, **117**, 126 (1999); N. Cretton, H.-W. Rix, T. de Zeeuw: ApJ, **536**, 319 (2000); K. Gebhardt *et al.*: AJ, **119**, 1157 (2000); M. Cappellari *et al.*: ApJ, **578**, 787 (2002); K. Gebhardt *et al.*: ApJ, **583**, 92 (2003)
28. R. Bender, R.P. Saglia, O. Gerhard: MNRAS, **269**, 785 (1994)
29. E. Verolme *et al.*: in Galaxies and Chaos, ed. by G. Contopoulos, N. Voglis, Lecture Notes in Physics, **626**, 279 (2003)
30. D. Syer, S.D. Tremaine: MNRAS, **282**, 223 (1996); N. Bissantz, V.P. Debattista, O.E. Gerhard: ApJ, **601**, L155 (2004); N. Sambhus, F. de Lorenzi, O.E. Gerhard: this proceedings; F. de Lorenzi, V.P. Debattista, O.E. Gerhard: this proceedings
31. O.E. Gerhard *et al.*: AJ, **121**, 1936 (2001)
32. H.-W. Rix, M. Carollo, K.C. Freeman: ApJ, **513**, L25 (1999)
33. M. Persic, P. Salucci, F. Stel: MNRAS, **281**, 27 (1996)
34. M. Baes, H. Dejonghe: ApJ, **563**, L19 (2001); MNRAS, **335**, 441 (2002)
35. M. Capaccioli *et al.*: ApJ, **371**, 535 (1991); T.S. Statler, T. Smecker-Hane: AJ, **117**, 839 (1999)
36. A. Matthieu, H. Dejonghe, X. Hui: A&A, **309**, 30 (1996)
37. T.H. Puzia *et al.*: A&A, **415**, 123 (2004)
38. T. Richtler *et al.*: AJ, **127**, 2094 (2004)
39. Y. Ikebe *et al.*: Nature, **379**, 427 (1996); M. Paolillo *et al.*: ApJ, **565**, 883 (2002)
40. P. Côté *et al.*: ApJ, **559**, 828 (2001)
41. M. Franx, J.H. van Gorkom, T. de Zeeuw: ApJ, **436**, 642 (1994)
42. F. Bertola *et al.*: ApJ, **416**, L45 (1993); T.A. Oosterloo, R. Morganti, E.M. Sadler: AJ, **123**, 729 (2002)
43. H. Awaki *et al.*: PASJ, **46**, L65 (1994); K. Matsushita *et al.*: ApJ, **499**, L13 (1998); M. Paolillo *et al.*: ApJ, **586**, 850 (2003); D. Buote: ApJ, **577**, 183 (2002)
44. M. Loewenstein, R.E. White: ApJ, **518**, 50 (1999); S. Sato *et al.*: ApJ, **537**, L73 (2000)
45. T.A. McKay *et al.*: ApJ, **571**, L85 (2002); D. Rusin, C.S. Kochanek, C.R. Keeton: ApJ, **595**, 29 (2003)

Dynamics of Rotating Elliptical Galaxies

Flavio De Lorenzi[1], Victor P. Debattista[2], and Ortwin E. Gerhard[1]

[1] Astronomisches Institut, Universität Basel, CH-4102, Switzerland
[2] Institut für Astronomie, ETH Zürich, CH-8093, Zürich, Switzerland

Abstract. We present preliminary results from an ongoing study of rotation in triaxial elliptical galaxies. Using a newly developed adaptive particle method, we first construct a model of a rotating triaxial galaxy. We then attempt to build a non-rotating model with the same density and projected velocity and dispersion maps. Finally we compare the projected kinematic data to see whether we can distinguish between the rotating and non-rotating models.

1 Introduction

Some elliptical galaxies (e.g. NGC 4697 [1,2], and NGC 5128, [3,4]) show a clear rotation in their stellar and outer planetary nebulae kinematics. From a theoretical point of view it is unclear what the source of this rotation is, because we cannot easily recognize whether such rotation is entirely due to internal streaming of stars, or also includes a figure rotation of the galaxy. We wish to investigate whether a clear signature of figure rotation exists. To tackle this problem we construct a rotating triaxial Hernquist particle model, compute its two dimensional projected velocity and dispersion fields, and then attempt to build a non-rotating model which exhibits the same kinematics.

2 Method

The particle models are built using a new three-dimensional adaptation of the code described in [5] which incorporates a made-to-measure algorithm (M2M) invented in [6]. The main idea of the M2M algorithm is to assign each particle an individually adaptable weight, which is adjusted to minimize the deviations of a model from target observables (e.g. projected velocities) while the particles move in the gravitational potential generated by the whole system. The observables y_{j} of a system with N particles having weights w_{i} and phase-space positions $\boldsymbol{z}_{\mathrm{i}}$ ($\mathrm{i} \leq \mathrm{N}$) are

$$y_{\mathrm{j}}(t) = \sum_{\mathrm{i}=1}^{\mathrm{N}} w_{\mathrm{i}} K_{\mathrm{j}} \left[\boldsymbol{z}_{\mathrm{i}}(t)\right], \tag{1}$$

where the function K_{j} is a kernel. The heart of the algorithm is the prescription for changing the weights and is given by the "force of change"

$$\frac{\mathrm{d}w_{\mathrm{i}}(t)}{\mathrm{d}t} = -\varepsilon w_{\mathrm{i}}(t) \sum_{\mathrm{j}} \frac{K_{\mathrm{j}}\left[\boldsymbol{z}_{\mathrm{i}}(t)\right]}{Z_{\mathrm{j}}} \Delta_{\mathrm{j}}(t) \tag{2}$$

where $\Delta_j = y_j(t)/Y_j - 1$, ε is small and positive, and $\boldsymbol{Z}$ is an arbitrary constant usually used for normalization. The Y_j are a given set of target observables. The algorithm also includes temporal smoothing and some profit function to remove the ill-conditioning in the solutions of (2) when the number of particles exceeds the number of observables.

3 Models

First we generated a truncated spherical Hernquist particle model [7] of unit mass from a distribution function, using the methods described in [8]. The model consists 250000 particles. We use units in which the scale-length of the model, $a = 1$; then the effective radius is $R_{\rm eff} \approx 1.42$. Using the M2M algorithm, we deform the spherical particle distribution to a triaxial one having a density

$$\varrho(s) \propto \frac{1}{s(s+a)^3}, \quad s^2 = x^2 + (\frac{y}{y_0})^2 + (\frac{z}{z_0})^2, \tag{3}$$

where $y_0 = 0.6$ and $z_0 = 0.8$. The targets used are the coefficients of the spherical harmonic decomposition of (3), binned on radial shells and normalized: We consider concentric shells at a set of radii $r_{\rm k}$ with $\rm k = 1, \cdots, n$ and $r_{\rm k} < r_{\rm k+1}$. The spherical harmonic decomposition of a density $\varrho(r, \vartheta, \varphi)$ is given by

$$\varrho(r, \vartheta, \varphi) = \sum_{\rm lm} A_{\rm lm}(r) Y_{\rm l}^{\rm m}(\vartheta, \varphi). \tag{4}$$

We define the contribution of shell $r_{\rm k-1}$ to shell $r_{\rm k}$ as

$$A_{\rm lm}^{-}(k) = \int_{r_{\rm k-1}}^{r_{\rm k}} A_{\rm lm}(r) \left(1 - \frac{r_{\rm k} - r}{r_{\rm k} - r_{\rm k-1}}\right) r^2 {\rm d}r. \tag{5}$$

A similar expression for $A_{\rm lm}^{+}(k)$, the contribution of shell $r_{\rm k+1}$ to shell $r_{\rm k}$, can easily be found. The $A_{\rm lm}^{\pm}(k)$ are obtained by adding up the contributions of the adjacent shells

$$A_{\rm lm}^{\pm}(k) = A_{\rm lm}^{-}(k) + A_{\rm lm}^{+}(k). \tag{6}$$

The innermost (outermost) shell is an exception because only the shell immediately exterior (interior) contributes. If we define $M_r(k) = 4\pi A_{00}^{\pm}(k)$ then $\sum_{\rm k} M_r(k)$ equals the mass enclosed within r_0 and $r_{\rm n}$. For $\rm l > 0$ we finally introduce the normalized $A_{\rm lm}(k)$

$$A_{\rm lm}(k) = \frac{A_{\rm lm}^{\pm}(k)}{M_r(k)}. \tag{7}$$

We then spun up the triaxial model by evolving it in a rotating frame such that the corotation radius, $R_{\rm cor} \approx 3.4 R_{\rm eff}$, while retaining its shape again using the M2M code. Once we had this rotating triaxial model, we computed its projected velocity and dispersion maps with respect to the rest-frame. Finally we built a non-rotating particle model with the same density and projected kinematics as the rotating model, also using the M2M code.

4 Results

In the upper plot of Fig. 1 we show the radial mass binned on shells. The lower panel shows the amplitude of Y_2^2, A_{22}, binned the same way. The targets are shown as solid lines, whereas the observables of the rotating particle model are represented by dashed lines. The match between the target and the model observables is excellent.

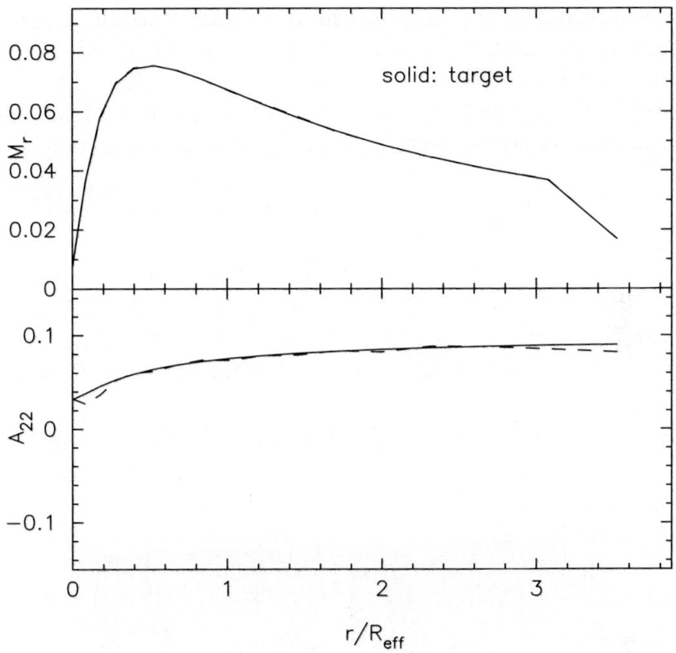

Fig. 1. Radial mass and A_{22} component in radial shells.

The left plots of Fig. 2 show the projected velocities (upper plot) and dispersions (lower plot) along the major axis. The right ones show the same quantities but along a slit parallel to the major axis offset from the center by $1.53R_{\rm eff}$. The model observables fit the targets quite well for radii $R \lesssim R_{\rm eff}$, but start to differ at $R \gtrsim R_{\rm eff}$.

Fig. 3 shows the projected velocity and dispersion maps for the rotating (upper maps) and non-rotating models (lower maps). The rotating model shows a cylindrical rotation in its velocity map. It seems that inside a circle of radius $R_{\rm eff}$ this rotation is well reproduced by the non-rotating model. Outside that region the velocity field of the non-rotating model departs from a cylindrical rotation, showing a drop in velocity with increasing height. The velocity dispersion fields differ in the outer regions: the contours of the non-rotating model are more elongated along the major axis.

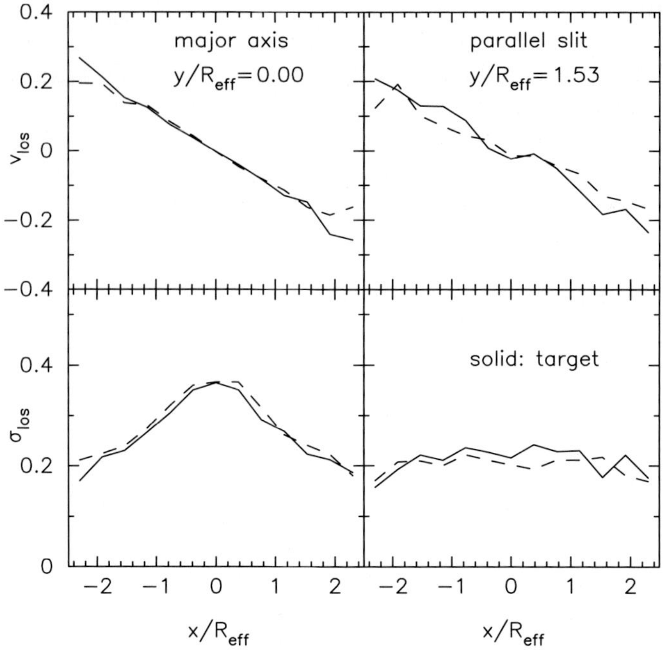

Fig. 2. Projected velocities and dispersions along the major axis and a parallel slit.

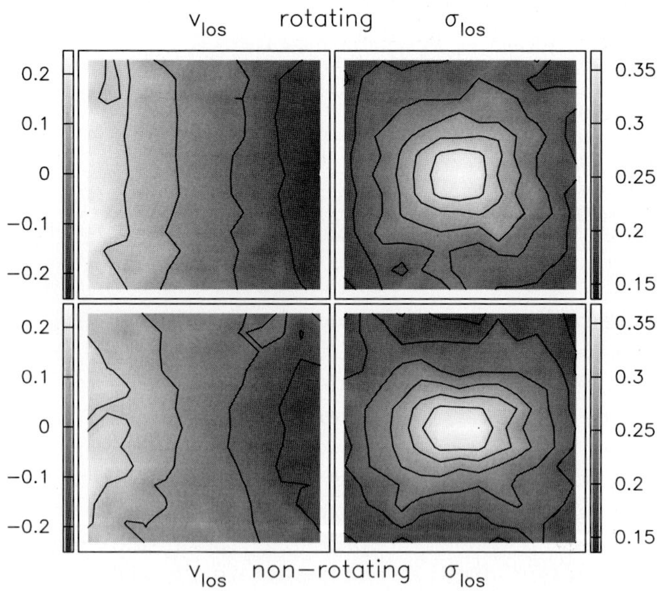

Fig. 3. Maps of the projected velocity and velocity dispersion for the rotating model (upper panels) and the non-rotating model (lower panels). A single panel extends approximately $-2.25R_{\rm eff}$ to $2.25R_{\rm eff}$ along each direction.

5 Conclusions

The projected velocity and dispersion fields of the rotating triaxial system are well reproduced by the stationary model, inside a circle of radius $R_{\rm eff}$. Outside that region, $R \gtrsim R_{\rm eff}$, the velocity field of the non-rotating model starts to depart from the linear, cylindrical rotation of the tumbling triaxial system. Likewise the dispersion fields differ in the outer regions in the sense that the contours of the non-rotating model are more elongated along the major axis than in the rotating case. This is for a corotation radius of $3.4R_{\rm eff}$. We expect a slower rotating triaxial galaxy, or a system with a non-linear projected rotation curve to become even more difficult to distinguish from a non-rotating galaxy.

It is possible that enough curvature in the projected velocity field of a triaxial elliptical galaxy may be a signature of internal stellar streaming. In future work we will investigate this further.

References

1. J. J. Binney, R. L. Davies, G. D. Illingworth: ApJ **361**, 78–97 (1990)
2. R. H. Méndez, A. Riffeser, R.-P. Kudritzki, M. Matthias, K. C. Freeman, M. Arnaboldi, M. Capaccioli, O. E. Gerhard: ApJ **563**, 135–150 (2001)
3. X. Hui, H. C. Ford, K. C. Freeman, M. A. Dopita: ApJ **449**, 692–615 (1995)
4. E. Peng, H. C. Ford, K. C. Freeman: ApJ **602**, 685–704 (2004)
5. N. Bissantz, V. P. Debattista, O. Gerhard: ApJ **601**, L155–L158 (2004)
6. D. Syer, S. Tremaine: Mon. Not. R. Astron. Soc. **282**, 223–233 (1996)
7. L. Hernquist: ApJ **356**, 359–364 (1990)
8. V. P. Debattista, J. A. Sellwood: ApJ **543**, 704–721 (2000)

Dynamics of the NGC 4636 Globular Cluster System

Ylva Schuberth[1], Tom Richtler[2], Boris Dirsch[2], Michael Hilker[1], Leopoldo Infante[3], Søren Larsen[4], and Markus Kissler-Patig[4]

[1] Sternwarte der Universität Bonn, Auf dem Hügel 71, D-53121 Bonn, Germany
[2] Universidad de Concepción, Departemento de Física, Casilla 160-C, Concepción, Chile
[3] Departamento de Astronomía y Astrofísica, P. Universidad Católica, Casilla 104, Santiago 22, Chile
[4] European Southern Observatory, Karl-Schwarzschild-Str. 2, D-85748 Garching, Germany

1 Introduction

NGC 4636, a giant elliptical in the Virgo Southern Extension at a distance of 15 Mpc [10] has some outstanding properties. Its specific frequency of globular clusters is unusually high [4],[2]. Far-infrared emission points to a considerable amount of warm dust, perhaps supplied in a recent merger event [9]. Moreover, it exhibits a bright X-ray halo [7],[3] which extends out to a radius of 300 kpc. Both ASCA [7] and Chandra [6] data have been used to infer an extremely dark matter dominated total mass. However, other work based on Chandra observations [3] interpret the strong X-ray emission as being due to a transient period of a high cooling rate, rendering the derivation of the mass profile doubtful.
Here, we present the first dynamical study of the NGC 4636 globular cluster system (GCS). We use the radial velocities of 176 globular clusters (GCs) in the range 4–70 kpc, the overwhelming majority being within 30 kpc, to derive a mass profile by stellar dynamical means.

2 The Dataset

Several hundred medium resolution ($\sim 3\,\text{Å}$) spectra were acquired at the VLT with the FORS2 Mask Exchange Unit (MXU) during four nights in May 2002 (ESO program ID 69.B.-0366(B)). The velocities for 176 GCs could be measured with uncertainties of the order $30\,\mathrm{km\,s^{-1}}$. The spectroscopy is complemented by wide field photometric data [2] which provides the spatial distribution of the blue (metal-poor) and the red (metal-rich) GC populations.
Their radial number densities are rather similar. Yet, while the profile of the blue GCs follows a power-law with an exponent of -1.5, the distribution of the red clusters is best described by two power-laws with exponents of -0.8 and -1.7 in the ranges of $1'$ –$3'$ and $3'$ –$7'$, respectively ($1' = 4.4\,\mathrm{kpc}$). At a radial distance of $9'$ ($7'$) for the blue (red) clusters, the power-law index drops to about -5, a

behavior which has not been observed in any GCS before. This combined spectroscopic and photometric dataset was then used to investigate the dynamics of the NGC 4636 GCS.

To obtain the line-of-sight velocity dispersion and study its dependence on radius, we used the maximum-likelihood dispersion estimator presented by Pryor & Meylan [8]. For the blue clusters, for which we did not detect any sign of rotation, we find a constant projected velocity dispersion of $200 \pm 15\,\mathrm{km\,s^{-1}}$ within a radial interval of 9 to 26 kpc. The red GCs show a completely different behavior. We observe a drop in the velocity dispersion from about 200 to about $170\,\mathrm{km\,s^{-1}}$ at $3'$. Moreover, rotation about the minor axis with an amplitude of $87 \pm 18\,\mathrm{km\,s^{-1}}$ was found, which together with the steepening of the number density profile might explain the drop in the velocity dispersion. Hence, only the blue GCs are considered for the mass models.

3 Mass Models

We assumed spherical symmetry and an isotropic velocity distribution. The non-rotating spherical Jeans equation (see [1]) was then used to infer the (constant) circular velocity from the power-law exponent and the line-of-sight velocity dispersion of the blue clusters: $v_{\mathrm{circ}} = 317 \pm 25\,\mathrm{km\,s^{-1}}$. Using the high stellar mass-to-light ratio from [5], scaled to the distance of 15 Mpc and converted to R-magnitudes, i.e. $\mathrm{M/L_R} = 7$, we find that a NFW dark matter halo with a dark matter fraction of 14 % at 10 kpc is sufficient to reproduce the circular velocity. Using a lower stellar mass-to-light ratio of $\mathrm{M/L_R} = 4.3$, requires a more massive dark component, with a dark matter fraction of almost 50 % at 10 kpc. In this case, however, the circular velocity of the galaxy itself [5] cannot be reproduced. We conclude that the NGC 4636 dark halo is not as massive as claimed in previous studies.

References

1. J. Binney, S. Tremaine: *Galactic Dynamics*, (Princeton University Press, Princeton, NJ 1987)
2. B. Dirsch, Y. Schuberth, T. Richtler: A&A, *submitted*
3. C. Jones, W. Forman, A. Vikhlinin, M. Markevitch, L. David, A. Warmflash, S. Murray, P. E. J. Nulsen: ApJ **567**, L115 (2002)
4. M. Kissler, T. Richtler, E. V. Held, E. K. Grebel, S. J. Wagner, M. Capaccioli: A&A **287**, 463 (1994)
5. A. Kronawitter, R. P. Saglia, O. Gerhard, R. Bender: A&AS **144**, 53 (2000)
6. M. Loewenstein, R. Mushotzky: Nuclear Physics B Proc. Suppl. **124**, 91
7. K. Matsushita, K. Makishima, Y. Ikebe, E. Rokutanda, N. Yamasaki, T. Ohashi: ApJ **499**, L13 (1998)
8. C. Pryor, G. Meylan: 'Velocity Dispersions for Galactic Globular Clusters'. In: *ASP Conf. Ser. 50: Structure and Dynamics of Globular Clusters* (1993) pp. 357–371
9. P. Temi, W. G. Mathews, F. Brighenti, J. D. Bregman: ApJ **585**, L121 (2003)
10. J. L. Tonry, A. Dressler, J. P. Blakeslee, E. A. Ajhar, A. B. Fletcher, G. A. Luppino, M. R. Metzger,C. B. Moore: ApJ **546**, 681 (2001)

Dynamics of the Elliptical Galaxy NGC 4697 from Integrated Light and PNe

Niranjan Sambhus, Flavio De Lorenzi, and Ortwin Gerhard

Astronomical Institute, University of Basel, Switzerland

Abstract. We present results from our ongoing calculations pertaining to the dynamics of the E6 galaxy NGC 4697. We construct a (self-consistent) N-body model of the galaxy by employing the deprojected photometric data and absorption line kinematics. Assuming that the PNe in the vicinity of NGC 4697 are drawn from the stellar distribution, we can use them as kinematic probes that trace the underlying mass distribution particularly in the outer regions of the galaxy. We construct a smoothed, non-parametric velocity field from the PNe radial velocities, and compare with the kinematic predictions of our model.

1 Introduction

The presence of Dark Matter (DM) has long been inferred around spiral galaxies from their flat rotation curves [1]. The situation has not been so clear for early type galaxies where a lack of bright mass tracers in their outer regions has limited the detection of DM to within $1-2\,R_{\rm eff}$. ΛCDM models for galaxy formation predict massive, extended halos around ellipticals. Mass estimates of ellipticals have been limited in the past to a few bright galaxies whose gravitational potential can be measured using gravitational lensing, satellite dynamics, or X-ray emission. Only in the last few years have astronomers been able to employ alternative probes such as Globular Clusters (GC) and Planetary Nebulae (PNe) to map the mass and angular momentum distribution in halos of ordinary ellipticals. GC's and PNe are individual bright objects detectable to large radii. Moreover, the PNe population is expected to arise from the underlying galactic stellar distribution. Large PNe surveys of individual galaxies have been reported in the recent literature [2–4].

One such galaxy is NGC 4697, a normal, oblate axisymmetric E6 galaxy ($R_{\rm eff}$ = 95") located along the direction of the Virgo southern extension. Previous X-ray emission studies [5] suggest a lack of substantial amount of DM in the halo of this galaxy. Radial velocities of 531 PNe extending up to 300" were measured by [4]. These authors could fit a spherical, isotropic Hernquist model with a constant B-band mass-to-light ratio ($\Upsilon_{\rm B} = 11$) to the PNe velocity dispersion data. This value was found to be consistent with a 10 Gyr stellar population with Salpeter mass function and slightly super solar metallicity. Thus they too concluded a lack of DM at least within $3\,R_{\rm eff}$. In the present analysis we employ integrated stellar light photometry and kinematics, along with the PNe radial velocities to study the halo dynamics of this galaxy. We deproject the galaxy surface brightness distribution to recover its luminosity density (j).

The mass density is obtained by multiplying j with an (as yet undetermined) Υ. Using an N-body code based on the Syer-Tremaine algorithm [6], and stellar line kinematic data as constraints, we construct a self-consistent galaxy model that reproduces the observed photometry. Line-of-sight (los) velocities predicted by the model are then compared with a smoothed, non-parametric velocity field derived from the observed PNe radial velocities.

2 Data and Analysis

R-band photometric observations of the galaxy using the Wide Field Imager at the ESO-MPIA 2.2m telescope were kindly provided to us by Roberto Saglia. The galaxy appears as an oblate axisymmetric elliptical, inclined almost edge-on. This makes it an easy candidate for deprojection. It is well-known that deprojection of less inclined ellipticals results in undetermined konus densities [7,8]. NGC 4697 does not suffer from this ill-condition. We use a code based on a Richardson-Lucy algorithm, described in [9] and improved by [10] to deproject the observed photometry. Figures 1(a) and (b) show the resulting luminosity density, deprojected assuming an exactly edge-on galaxy. Long-slit stellar absorption line spectroscopic observations of this galaxy were obtained by [9,11]; we use these kinematic observables to constrain our N-body models.

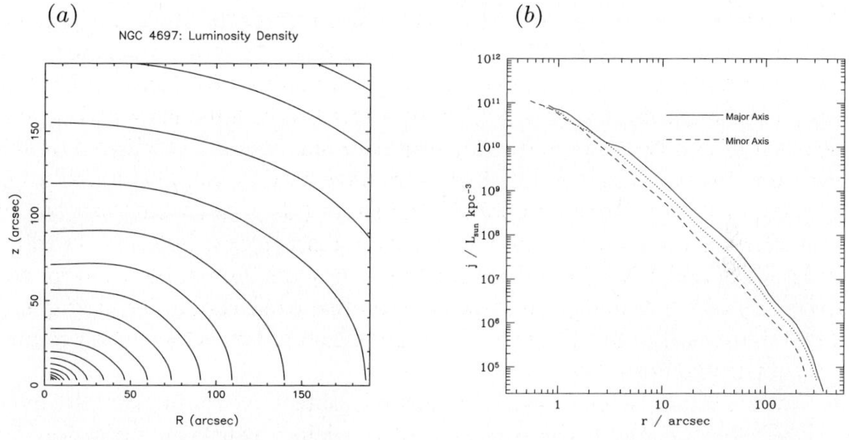

Fig. 1. NGC 4697 deprojected luminosity distribution: (a) Isocontours of density. (b) Radial cuts along the major, minor, and an intermediate axis.

Using apparent magnitudes m(5007) of 531 PNe, [4] determined the PN Luminosity Function (PNLF) for NGC 4697. They estimate a new distance of 10.5 ± 0.5 Mpc, contrary to the previous estimate of 24 Mpc, now placing NGC 4697 midway between the local group and the Virgo Cluster. A major difference between measured PNe emission line and stellar absorption line radial velocities is that the latter are obtained by modelling the observed los velocity profile, whereas each PN samples a single number from the (unobserved) underlying velocity distribution at their location. Hence some processing of the PNe data is needed to ensure a faithful deduction of the underlying kinematics. We use a thin plate spline interpolation routine that allows smoothing [12,13], to recover non-parametric mean velocity and velocity dispersion fields from the PNe data. Various tests were performed on the resulting velocity distribution to check its robustness. The interpolation routine was tested using fiducial data generated from random samplings of a parametric field similar to that of NGC 4697. The routine was constrained to recover a smooth distribution which was then compared with the underlying field from which the pseudo data were drawn. Further, we check for the Gaussianity of the difference between the PNe velocities and the recovered velocities. Monte Carlo tests were performed on the recovered velocities to find the statistical significance of the occurance of the data. The entire procedure was repeated for different levels of smoothing, and the best result (in the χ^2 sense) was accepted as a fair representation of the actual velocity field as traced by the PNe. Figure 2 shows the spatial distribution and measured velocities of 531 PNe, along with the recovered velocity and dispersion fields.

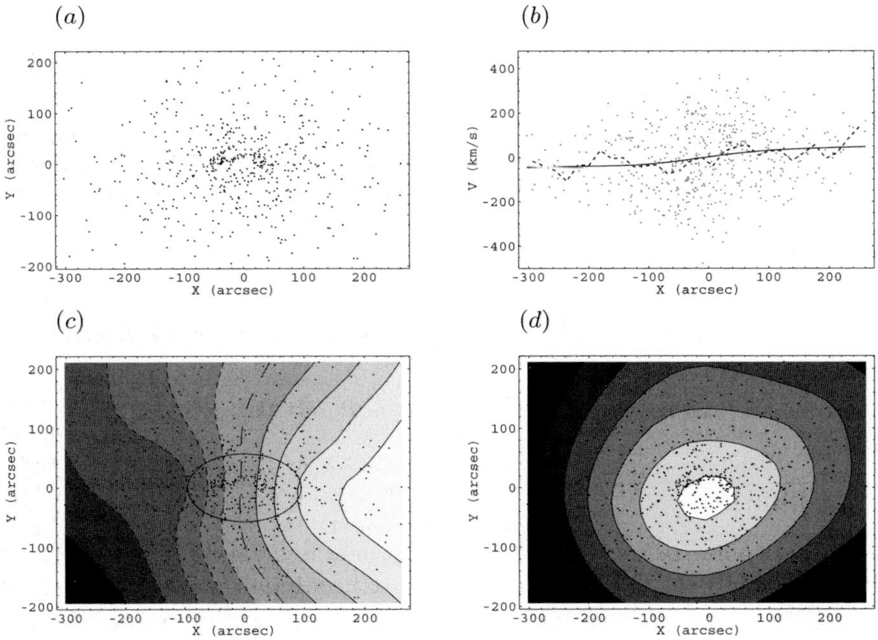

Fig. 2. NGC 4697 PNe velocity field: (a) Spatial distribution of PNe. (b) PNe velocities projected along the major axis. Dotted line is observed mean velocity binned along the major axis. Continuous line is the underlying smoothed, recovered mean velocity. Note a clear trend of rotation. (c) Recovered mean velocity field. The long dashed contour denotes the zero-velocity. Positive and negative velocities are denoted by continuous and thin dashed contours respectively. The contour spacing is $10\,\mathrm{km\,s^{-1}}$. The central ellipse corresponds to $1\,R_{\mathrm{eff}}$. (d) Recovered velocity dispersion field. Contours decline from a central value of $170\,\mathrm{km\,s^{-1}}$ and are spaced $10\,\mathrm{km\,s^{-1}}$ apart.

3 Model Construction

A N-body model of the galaxy is constructed using a three-dimensional version of a code that incorporates a made-to-measure algorithm (M2M) as described in [6,14]. Details of the code are described elsewhere in this volume (see De Lorenzi et al.). Starting from a spherical Hernquist model, the M2M code evolves 250, 000 particles so as to achieve the observed luminosity density distribution. A quantitative check of this convergence is provided in Fig 3, where radial mass distribution of the galaxy and the model are plotted. Except in the innermost region where the galaxy is known to have a stellar disk that we have not incorporated in our modelling, the match between the two is good.

The absorption line long-slit data consist of mean velocity and velocity dispersion along a few directions parallel and perpendicular to the galaxy major axis. Although these data exist only within the inner bright regions of the galaxy ($r \lesssim R_{\rm eff}$), they serve as a kinematic constraint for the model, as well as a handle to determine $\Upsilon_{\rm R}$. Results of our model predictions compared with the PNe kinematics are presented in the next section.

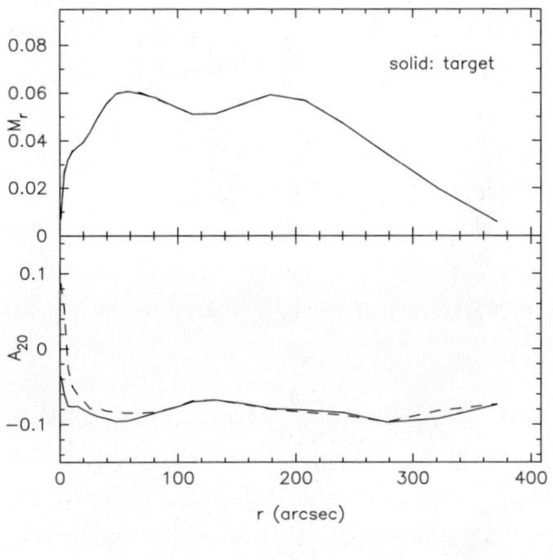

Fig. 3. NGC 4697 mass decomposition: Upper panel shows radial mass distribution (M_r), binned in spherical shells. Units are arbitrary. Lower panel shows the coefficient of Y_2^0, scaled by M_r. The solid lines correspond to the galaxy, and the dashed lines are results from the N-body model.

4 Results and Conclusion

In Figure 4 we show representative kinematic slit-cuts along major and minor axes. The match between the absorption line and model kinematics is excellent. Moreover, except for the major axis mean velocity, the PNe mean velocities and dispersions are in good agreement with our model values and stellar data. Similar comparisons of model and PNe kinematics with slit-data along other directions show an equally good fit. NGC 4697 is known to harbor a weak stellar disk inside $R_{\rm eff}$, lying along the equatorial plane of this galaxy [16,17]. Such a disk is not expected to produce large number of PNe. However, it is bound to affect stellar kinematics; in particular, increase the rotation velocity. Our present models do not include a stellar disk and we are working towards adding one.

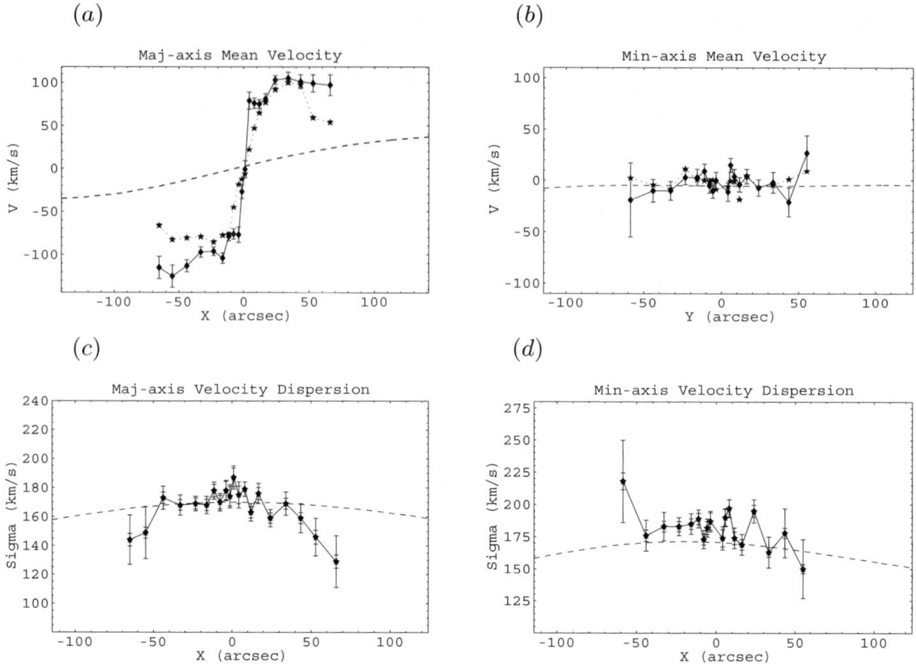

Fig. 4. NGC 4697 kinematic comparisons. Stellar long-slit data [9] is shown in continuous lines, the N-body model predictions are shown in small dotted lines, while the PNe velocities are drawn as long dashed lines. Except for mean velocity along the major axis, the model predictions are in good agreement with stellar and PNe kinematic data.

With only the luminous mass distribution as input, the M2M code is able to reproduce the observed kinematics, at least up to $3\,R_{\rm eff}$. This eliminates the need for having abundant DM in this galaxy, thus confirming earlier findings. Similar conclusion can be reached by noticing an apparent lack of large rotation velocities of PNe in the region beyond $1\,R_{\rm eff}$ (see Fig 2).

References

1. M. Persic, P. Salucci, F. Stel: Mon. Not. R. Astron. Soc. **281**, 27 (1996)
2. A. J. Romanowsky et al.: Science **301**, 1696 (2003)
3. E. Peng, H. C. Ford, K. C. Freeman: ApJ **602**, 685 (2004)
4. R. H. Méndez et al.: ApJ **563**, 135 (2001)
5. C. L. Sarazin, J. A. Irwin, J. N. Bregman: AJ **544**, L101 (2000)
6. D. Syer, S. Tremaine: Mon. Not. R. Astron. Soc. **282**, 223 (1996)
7. O. Gerhard, J. Binney: Mon. Not. R. Astron. Soc. **279**, 993 (1996)
8. C. Kochanek, G. B. Rybicki: Mon. Not. R. Astron. Soc. **280**, 1257 (1996)
9. J. Binney, R. L. Davis, G. D. Illingworth: ApJ **361**, 78 (1990)
10. W. Dehnen: Mon. Not. R. Astron. Soc. **274**, 919 (1995)

11. H. Dejonghe et al.: AA **306**, 363 (1996)
12. G. Wahba, J. Wendelberger: Monthly Weather Review **108**, 1122 (1980)
13. M. Arnaboldi et al.: ApJ **507**, 759 (1998)
14. N. Bissantz, V. P. Debattista, O. Gerhard: ApJ **601**, L155 (2004)
15. R. F. Peletier et al.: AJ **100**, 1091 (1990)
16. D. Carter: ApJ **312**, 514 (1987)
17. C. Scorza, R. Bender: AA **293**, 20 (1995)

Planetary Nebulae as Dynamical Tracers: Mass-to-Light-Ratio Gradients in Early-Type Galaxies

N.R. Napolitano[1], A.J. Romanowsky[2], M. Capaccioli[3], K. Kuijken[1,4], M.R. Merrifield[2], N.G. Douglas[1], M. Arnaboldi[5], K.C. Freeman[6], and O. Gerhard[7]

[1] Kapteyn Astronomical Institute, Groningen, The Netherlands
[2] School of Physics & Astronomy, University of Nottingham, UK
[3] INAF – Observatory of Capodimonte, Naples, Italy
[4] University of Leiden, Leiden, The Netherlands
[5] INAF – Observatory of Pino Torinese, Turin, Italy
[6] RSAA, Mt. Stromlo and Siding Spring Observatory, Canberra, Australia
[7] Astronomical Department, University of Basel, Basel, Switzerland

Abstract. Planetary Nebulae (PNe) have enabled mass-to-light ratios (M/L) in early-type galaxies to be constrained to unprecedented distance from the center, showing in some cases clear evidence of increasing M/L, in other cases fairly constant M/L. We combine the information obtained from PN kinematics with radially extended long-slit spectroscopy data in order to constrain the M/L trends in a heterogeneous sample of early-type galaxies. We discuss whether these trends are expected in the ΛCDM framework of galaxy formation.

1 Introduction

N-body cosmological simulations of galaxy formation predict that galaxies are surrounded by dark matter (DM) halos with a rather universal density profile [20,21] (NFW hereafter). The inner regions of galaxies, however, are dominated by the effects of baryon dissipation, which are only approximately allowed for in the semi-analytical or hydrodynamical cosmological simulations [23,27]. In ellipticals, observations of stellar kinematics from integrated-light spectra, mostly confined within $2R_e$ (R_e is the effective radius which contains half of the total galaxy light), show no strong indication for an increasing M/L [26,11]. At these distances from the galaxy centers, messy baryonic physics can still affect the dark matter distribution and cause its deviation from the behavior expected from N-body simulations. A clearer view on the dark-matter properties of early-type galaxies must come from their outskirts, where it should be easier to disentangle the luminous and dark matter mass contributions.
Different methods have been developed to study the DM distribution in the halo regions of ellipticals: X-ray emission from hot gas and gravitational lensing indeed provide constraints on the mass distribution at large galaxian radii and favor the occurrence of a significant amount of DM in the halos [15,14]. However, X-ray emission occurs mostly in the brightest objects, and the results may

be affected by this luminosity bias; on the other hand gravitational lensing is mostly observed in high density environments (cluster cores) and only relatively few events have been observed in isolated galaxies. Neither of these methods are optimal for systematic studies of DM properties in a selection-free sample of galaxies.

Planetary Nebulae (PNe) provide the natural probes to unveil the stellar kinematics in the outskirts of elliptical and S0 galaxies [17]. They have been used successfully to gauge the mass distribution in the outer regions of early-type galaxies, either confirming [13,1,8,18,24], or ruling out [16,25], the presence of a significant dark halo. Following a preliminary study [7], here we investigate the possible connections between the dark and luminous properties for a heterogeneous sample of early-type galaxies for which extended estimates of M/L were available in the literature (both from PNe discrete radial velocity fields and inner stellar kinematics – long slit spectroscopy). The DM distribution in these objects is characterized via the radial behavior of M/L: the outermost M/L estimates span a range from $\sim 2R_e$ to $15R_e$.

2 M/L Gradients and Correlation with Structural Parameters

Structural parameters (total luminosity, isophote shape, inner surface brightness slope) have been used to characterize the evolution of early-type galaxies [22,6,10]. Here we want to look for any correlation between such parameters and dark-to-luminous galaxies' properties, via M/L gradients.
Structural parameters are computed simply as $\Delta\Upsilon_B \Delta R^{-1} \Upsilon_*^{-1} = \Delta\Upsilon_{R,\Upsilon_*}$, where $\Delta R = (R_{\rm out} - R_{\rm in})/R_e$, $\Upsilon_B = M/L_B$[1] and $\Upsilon_* = M/L_{B,*}$ is the global mass-to-light ratio of the stellar component in B band (deduced from stellar population synthesis models) assumed to be constant with radius. This is a global estimate, which can easily be related to the radial behavior of the dark mass distribution via the equation:

$$\frac{\Delta\Upsilon}{\Delta R} = \frac{\Upsilon_{\rm out} - \Upsilon_{\rm in}}{R_{\rm out} - R_{\rm in}} = \frac{\Upsilon_*}{\Delta R}\left[\left(1+\frac{M_d}{M_*}\right)_{\rm out} - \left(1+\frac{M_d}{M_*}\right)_{\rm in}\right] = \frac{\Upsilon_*}{\Delta R}\left[\left(\frac{M_d}{M_*}\right)_{\rm out} - \left(\frac{M_d}{M_*}\right)_{\rm in}\right] \quad (1)$$

Thus we can define the *normalised* M/L *gradient* as

$$\frac{\Delta\Upsilon}{\Upsilon_* \, \Delta R} = \frac{1}{\Delta R}\left[\left(\frac{M_d}{M_*}\right)_{\rm out} - \left(\frac{M_d}{M_*}\right)_{\rm in}\right] \quad (2)$$

where M_* and M_d are respectively the luminous (stellar) and the dark masses enclosed within a certain radius. As M_d/M_* can be reasonably assumed to be a radially linear quantity [9], i.e. without strong local gradients, $\Delta\Upsilon_{R,\Upsilon_*}$ allows us to evaluate the relative trend in the radial M/L distribution among galaxies in our catalog independently of the radial coverage of the M/L estimates, under the assumption of constant Υ_*. From Eq. 2, the quantity $\Delta\Upsilon_{R,\Upsilon_*}$ is independent

[1] In the following, R and Υ_B will be given respectively in R_e and solar units.

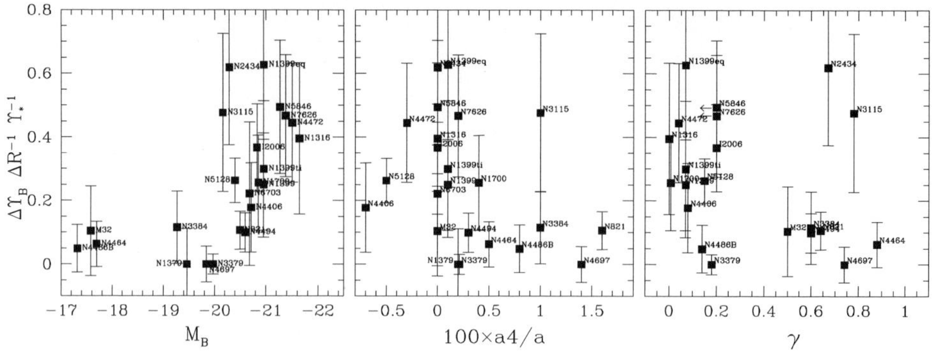

Fig. 1. The normalised M/L gradients are plotted as a function of the intrinsic parameters: the total luminosity, $M_{\rm B}$ (left); the isophotal shape, a_4 (center); and the central surface brightness slope, γ (right). Error bars are 1-σ level.

of Υ_*, and only relates to the dark-to-luminous matter properties of the galaxies. Galaxies with fairly constant M/L should have M/L gradients nearly equal to zero. Galaxies with increasing M/L due to extra mass residing in the halo, should show non-zero gradients, roughly increasing with the dark mass fraction.

The M/L gradients are plotted in Fig. 1 versus the total luminosity (from the LEDA catalogue, corrected for our adopted distance), the isophotal shape, a_4, and γ parameters. As already discussed in a series of conference papers [7,19], there is a general correlation of the $\Delta\Upsilon_{R,\Upsilon_*}$ values with the structural parameters, except for some outliers. These outliers are systems which possibly live in a peculiar environment (the center of the galaxy cluster NGC 1399, for instance) or they are thought to be interacting systems (NGC 2434). These environments could have affected their mass estimates (see [20] for a detailed discussion on NGC 1399).

Amongst others, the correlation of $\Delta\Upsilon_B\Delta R^{-1}\Upsilon_*^{-1}$ with the total luminosity is fairly convincing. Here we want to compare such a trend with that expected from ΛCDM.

3 How Do Observed M/L Gradients Compare with ΛCDM Predictions?

We want to verify whether this trend with luminosity is actually consistent with the expectations from the ΛCDM. In order to do that, we build spherical representations of early-type galaxy mass profiles, using a constant-M/L model for the stellar distribution plus a ΛCDM model of the dark halo [3].

We combine these components and derive the M/L predictions for various model parameters. From Eq. 2, it is easily understood that this quantity is strongly dependent on the ratio between the total dark mass in the halo, $M_{\rm d,tot}$,

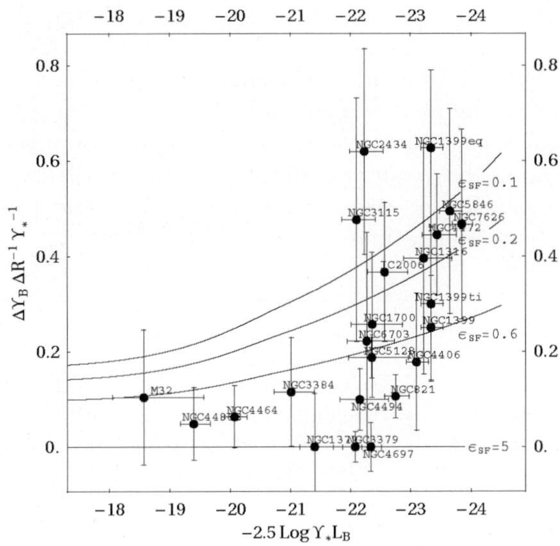

Fig. 2. Normalised M/L gradient versus logarithmic total stellar mass. Solid curves are predictions from ΛCDM parametrised with different stellar formation efficiencies. The observed M/L trends with luminosity (and then with $M_{*,\mathrm{tot}}$) are considered in the ΛCDM framework, but low luminosity systems seem to require an unrealistic star formation efficiency (see text, Sect.3).

and the total stellar mass , $M_{*,\mathrm{tot}}$. These can be related to the net star formation efficiency, ϵ_{SF} (including stellar mass loss):

$$f_{\mathrm{b}}^{-1} \approx \frac{M_{\mathrm{vir}}}{M_{\mathrm{b}}} = 1 + \frac{\epsilon_{\mathrm{SF}}\ M_{\mathrm{d,tot}}}{M_{*,\mathrm{tot}}}, \tag{3}$$

where M_{b} is the baryonic mass, f_{b} =0.17 [2] is the cosmic baryon fraction, and we used the relation $M_{\mathrm{vir}} = M_{\mathrm{b}} + M_{\mathrm{d,tot}}$ and $M_{*,\mathrm{tot}} = \epsilon_{\mathrm{SF}} M_{\mathrm{b}}$. In this way we can parametrise the modeled M/L with the star formation efficiency. The latter does not vary arbitrarily in galaxies and recent estimates place its limit between 0.1 and 0.6 in early type galaxies. In Fig. 2, the observed and modeled gradients are shown as a function of the logarithmic total stellar mass ($LSM = -2.5 \mathrm{Log} M_{*,\mathrm{tot}}$, where $M_{*,\mathrm{tot}} = \Upsilon_* \times L_{\mathrm{B},\odot}$ for the galaxy sample). There seems to be a clear dichotomy: galaxies with $LSM > -22.5$ (corresponding to $M_{*,\mathrm{tot}} \sim 1.6 \times 10^{11} M_\odot$ and $M_{\mathrm{B}} \sim -20.5$) are generally consistent with $\epsilon_{\mathrm{SF}} > 0.6$ (which violates the observed baryon budget), while galaxies with $LSM < -22.5$ (i.e. $M_{*,\mathrm{tot}} > 1.6 \times 10^{11} M_\odot$) are all consistent with ϵ_{SF} between 0.1 and 0.6. Such a dichotomy is reminiscent of previous evidence found with respect to galaxy structural properties [22,10,7,19,4,5]. It is particularly intriguing that this dichotomy manifests itself at the same luminosity, $M_{\mathrm{B}} \sim -20.5$, (and mass) scale in both dark matter and luminous properties [12]. This possibly suggests a connection between dark halo properties and galaxy formation mechanisms.

More investigations are needed in order to solve the conflict raised by the intermediate/low luminosity systems (see Romanowsky, these proceedings). This conflict can be alleviated only assuming halo concentration parameters, $c_{\rm vir}$[2] $= 7 \pm 2$, i.e. well below the expected concentration values for DM halos [3]. Here the critical ingredients could be the treatment of baryon physics and feedback processes which could have affected the dark matter distribution. This would explain the discrepancies between observed properties and those predicted by "collisionless" N-body simulations.

References

1. Arnaboldi, M., Freeman, K. C., Gerhard, O., et al.: ApJ **507**, 759 (1998)
2. Bennett, C.L., Halpern, M., Hinshaw, G., et al.: ApJs **148**, 1 (2003)
3. Bullock, J.S., Kolatt, T. S., Sigad, Y., et al.: MNRAS **321**, 559 (2001)
4. Capaccioli, M. & Caon, N.: MNRAS **248**, 523 (1991)
5. Capaccioli, M., Caon, N., D'Onofrio, M., & Trevisani, S.: Memorie della Societa Astronomica Italiana **63**, 509 (1992)
6. Capaccioli, M., Caon, N. & D'Onofrio, M.: *ESO ESP/EIPC Workshop on Structure, Dynamics and Chemical Evolution of Early-type Galaxies*, Isola d'Elba , eds. J. Danziger, W. W. Zeilinger, and K. Kjar, ESO: Garching, **43** (1991)
7. Capaccioli, M., Napolitano, N. R., Arnaboldi, M.: *Sakharov Conference of Physics*, Moskow, June 2002, [preprint:astro-ph/0211323] (2003)
8. Douglas, N., Gerssen, J., Kuijken, K., & Merrifield, M. R.: MNRAS **316**, 795 (2000)
9. Dubinski, J.: ApJ **502**, 141 (1998)
10. Faber, S.M., Tremaine, S., Ajhar, E.A., et al.: AJ **114**, 1771 (1997)
11. Gerhard O., Kronawitter, A., Saglia, R. P., & Bender, R.: AJ **121**, 1936 (2001)
12. Graham, A.W., Erwin, P., Trujillo, I., & Asensio Ramos, A.: AJ **125**, 2951 (2003)
13. Hui, X., Ford, H., Freeman, K.C., & Dopita, M.A.: ApJ **449**, 5 (1995)
14. Keeton, C.R.: ApJ **561**, 46 (2001)
15. Loewenstein, M., & White, R.E.: ApJ **518**, 50 (1999)
16. Méndez, R.H., Riffeser, A., Kudritzki, R.P., et al.: ApJ **563**, 135 (2001)
17. Napolitano, N.R., Arnaboldi, M., Freeman, K.C., & Capaccioli, M.: A&A **377**, 784 (2001)
18. Napolitano, N.R., Arnaboldi, M., & Capaccioli, M.: A&A **383**, 791 (2002)
19. Napolitano, N.R., Capaccioli, M., Arnaboldi, M., et al.: IAUS **220**, 93 [preprint astro-ph/0310798] (2003)
20. Navarro, J.F., Frenk, C.S., & White, S.D.: ApJ **462**, 563 (1996)
21. Navarro, J.F., Frenk, C.S., & White, S.D.: ApJ **490**, 493 (1997)
22. Nieto, J.L., & Bender, R.: A&A **215**, 266 (1989)
23. Pearce, F.R., Jenkins, A., Frenk, C.S., White, S.D.M.: MNRAS **326**, 649 (2001)
24. Peng, E.W., Ford, H.C., & Freeman, K.C.: ApJ **602**, 685 (2004)
25. Romanowsky, A.J., Douglas, N.G., Arnaboldi, M., et al.: Science **301**, 1696 (2003)
26. Saglia, R.P., et al.: ApJ **403**, 567 (1993)
27. Springel, V., White, S.D.M., Tormen, G., Kauffmann, G.: MNRAS **328**, 726 (2001)

[2] The halo concentration parameter is defined as $c_{\rm vir} = R_{\rm vir}/r_{\rm s}$, where $R_{\rm vir}$ is the virial radius and $r_{\rm s}$ is a scale length [3].

Part VIII

PNe in the Intracluster Environment

Intracluster Planetary Nebulae as Probes of Intracluster Starlight

John Feldmeier

Case Western Reserve University, Cleveland OH, 44106, USA

Abstract. I review the progress in research on Intracluster Planetary Nebulae (IPN). Hundreds of IPN candidates have now been found in the Virgo and Fornax galaxy clusters, and searches of two nearby galaxy groups have been made. From the results thus far, approximately 10–20% of all stars in Virgo and Fornax are in an intracluster component, but there are few such stars in galaxy groups. From the spatial distribution of IPN, it appears that the intracluster stars are clustered, in agreement with tidal-stripping scenarios. In Virgo, the IPN have a large line-of-sight depth, which implies that the bulk of intracluster stars in this cluster derive from late-type galaxies and dwarfs. I also discuss other important developments in IPN research such as the detection of intracluster H II regions, a possible detection of IPN in the Coma Cluster, and future observational and theoretical developments.

1 Introduction

Why do planetary nebulae (PN), which are a late phase of stellar evolution, make excellent tracers of intracluster stars, stars between the galaxies inside a galaxy cluster? An example of the utility of PN under similar circumstances can be seen in Figure 1. On the left is a standard broad-band image of the famous interacting galaxy pair M 51. Although there are numerous signs of the interaction at high surface brightnesses, there are many lower surface-brightness features [6] that are barely visible in this representation.

However, the distribution of PN found in M51 [21] displayed on the right gives additional information on this interacting system. There is a clear tidal tail structure to the west of the secondary galaxy, and there are also signs of an extension of the spiral arm to the south. The lack of PN in the central regions of the galaxies is due to crowding effects. What this illustration reminds us is that *planetary nebulae trace stellar luminosity, not stellar surface brightness.* Because PN are an end-phase of stellar evolution for most stars between 1–8 solar masses, the distribution of planetary nebulae closely follows the distribution of stars in galaxies [10]. If there is enough luminosity in a stellar population for sufficient numbers of PN to be present, then we can detect and study that population, regardless of its surface brightness. Intracluster stars, which are believed to be quite luminous (anywhere between 10% and 70% of the cluster's total stellar luminosity; [52,42]), but have a very low surface brightness (less than 1% of the night sky in the optical bands; [31,23]) are therefore an ideal target for PN searches.

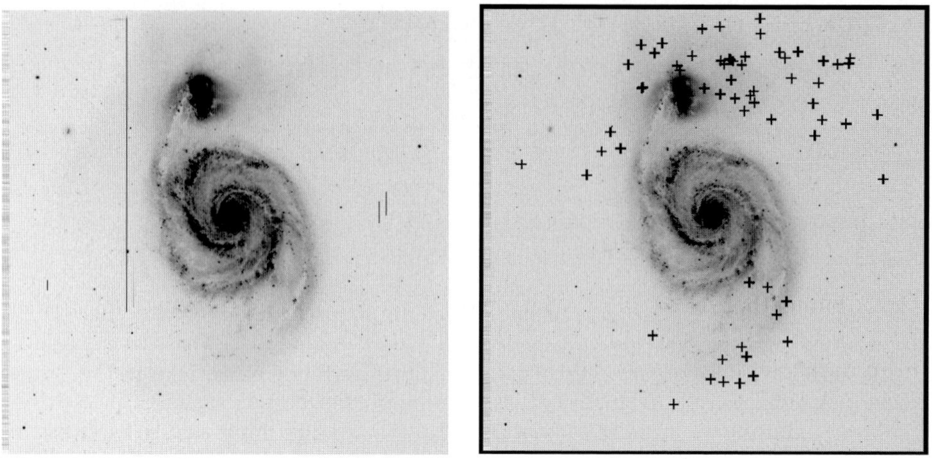

Fig. 1. Two images of the interacting galaxy pair M51. On the left is a broad-band image taken with the KPNO 4-m telescope showing the classical high-surface brightness features of this interacting system. On the right shows the location of the 64 planetary nebula candidates found by [21]. Note the clear tidal tail structures that are extremely difficult to detect in the broad-band image, but can be seen easily in the PN distribution.

PN have additional features that make them useful probes of intracluster starlight. Because PN are emission-line objects, any bright PN that can be detected photometrically can also be observed spectroscopically to high precision ($\sigma \approx 12$ km s^{-1}). In the case of M 51, spectroscopic follow-up of the PN [19] found that the western tidal tail consists of two discrete structures that overlap in projection, one from each galaxy. With the high velocity precision of the PN observations, the separation of the two kinematic structures was trivial. Although I will not focus on the dynamical aspects of IPN/PN in this review (see reviews by Arnaboldi, Douglas, Gerhard, & Peng this conference), it is important to state that IPN are the easiest (and perhaps, the only) way to obtain dynamical information about the intracluster light.

Finally, through the [O III] λ 5007 Planetary Nebulae Luminosity Function (PNLF), extragalactic PN make excellent distance indicators (see Ciardullo, this conference). The distinctive PNLF can be used in the intracluster environment to gain information on the line-of-sight distribution of the intracluster stars.

Why study intracluster stars in the first place? Once a curiosity proposed by Zwicky ([63]), intracluster light is potentially of great interest to studies of galaxy and galaxy cluster evolution. The dynamical evolution of cluster galaxies is complex, and involves the poorly understood processes of cluster accretion and tidal stripping [15]. The intracluster light provides a unique way to study these mechanisms. Modern numerical simulations show that the intracluster light [16,45,56,60,40] has a complex structure, and can be used to gain information on the dynamical evolution of galaxies and galaxy clusters.

2 History of Intracluster Planetary Nebula Research

The history of IPN research begins over a decade ago with the first PN survey of the Virgo cluster [33] (hereafter JCF). In this survey of elliptical galaxies, JCF found 11 PN that were much brighter than the expected [O III] λ 5007 PNLF cut-off magnitude. JCF attempted to explain these "overluminous" PN with a number of hypotheses, but none was entirely satisfactory.

The next step involved spectroscopic follow-up of objects from the JCF survey. During a radial velocity survey of PN in the Virgo elliptical galaxy M 86 [1], it was found that 16 of the 19 detected PN velocities were consistent with the galaxy's mean velocity (v_{radial} = -227 km s^{-1}). The other three planetaries had mean radial velocities of $\sim$ 1600 km s^{-1}, more consistent with the Virgo cluster's mean velocity. It was argued convincingly that these objects were intracluster planetary nebulae [1], and it is here that the term first enters the literature. The "overluminous" PN candidates were thus naturally explained as a population of intracluster stars in front of the target galaxies. Almost simultaneously, the first search for IPN candidates in the Fornax cluster was published [58], and more detections of IPN candidates in Virgo quickly followed [37,11,22].

However, a surprise was in the works. Spectroscopic follow-up of the IPN candidates revealed that some were not IPN, but instead background emission-line objects with extremely high equivalent width [28,35]. This was unexpected, because previous deep emission-line surveys had found very few such objects [49], though many have now been detected at fainter magnitudes [51]. The most likely source of the contamination was found to be Lyman-α galaxies at redshifts 3.12–3.14, where the Lyman-α λ 1215 line has been redshifted into the [O III] λ 5007 filters used in IPN searches. However other types of contaminating objects may also exist [57,47].

Although these contaminants caused some consternation at first, a number of lines of evidence quickly showed that the majority of IPN candidates are in fact, actual IPN. Observations of blank control fields with identical search procedures as the IPN surveys [12,8] have found that the contamination fraction is significant, but was less than the observed IPN surface density. The surface densities found correspond to a contamination rate of $\approx$ 20% in the Virgo cluster and $\approx$ 50% in Fornax (Fornax is more distant than Virgo, so its PNLF is fainter, and therefore further down the contaminating sources luminosity function). There are still significant uncertainties in the background density due to large-scale structure, and to the small numbers of contaminating objects found thus far. However, deeper and broader control fields are forthcoming. Spectroscopic follow-up of IPN candidates [28,12,3], and Arnaboldi these proceedings, clearly show large numbers of IPN candidates having the expected [O III] λ 5007 and 4959 emission lines, with a contamination rate similar to the blank field surveys. Finally, there is independent evidence of individual intracluster stars in Virgo from observations of individual Intracluster Red Giant stars (IRGs; [27,18]) from the *Hubble Space Telescope*.

It is worth reiterating that *all* deep [O III] λ 5007 emission-line searches will have such contamination at fainter magnitudes. In fact, such objects have already been seen in conventional PN surveys ([38,19] and Romanowsky, private communication). Extragalactic PN researchers should take this into account. The brightest of such objects have been detected with a m_{5007} magnitude of 25.5 [28], though the luminosity function of the contaminants is still poorly known.

Currently, with the widespread use of mosaic CCD detectors, and automated detection methods derived from DAOPHOT and SExtractor, over a hundred IPN candidates can be found in a single telescope run [48,24,3,26]. IPN candidates are easily identified as stellar sources that appear in a deep [O III] λ 5007 image, but completely disappear in an image through a filter that does not contain the [O III] line. Currently, over 400 IPN candidates have been detected in the Virgo cluster, and over 100 IPN candidates have been found in the Fornax cluster. Figure 2 summarizes the status of the different surveys. Despite all of the effort, to date, only a few percent of the total angular area of Virgo and Fornax have been surveyed. Literally thousands of IPN wait to be discovered by 4-meter class telescopes.

3 The Spatial Distribution of the Intracluster Light

If we want to understand the mechanisms that produce intracluster stars, it is useful to compare the intracluster stars' spatial distribution with that of the better-studied components of galaxy clusters: the cluster galaxies, the hot intracluster gas, and the invisible dark matter. Theoretical work predicts very different spatial distributions for the intracluster light depending on its production time. For example, if most intracluster stars are removed early in a cluster's lifetime [39], the distribution of this diffuse component will be smooth and follow the cluster potential. However, if a significant portion of the stars are removed at late epochs via galaxy encounters and tidal stripping [52,43], then the intracluster light should be clumpy and have a non-relaxed appearance. In particular, the "harassment" models [43,44] first predicted that many intracluster stars will exist in long ($\sim$ 2 Mpc) tidal tails, which may maintain their structure for Gyrs.

With regards to angular clustering, there is good evidence that the IPN follow a non-random distribution. Evidence for clumps of IPN within their survey fields was found [48,26] and also a strong clustering signal using a standard two-point correlation function analysis [4]. The scale of the angular clustering ranges from one to ten arcminutes, (corresponding 5 to 50 kpc assuming a mean Virgo distance of 15 Mpc), depending on the field studied. However, no tail-like features have yet been seen in the IPN distribution, although tidal arc features have been seen in broad-band images of other galaxy clusters [59,32,7].

With the use of the [O III] λ 5007 PNLF, we can also obtain line-of-sight information on the intracluster stars, a property unique to IPN. Although a full analysis of the observed luminosity functions has not yet been attempted, we can already learn something useful from the brightest IPN candidates in each field. Given the empirical PNLF, there is a maximum absolute magnitude M^*, that a PN may have in the light of [O III]. If we assume that the brightest observed IPN candidate in each field has this absolute magnitude, we can derive

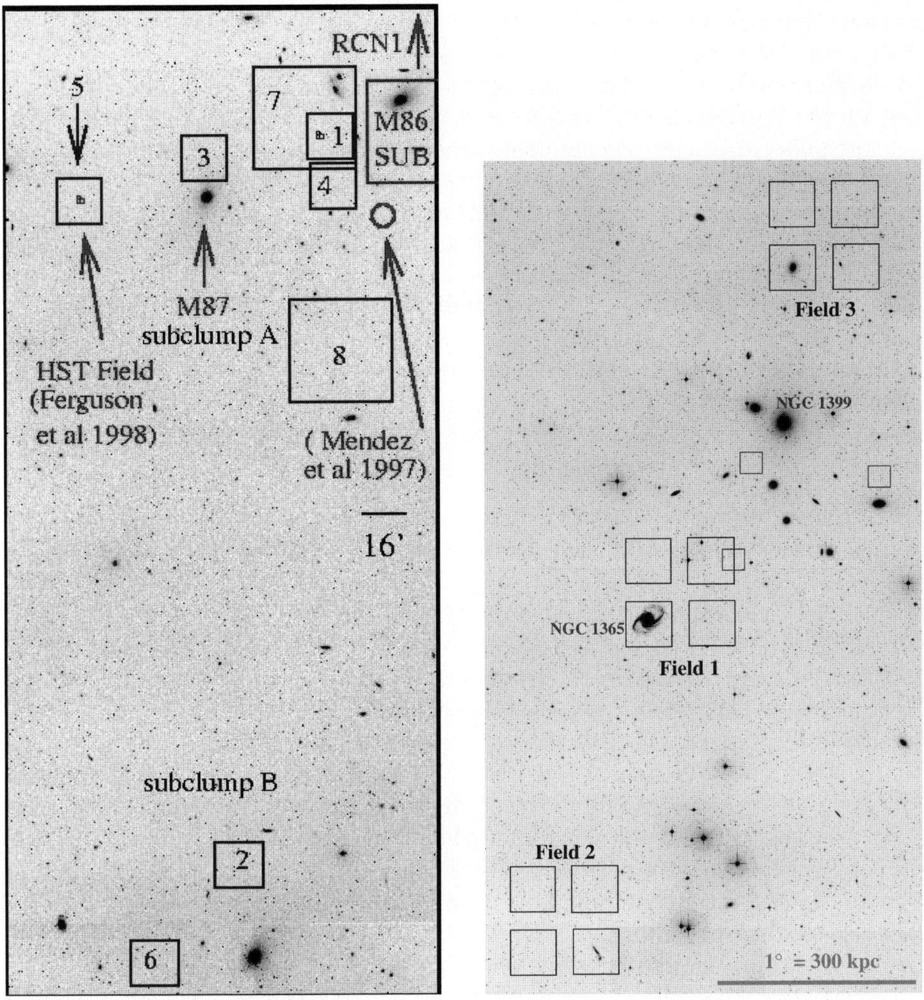

Fig. 2. Images of the Virgo cluster (left), and the Fornax cluster (right), with the various published IPN survey fields marked, as well as the two *HST* IRG fields. Soon to come are additional fields from Aguerri et al. (2004; Virgo), Ciardullo et al. (2004; Fornax), Feldmeier et al. (2005; Virgo) and Ciardullo et al. (2005; IRG Virgo)

an upper limit on its distance, and therefore estimate the distance to the front edge of the Virgo intracluster population. This is plotted in Figure 3. As can be clearly seen, there is an offset between the IPN of subclusters A and B: this is interpreted as due to the differing distances of these two subclusters. However, the most revealing feature of the distribution of the IPN is in subclump A. The upper limit distance for most of the fields is well in front of the cluster core. This is partially due to a selection effect: it is easier to detect an IPN on the near side of Virgo than the far side. Nevertheless, the depth implied from the measurements is remarkable. If we take the data at face value, then the IPN distribution has a line-of-sight radius of over 4 Mpc. If we compare this radius to the classical radius of Virgo on the sky (six degrees, or 1.6 Mpc), we find that the Virgo cluster is more than 2.6 times as deep as it is wide. Virgo is nowhere near a spherical cluster: it contains considerable substructure, and is elongated significantly along our line of sight.

The great depth derived for the IPN distribution gives an important clue to the parent galaxies of the intracluster stars. Numerical simulations [44,17] show that the majority of intracluster stars are ejected into orbits similar to that of their parent galaxies. In Virgo, the cluster ellipticals are clustered within a radius of ≈ 2 Mpc from the cluster core [33]. In contrast, the hydrogen deficient spirals found in the Virgo cluster have a radius of 4 Mpc or larger [55], and the dwarf ellipticals have a depth of 6 Mpc [34]. Therefore, it seems clear that a significant portion of Virgo's intracluster stars originate from late-type galaxies whose highly radial orbits take them in and out of the cluster core.

4 Converting IPN Densities to Luminosity Densities

In principle, determining the amount of intracluster luminosity from the observed numbers of IPN is straightforward. Theories of simple stellar populations [50] have shown that the bolometric luminosity-specific stellar evolutionary flux of non-star-forming stellar populations should be $\sim 2 \times 10^{-11}$ stars-yr^{-1}-$L_{\odot}^{-1}$, (nearly) independent of population age or initial mass function. If the lifetime of the planetary nebula stage is $\sim$25,000 yr, and if the empirical PNLF is valid to $\sim$ 8 mag below the PNLF cutoff, then every stellar system should have $\alpha \sim 50 \times 10^{-8}$ PN $L_{\odot}^{-1}$. According to the empirical PNLF, approximately one out of ten of these PNe will be within 2.5 mag of M^*. Thus, under the above assumptions, most stellar populations should have $\alpha_{2.5} \sim 50 \times 10^{-9}$ PN-$L_{\odot}^{-1}$. The observed number of IPN, coupled with the PNLF, can therefore be used to deduce the total luminosity of the underlying stellar population.

In practice, there are a number of systematic effects that must be accounted for before we can transform the numbers of IPN to a stellar luminosity [26], which we briefly summarize here.

First, although stellar evolution theory originally predicted a constant $\alpha_{2.5}$ value for all non star-forming populations, observations present a more complicated picture. It was found that in a sample of 23 elliptical galaxies, lenticular galaxies, and spiral bulges, the observed value of $\alpha_{2.5}$ never exceeded $\alpha_{2.5} = 50 \times 10^{-9}$ PN-$L_{\odot}^{-1}$ but was often significantly less, with higher luminosity galaxies having systematically smaller values of $\alpha_{2.5}$ [9]. Since the amount of intracluster starlight derived is inversely proportional to the $\alpha_{2.5}$ parameter, a large error in the amount of intracluster light can result if this is not accounted for.

By comparing the numbers of IPN in a field surrounding an *HST* WFPC2 field, with RGB and AGB star counts, a value of $\alpha_{2.5} = 23^{+10}_{-12} \times 10^{-9}$ PN-$L_{\odot}^{-1}$ was found for Virgo's intracluster population [18]. This observational measurement is bolstered by the new theoretical work of Buzzoni & Arnaboldi (these

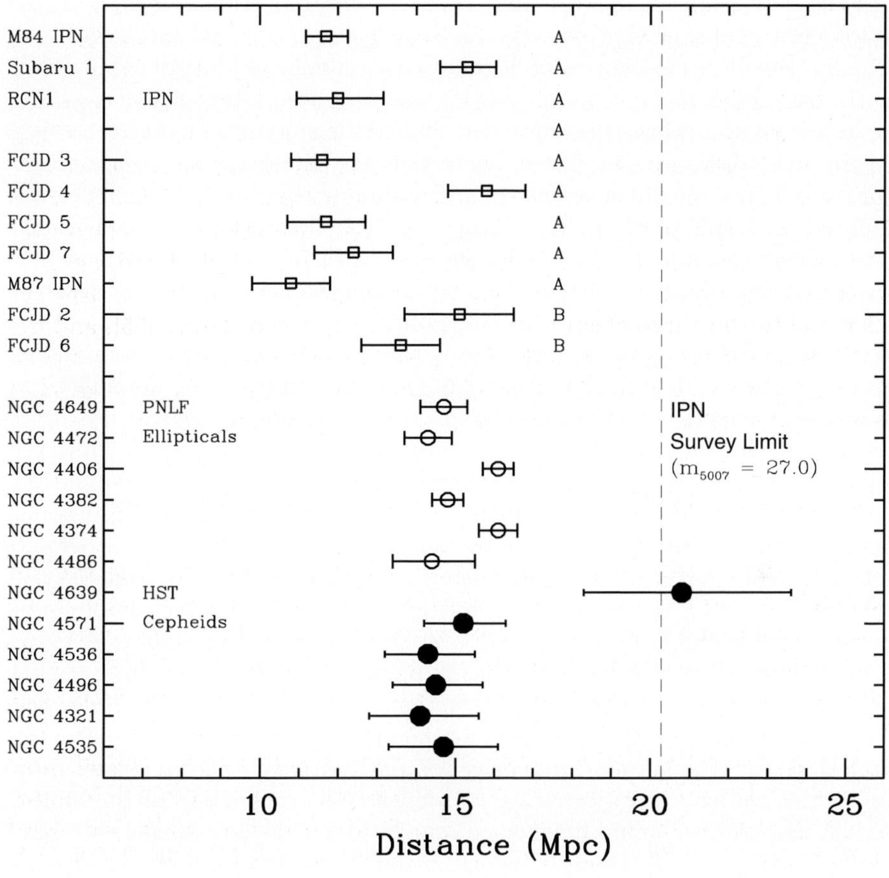

Fig. 3. A comparison of the upper limit distances obtained from the intracluster planetary nebulae to direct distances to Virgo Cluster galaxies. At the top are the upper limit distances (denoted by the open squares) from IPN observations by Okamura et al. (2002) and Arnaboldi et al. (2002). Below that are the distances derived from as the overluminous IPN found in front of M87 (Ciardullo et al. 1998). These upper limit distances are compared to the PNLF distances of Virgo ellipticals (denoted by the open circles; Jacoby, Ciardullo, & Ford 1990; Ciardullo et al. 1998), and Cepheid distances to spiral galaxies (denoted by the filled circles; Pierce et al. 1994; Saha et al. 1997; Freedman et al. 2001). The subcluster of Virgo in which each intracluster field resides is noted. Note the great depth of IPN, compared to that of the elliptical galaxies.

proceedings), which show that the observational value for $\alpha_{2.5}$ is well within the range for models of moderate luminosity galaxies.

Second, as we have discussed previously, IPN surveys are not pristine: approximately 20% of Virgo IPN candidates, and 50% of Fornax IPN candidates, are likely to be unrelated background sources. Although these objects are subtracted out statistically from all modern IPN surveys, the importance of field-to-field variations in the background due to large-scale structure is still unknown.

Finally, the IPN candidates of Virgo have a significant line-of-sight depth. Since the conversion between number of PN and luminosity depends on the shape of the luminosity function, this depth can change the amount of intracluster light found from the data. Models [22] indicate that the difference between a single-distance model (the most conservative) and assuming that the IPN are uniformly distributed in a sphere of radius 3 Mpc (the least conservative), changes the derived intracluster star luminosity by up to a factor of three. Thus far, all IPN researchers have adopted a single distance model in order to be conservative, but this effect is the least studied at this point.

After applying all of these corrections, the intracluster stellar fractions for Virgo vary between 10 and 20% [2,48,3,26], with the errors being dominated by the systematic effects. The IRG measurements [27,18] find somewhat less intracluster light (10–15%), but the various results agree within the errors. However, it is important to note that each of these studies makes different assumptions concerning the calculation of intracluster luminosity, the amount of light bound to galaxies, and the contamination of background sources. For Fornax, there are only two measurements of the intracluster star fraction thus far from IPN. A fraction of up to 40% is reported [58], but this is before the detection of contaminating sources, and is likely to be an overestimate. Ciardullo (2004, in prep), reports an approximate fraction of 20%, with similar errors to the Virgo results. Regardless of the uncertainties, it is clear that a significant fraction of all stars in Virgo and Fornax are in an intracluster component.

How does the IPN luminosity density compare with well-known cluster properties such as radius, or projected galaxy density? A comparison of these properties in Virgo [26] has found little or no correlation. This may be due to Virgo's status as a dynamically young cluster, or due to a selection effect (IPN researchers must observe where the galaxies are not, to avoid confusion with normal extragalactic PN). More data will be needed to confirm this result.

5 Intracluster H II Regions

A recent discovery stemming from IPN searches is the detection of H II regions in the far halos of galaxies or in intracluster space. In the course of spectroscopic follow-up of a Virgo IPN field [30], an emission-line source was found that has the properties of a $\sim$ 400 solar mass, 3 Myr compact H II region, over 17 kpc away from the nearest Virgo galaxy (see also [36] for a similar, earlier, underappreciated example). There are now a number of other examples of intergalactic star formation in Virgo and in other galaxy clusters and groups [61,54,53,14,20].

These discoveries have two important implications. First, star formation can occur at large distances from galaxies, and environments quite different than that normally studied. This has implications for a number of fields of astrophysics, including the origin of metallicity in the intergalactic and intracluster medium, and the possible in situ origin of B stars in the Galactic halo. At this time, it does appear that intracluster star formation is only a small fraction of the total intracluster star production, but the exact amount is uncertain. Second, these objects are another form of contamination to pure IPN surveys. Discouragingly, the only way to separate these objects from IPN is through deep spectroscopy. Luminous PN have [O III] / Hα ratios of two or greater [13], where luminous H II regions do not. From the surveys thus far, these objects appear to be a relatively rare component of IPN surveys ($\sim$ 3%), but more study is needed in this area.

6 Intra-Group Starlight

Although the presence of intracluster starlight in clusters such as Virgo and Fornax is now well established, the amount of 'intra-group' starlight is still uncertain. Theoretical studies predict that if most intracluster stars are removed by galaxy collisions [52,43], the fraction of intra-group stars, to first order, should be a smooth function of galaxy number density ($L_{ICL} \sim N^2_{Gal}$). To test this hypothesis, two different groups have surveyed the nearby M 81 and Leo groups of galaxies [8,20], using similar detection methods as the cluster searches. In both cases, no genuine intra-group PN were found. These non-detections strongly imply there is substantially fewer (4–15 times) intergalactic stars in groups than there are in clusters.

When compared with other measurements of intracluster star fractions through modern deep imaging [5,31,23,25], or through the detection of intracluster supernovae [29], an interesting pattern emerges, plotted in Figure 4. For galaxy clusters, the data is consistent with an approximate fraction of 20%, albeit with large error bars or intrinsic scatter. When we move to the group environment, the fraction abruptly drops, with no sign of any smooth decline. This implies that there is something special about the cluster environment that promotes intracluster star production. More data will be needed to confirm and strengthen this result, especially IPN searches of additional galaxy groups.

7 IPN in the Coma Cluster?

At this conference, Gerhard et al. announced the possible detection of at least one IPN candidate in the Coma Cluster. If this candidate is confirmed, it will be the most distant individual PN ever detected, at a distance of over 100 Mpc. This object was found using a technique of filling the entire focal plane of the 8-m Subaru telescope with a slit mask through the appropriate [O III] λ 5007 filter, and looking for narrow-line emission sources. Deep broad-band imaging has implied that the Coma Cluster may have a very high intracluster star fraction, up to 50% [5], so it is plausible that a few IPN could be detected, despite the small area surveyed. The most exciting aspect of this observation is that it opens up a way to observe PN at much greater distances than previously thought possible, and makes it possible to place IPN density limits in more distant galaxy clusters.

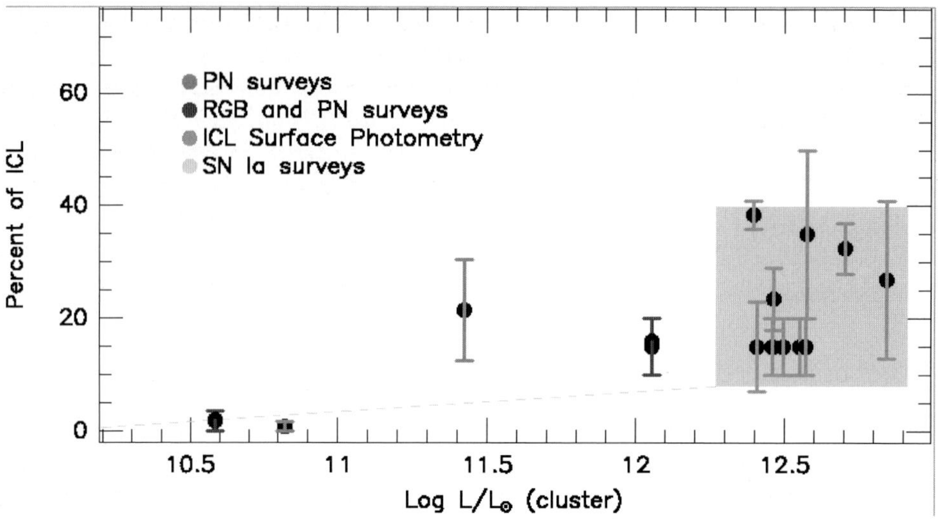

Fig. 4. A comparison of detected intracluster star fractions from modern measurements, as a function of the luminosity dispersion. Note the abrupt change from the cluster environment to the group environment.

8 The Future

Studies of IPN are not even a decade old, and much more work still needs to be done. The most crucial observations needed in the near term are spectroscopic follow-up of a large number of IPN candidates. This would allow us to better determine the contamination rate from background galaxies, avoid any other intracluster H II regions, and most importantly, gain information on the kinematics of intracluster stars. More IPN imaging is needed in Virgo and Fornax at differing radii and densities, in order to confirm the lack of correlations found thus far. IPN surveys of additional galaxy groups are needed to determine the lack of intra-group stars in these systems, and to better characterize the steepness of the intracluster production "cliff."

However, the future of IPN research will involve close comparisons of IPN properties to other observations of extragalactic PN and galaxy clusters and close comparison to the modern numerical simulations now being undertaken. Some examples are given below, but it is almost certain that many more will be added in the next few years.

Villaver and Stanghellini (these proceedings) presented the first numerical hydrodynamical simulations of IPN within a hot intracluster gaseous medium. The results imply that IPN do survive in the intracluster medium, but the nebula may be spatially distorted.

The first phase of a deep CCD imaging survey of the Virgo cluster core, using the techniques of ultra-deep surface photometry, has just been completed [41]. These new observations reach extremely faint surface brightnesses ($\mu_V \approx 29.6$)

over large angular areas (≈ 2 square degrees). Comparing these deep imaging maps with the IPN observations should be revealing, and allow us to better determine the reality of tidal features in both data sets, and to give clear targets for follow-up IPN observations.

HST+ACS broad-band observations are planned of an intracluster field in Virgo by Summer 2005, with the goal of obtaining the first color-magnitude diagram of the intracluster stars. By placing tighter constraints on the age and luminosity of the intracluster population, we should be able to better constrain the $\alpha_{2.5}$ parameter, and improve the accuracy of the total amount of intracluster starlight.

Finally, with the advent of advanced numerical simulations of galaxy clusters, we will be able to compare the IPN data to models of intracluster star production. Due to the large angular size of nearby clusters, it will be many years before the majority of the clusters will be surveyed for IPN. Numerical simulations will hopefully allow us to survey more efficiently, and to obtain better quality information on the properties of intracluster starlight.

I would like to thank the conference organizers for allowing me to give this review, and for running an excellent conference. I would also like to thank my collaborators for their years of effort on the science presented here. This work was supported in part by NSF grant AST 03-02030 and NASA grant NAG 5-9377.

References

1. Arnaboldi, M., et al. 1996, Ap. J., 472, 145
2. Arnaboldi, M., et al. 2002, A. J., 123, 760
3. Arnaboldi, M., et al. 2003, A. J., 125, 514
4. Aguerri, J. A. L., Iglesias-Paramo, J., et al.. 2004, A. J., 127, 1344
5. Bernstein, G. M., et al. 1995, A. J., 110, 1507
6. Burkhead, M. S. 1978, Ap. J. Suppl., 38, 147
7. Calcáneo-Roldán, C., et al. 2000, M.N.R.A.S., 314, 324
8. Castro-Rodríguez, N., et al. 2003, Astr. Ap. 405, 803
9. Ciardullo, R. 1995, in IAU Highlights of Astronomy, 10, ed. I. Appenzeller (Dordrecht: Kluwer), p. 507
10. Ciardullo, R., et al. 1989, Ap. J., 339, 53
11. Ciardullo, R., et al. 1998, Ap. J., 492, 62
12. Ciardullo, R., et al. 2002a, Ap. J., 566, 784
13. Ciardullo, R., et al. 2002b, Ap. J., 577, 31
14. Cortese, L., et al. 2004, A.& A., 416, 119
15. Dressler, A. 1984, A.R.A.& A., 22, 185
16. Dubinski, J. 1998, Ap. J., 502, 141
17. Dubinski, J., Murali, C.,& Ouyed, R. 2000, unpublished preprint
18. Durrell, P. R., et al. 2002, Ap. J., 570, 119
19. Durrell, P. R., et al. 2003, Ap. J., 582, 170
20. Durrell, P. R., et al. 2004, IAU Symposium, 217, 90
21. Feldmeier, J. J., Ciardullo, R., & Jacoby, G. H. 1997, Ap. J., 479, 231

22. Feldmeier, J. J., Ciardullo, R., & Jacoby, G. H. 1998, Ap. J., 503, 109
23. Feldmeier, J. J., et al. 2002, Ap. J., 575, 779
24. Feldmeier, J. J., et al. 2003a, Ap. J.Suppl., 145, 65
25. Feldmeier, J. J., et al. 2004a, Ap. J., in press
26. Feldmeier, J. J., et al. 2004b, Ap. J., in press
27. Ferguson, H. C., Tanvir, N. R., & von Hippel, T. 1998, Nature, 391, 461
28. Freeman, K. C. et al. 2000, ASP Conf. Ser. 197: Dynamics of Galaxies: from the Early Universe to the Present, ed. F. Combes, G. A. Mamon, & V. Charmandaris (San Francisco: ASP), 389.
29. Gal-Yam, A., et al. 2003, A J., 125, 1087
30. Gerhard, O., et al. 2002, Ap. J. L., 580, L121
31. Gonzalez, A. H., et al. 2000, Ap. J., 536, 561
32. Gregg, M. D., & West, M. J. 1998, Nature, 396, 549
33. Jacoby, G. H., Ciardullo, R., & Ford, H. C. 1990, Ap. J., 356, 332
34. Jerjen, H., Binggeli, B. & Barazza, F.D. 2004, A.J., 127, 771
35. Kudritzki, R.-P., et al. 2000, Ap. J., 536, 19
36. Lee, H., Richer, M. G., & McCall, M. L. 2000, Ap. J.L., 530, L17
37. Méndez, R. H., et al. 1997, Ap. J. L., 491, 23
38. Méndez, R. H., 2001, Ap. J., 563, 135
39. Merritt, D. 1984, Ap. J., 276, 26
40. Mihos, J.C., et al. 2004a, in preparation
41. Mihos, J.C., et al. 2004b, in preparation
42. Miller, G.E. 1983, Ap. J., 268, 495
43. Moore, B., et al., 1996, Nature, 379, 613
44. Moore, B., Lake, G. & Katz, N. 1998, Ap. J., 495, 139
45. Murante, G., et al. 2004, Ap. J. L., 607, L83
46. Napolitano, N. R., et al. 2003, Ap. J., 594, 172
47. Norman, C., et al. 2002, Ap. J., 571, 218
48. Okamura, S., et al. 2002, P.A.S.J., 54, 883
49. Pritchet, C. J. 1994, P.A.S.P., 106, 1052
50. A. Renzini, A. Buzzoni: In: *Spectral Evolution of Galaxies*, ed. by C. Chiosi, A. Renzini (Reidel, Dordrecht), p. 195 (1986)
51. Rhoads, J. E., et al. 2000, Ap. J. L., 545, L85
52. Richstone D. O., & Malumuth, E. M. 1983, Ap. J., 268, 30
53. Ryan-Weber, E. V., et al. 2004, A. J., 127, 1431
54. Sakai, S., Kennicutt, R. C., van der Hulst, J. M., & Moss, C. 2002, Ap. J., 578, 842
55. Sanchis, T., et al. 2002, Ap J., 580, 164
56. Sommer-Larsen, J., Romeo, A.D., Portinari, L. 2004, M.N.R.A.S., submitted, available as astro-ph/0403282
57. Stern, D., Bunker, A., Spinrad, H., & Dey, A. 2000, Ap. J., 537, 73
58. Theuns, T., & Warren, S. J. 1997, M.N.R.A.S., 284, L11
59. Trentham, N. & Mobasher, B. 1998, M.N.R.A.S., 293, 53
60. Willman, B., et al. 2004, M.N.R.A.S., submitted - available as astro-ph/0405094
61. Xu, C., Sulentic, J. W., & Tuffs, R. 1999, Ap. J., 512, 178
62. Yun, M.S., Ho, P.T.P., & Lo, K.Y. 1994, Nature, 372, 530
63. Zwicky, F. 1951, P.A.S.P., 63, 61

Planetary Nebulae as Tracers of Galaxy Clusters

Magda Arnaboldi

INAF, Osservatorio Astronomico di Torino
Strada Osservatorio 20
I-10025, Pino Torinese

Abstract. Stars are usually observed to form in galaxies (disks, dwarfs and starbursts). In galaxy clusters, however, a diffuse intracluster stellar component has been detected from deep imaging and observations of individual intracluster stars.

Intracluster light (ICL) is potentially of great interest for studies of galaxy and galaxy cluster evolution. The dynamical evolution of cluster galaxies involves complex and imperfectly understood processes such as galactic encounters, cluster accretion, and tidal stripping. Various studies have suggested that between 10% and 50% of a cluster's total luminosity may be contained in the ICL, with a strong dependence on the dynamical state of the cluster. The properties of the ICL may also be sensitive to the distribution of dark matter in cluster galaxies, as simulations have shown that the structure of dark matter halos in galaxies plays a central role in the formation and evolution of tidal debris.

Recently some progress has been made in the study of intracluster star light on several fronts. Individual intracluster stars, including planetary nebulae detected from the ground and red giants detected using HST, have been discovered in the Virgo cluster. These intracluster stars give the promise of studying in detail the kinematics, metallicity and age of the intracluster stellar population in nearby galaxy clusters and thereby learning about the origin of this diffuse stellar component and the details of the cluster origin.

1 Properties of Diffuse Light in Clusters of Galaxies

The first studies on this subject date back to the work of Zwicky (1951) on the luminosity function of galaxies in the Coma cluster. In his work, Zwicky pointed out that vast and irregular *swarms of stars* existed in the spaces between the standard galaxies in Coma and commented whether they could be incorporated into the distribution function of known galaxy types. Zwicky's efforts were then followed by photographic surveys for diffuse light in Coma and other rich clusters in the 1970s. In the 1990s, CCD photometry provided the first accurate measurements in Coma, as described by Bernstein et al. (1995).

Two kind of problems affect these experiments: (a) the typical surface brightness of intracluster light (ICL) is less than 1% of the dark sky brightness; (b) it is difficult to disentangle the diffuse light associated with the halo of the central dominant (cD) galaxy at the cluster center and the diffuse light component.

Since 1995, wide-field cameras equipped with a CCD mosaic have allowed accurate measurements of diffuse light in Abell clusters. This stellar component is traced by tails, arcs and/or plumes with typical $\mu_B = 27.8$ mag arcsec^{-2}, very narrow ($\sim$ 2 kpc) and extended ($\sim$ 100 kpc) in Coma and Centaurus (Gregg & West 1998, Threntam & Mobasher 1998, Calcáneo-Roldán et al. 2000). In addition, different groups have measured the radial surface brightness profile of the extended cD halos out to very large cluster radii. These measurements were carried out for the Abell cluster 1651 (Gonzalez et al. 2000), Abell 1413 & MKW7 (Feldmeier et al. 2002), in four Abell type II-III (non-cD) clusters (Feldmeier et al. 2004b) and for the compact group HGC90 (White et al. 2003). Quoting from Uson et al. (1991), " *...whether this diffuse light is called the cD envelope or diffuse intergalactic light is a matter of semantics: it is a diffuse component which is distributed with elliptical symmetry about the center of the cluster potential*". All these independent measurements place the lower limit to the fraction of diffuse light in clusters with respect to the amount of light in individual galaxies at 10% – 20%. An alternative method for probing ICL is through the direct detection and measurements of the stars themselves.

The detection of intergalactic supernovae was first reported by Smith (1981): a SNIa was observed in the Virgo cluster, in the region between M86 and M84. In 2003, Gam-Yam et al. observed two SN1a in Abell 403 ($z = 0.10$) and in Abell 2122/4 ($z = 0.066$) . Both events appear projected on the halos of the central cDs, no other obvious hosts are present, but these stars have a substantial velocity offset (750 - 2000 km s^{-1}) from the cD systemic velocity suggesting that they are not bound to it, but are free-flying in the cluster potential. Gam-Yam and collaborators estimate that 20% of the SN1a parent stellar population in clusters is intergalactic (Gam-Yam et al. 2003).

In 1995, West and collaborators argued that a population of intergalactic globular clusters (IGCs) exists in all clusters and are concentrated towards the center (West et al. 1995). High values of the GC frequency (S_N) in cDs is then the results of the accretion of a number of IGCs. Independent evidence was acquired by Coté et al. (2001) around M87, where they concluded that the metal poor GCs were not formed in situ, but stripped from the Virgo cluster dwarfs. In the cluster Abell 1185, Jordàn et al. (2003) searched for a population of IGCs, in a field centered on the peak of the cluster's X-ray emission, which contains no bright galaxies. An excess of point like sources is found with respect to HDF North, which Jordàn and collaborators associate with a population of IGCs. Bassino et al. (2003) also reported on the discovery of IGC candidates in the Fornax cluster.

In the intracluster (IC) regions, an additional kind of stellar cluster is found. Drinkwater et al. (2003) discovered Ultra-compact dwarfs (UDCs) in the Fornax cluster; they are nucleated dwarf galaxies whose outer envelopes were stripped by interactions with the cD, at the cluster center.

Direct observations of stars in Virgo IC fields were carried out by Ferguson, Tanvir & von Hippel in 1998 with HST. The presence of intracluster red giant stars (IRGBs) was inferred from the excess of red number counts in a Virgo IC field with respect to the HDF North. By comparing the I-band star counts into two Virgo cluster fields with similar observations for a metal-poor nucleated dwarf elliptical, Durrell et al. (2002) found an offset between the RGB tip of the

IRGBs and that in the dwarf. Their interpretation is that the bulk of the IRGBs are moderately metal rich ($-0.8 < [Fe/H] < -0.2$). The surface brightness associated with the IRGB counts is $\mu_L = 27.9$ mag arcsec^{-2}, and it amounts to 15% of the Virgo cluster galaxy I band luminosity. The recent discovery of an isolated compact HII region in the Virgo cluster (Gerhard et al. 2002) has shown that some star-formation activity can indeed take place in the outskirts of galaxy halos if not already in Virgo IC space.

2 Intracluster Light in Cosmological Simulations

The presence of diffuse 'intracluster light' in galaxy groups and clusters is now well established. The fraction of stars contained in this space-filling component seems to increase strongly with the density of the environment: from loose groups ($< 2\%$, Castro-Rodriguez et al. 2003; Durrell et al. 2003) to Virgo-like ($\sim 10\%$; Feldmeier et al. 2003; Arnaboldi et al. 2003) or Abell type II-III (non-cD) galaxy clusters ($\sim 10\%$; Feldmeier et al. 2004b), and rich clusters ($\sim 20\%$ or higher; Gonzalez et al. 2000, Feldmeier et al. 2002; Gal-Yam et al. 2003). This correlation may represent an important clue for understanding the mechanisms that produce ICL and drive its evolution in the cluster environment.

Cosmological simulations of structure formation facilitate studies of the diffuse light and its expected properties. Dubinski (1998) constructed compound models of disk galaxies and placed them into a partially evolved simulation of cluster formation, allowing an evolutionary study of the dark matter and stellar components independently. Using an empirical method to identify stellar tracer particles in high-resolution cold dark matter (CDM) simulations, Napolitano et al. (2003) studied a Virgo-like cluster, finding evidence of a young dynamical age of the intracluster component. The main limitations in these approaches is the restriction to collisionless dynamics.

Murante et al. (2004) analyzed for the first time the ICL formed in a cosmological hydrodynamical simulation including a self-consistent model for star formation. In this method, no assumptions about the structural properties of the forming galaxies need to be made, and the gradual formation process of the stars, as well as their subsequent dynamical evolution in the non-linearly evolving gravitational potential can be seen as a direct consequence of the ΛCDM initial conditions. Murante et al. (2004) identified 117 clusters in a large volume of $192^3\, h^{-3}\mathrm{Mpc}^3$ and analyze the correlations of the properties of diffuse light with, e.g., cluster mass and X-ray temperatures. These predictions can be tested against known properties of cD halos and used to plan observational tests to understand the physical properties of ICL.

The presence of the IC component is evident when the whole distribution of stars in the simulated clusters is analysed in a way similar to Schombert's (1986) photometry of brightest cluster galaxies (BCGs). Galaxies at the centers of Murante et al. (2004) simulated clusters have surface-brightness profiles which turn strongly upward in a $(\mu, R^{1/\alpha})$ plot. This light excess can be explained as IC stars orbiting in the cluster potential. Integrating its density distribution along the LOS, the slopes from the simulations are in agreement with those observed for the surface brightness profiles of the diffuse light in nearby clusters. In the Coma cluster, Bernstein et al. (1995) parameterize the surface brightness as r^β and find that the diffuse light is best fit by $\beta = -1.3 \pm 0.1$. In the Fornax cluster, the surface brightness profile of the cD envelope of NGC 1399 follows a power law of the form $\propto r^\beta$ with $\beta = -1.5$.

At large cluster radii, the surface brightness profile of the ICL appears more centrally concentrated than the surface brightness profile of cluster galaxies (see Fig. 1). In the simulations carried out by Murante et al. (2004), they also obtained the redshifts z_{form} at which the stars formed: those in the IC component have a z_{form} distribution which differs from that in cluster galaxies, see Fig. 2. The "unbound" stars are formed earlier than the stars in galaxies. The prediction for an old star age in the diffuse component agrees with the HST observation of the IRGB stars in the Virgo IC field, e.g. $t > 2$Gyr (Durrell et al. 2002), and points toward the early tidal interactions as the preferred formation process for the ICL. The different age and spatial distribution of the stars in the diffuse component indicate that it is a stellar population that is not a random sampling of the stellar populations in cluster galaxies.

Murante et al. (2004) studied the correlation between the fraction of stellar mass in the diffuse component and the clusters' total mass in stars, based on their statistical sample of 117 clusters. This fraction is ~ 0.1 for cluster masses $M > 10^{14} h^{-1} M_\odot$ and it increases with cluster mass: the more massive clusters have the largest fraction of diffuse light (Fig. 2). For $M \sim 10^{15} h^{-1} M_\odot$, simulations predict as many stars in the diffuse component as in cluster galaxies.

2.1 Dynamics of the ICL

In the currently favored hierarchical clustering scenario, fast encounters and tidal interactions within the cluster potential are the main players affecting the morphological evolution of galaxies in clusters. Fast encounters and tidal stirring cause a significant fraction of the stellar component in individual galaxies to be stripped and dispersed within the cluster in a few dynamical times. If the timescale for significant phase-mixing is on the order of a few cluster internal dynamical times, then a fraction of the ICL should still be located in long streams along the orbits of the parent galaxies. Detection of substructures in phase space would be a clear sign of late infall and harassment as the origin of the ICL.

A high resolution simulation of a Virgo-like cluster in a ΛCDM cosmology was used to predict the velocity and the clustering properties of the diffuse stellar component in the intracluster region at the present epoch (Napolitano et al. 2003). The simulated cluster builds up hierarchically and tidal interactions between member galaxies and the cluster potential produce a diffuse stellar component free-flying in the intracluster medium. The simulations are able to predict the radial velocity distribution expected in spectroscopic follow-up surveys (see Fig. 3): it is found that at $z = 0$ the intracluster stellar light is mostly dynamically unmixed and clustered in structures on scales of about 50 kpc at a radius of $400 - 500$ kpc from the cluster center.

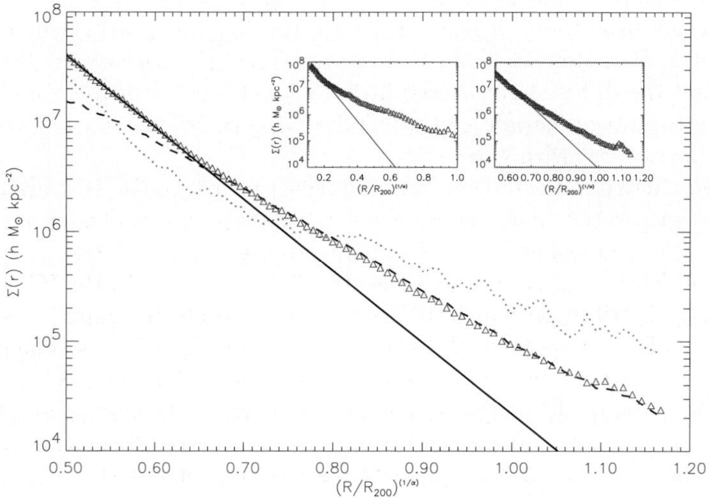

Fig. 1. Schombert–like analysis on the *stacked* 2D radial density profile (BCG + ICL) of clusters in the Murante et al (2004) simulation (triangles). The light excess is evident at large cluster radii. The solid line shows the function $\log \Sigma(r) = \log \Sigma_e - 3.33[(r/r_e)^{1/\alpha} - 1]$, with best–fit parameters $\log \Sigma_e = 20.80$, $r_e = 0.005$, $\alpha = 3.66$ to the BCG inner stellar light. Also shown are the averaged 2D density profile of stars in galaxies (dotted line) and in the field (dashed line). In the inserts, the results are shown from the same analysis for the most luminous clusters with X-ray temperature $T > 4$ keV (left panel), and for less luminous ones with $0 < T < 2$ keV (right panel). The resulting best–fit parameters are respectively $\log \Sigma_e = 16.47$, $r_e = 0.11$, $\alpha = 1.24$ and $\log \Sigma_e = 23.11$, $r_e = 0.00076$, $\alpha = 4.37$. In the main plot and in the inserts the unit $(R/R_{200})^{1/\alpha}$ refers to the α values given by each Sersic profile, where R_{200} is the radius which encloses an average density of 200 times the critical density; see Murante et al. (2004).

Willman et al. (2004) and Sommer-Larsen et al. (2004) have studied the dynamics of the ICL in cosmological hydrodynamical simulations. Willman et al. (2004) finds that the ICL show significant substructure in velocity space, tracing separate streams of stripped IC stars. Evidence is given that despite an un-relaxed distribution, IC stars are useful mass tracers, when several fields at a range of radii have measured line-of-sight (LOS) velocities. According to Sommer-Larsen et al. (2004) IC stars are colder than cluster galaxies. This is to be expected given that they are more centrally concentrated than cluster galaxies (Murante et al. 2004) and are both in equilibrium with the same cluster potential.

3 Planetary Nebulae as Tracers of Cluster Evolution

Intracluster planetary nebulae (ICPNe) have several unique features that make them ideal for probing the ICL. The diffuse envelope of a PN re-emits ~15%

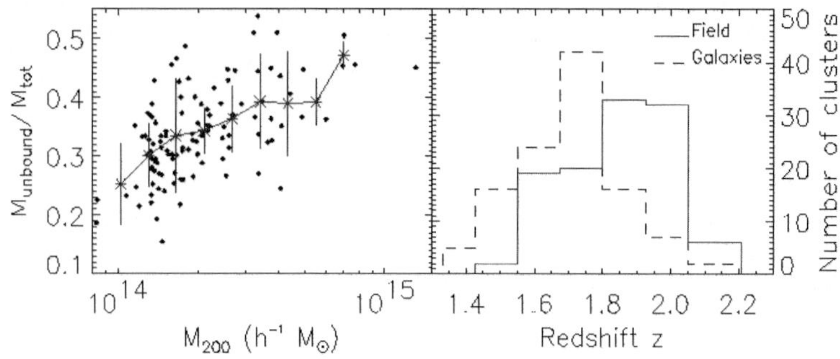

Fig. 2. Left: Fraction of stellar mass in diffuse light vs. cluster mass. Dots are for clusters in the simulated volume; asterisks show the average values of this fraction in 9 mass bins with errorbars. Right panel: histograms of clusters v. mean formation redshift, for bound (dashed) and IC star particles (solid line). Mean formation redshifts are evaluated for each cluster as the average on the formation redshift of each star particle. From Murante et al. (2004).

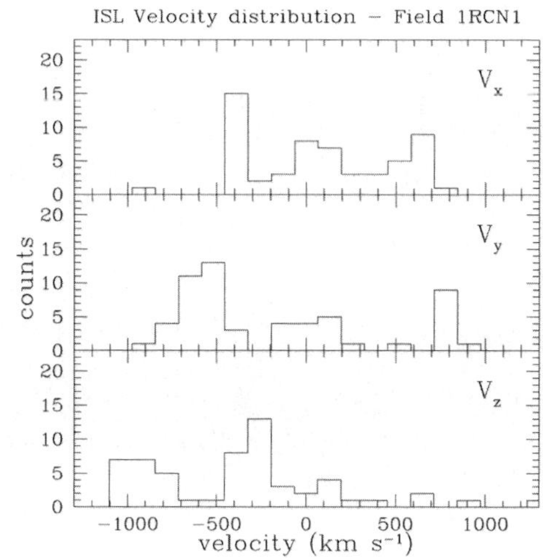

Fig. 3. Velocity distribution of a simulated ICPNe sample in a N-body simulation of a Virgo-like cluster. From Napolitano et al. (2003).

of the UV luminosity of the central star in one bright optical emission line, the green [O III]5007 Å line. PNe can therefore readily be detected in external galaxies out to distances of 25 Mpc and their velocities can be determined from moderate resolution ($\lambda/\Delta\lambda \sim 5000$) spectra: this enables kinematical studies of the IC stellar population.

PNe trace stellar luminosity and therefore provide an estimate of the total IC light. Also, through the [O III]5007 Å planetary nebulae luminosity function (PNLF), PNe are good distance indicators and the observed shape of the PNLF provides information on the line-of-sight distribution of the IC starlight. Therefore ICPNe are useful tracers to study the spatial distribution, kinematics, and metallicity of the diffuse stellar population in nearby clusters.

3.1 Current Narrow Band Imaging Surveys

Several groups (Arnaboldi et al. 2002, 2003; Aguerri et al. 2004; Feldmeier et al. 2003, 2004a) have embarked on narrow-band [O III] imaging surveys in the Virgo cluster, with the aim of determining the radial density profile of the diffuse light and gaining information on the velocity distribution via subsequent spectroscopic observations of the obtained samples. Given the use of the PNLF as distance indicators, one can also obtain valuable information on the 3D shape of the Virgo cluster from these ICPN samples. Wide-field mosaic cameras, such as the WFI

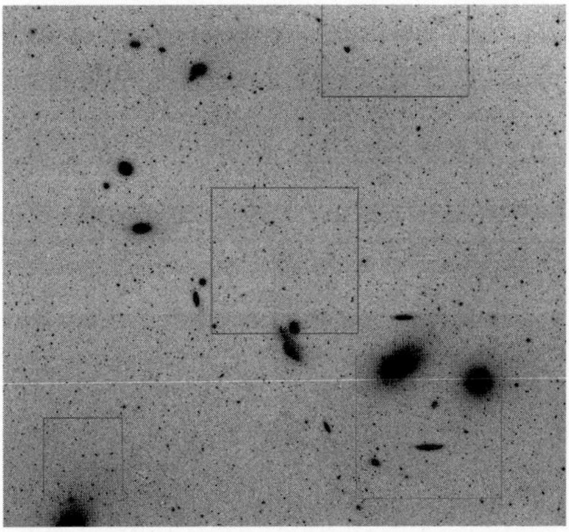

Fig. 4. Aguerri et al. (2004) surveyed several fields in the Virgo cluster core. The two upper fields were obtained at the ESO MPI 2.2m telescope, and the lower right field with the Suprime Cam at the 8.2m Subaru telescope. The lower left field was imaged by the prime focus camera of the Kitt Peak 4m telescope.

on the ESO MPI 2.2m telescope and the Suprime Cam on the Subaru 8.2m, allow us to identify the ICPNe associated with the extended ICL (Arnaboldi et al. 2002, 2003; Okamura et al. 2002). These surveys require the use of data reduction techniques suited for mosaic images, and also the development and refinement of selection criteria based on color-magnitude diagrams produced with SExtractor.

The data analysed by Aguerri et al. (2004) constitute a sizeable sample of ICPNe in the Virgo core region, constructed homogeneously and according to rigorous selection criteria; a layout of the pointings is shown in Fig. 4. From the study of four wide-fields they conclude that the mean contribution of the ICL to the total light in the CORE region of the Virgo cluster is 5-6%, the mean surface luminosity 2.7×10^6 $L_{B\odot}$ arcmin^{-2}, and the mean surface brightness $\mu_B = 29.1$ mag arcsec^{-2}. These values are in good agreement with those obtained from excess red giant counts in two deep HST images in the Virgo core. In addition, significant field-to-field variations are observed. The fraction of ICL versus total light ranges from 10% in the CORE and FCJ fields to less than 1% in the LPC field, which in its low ICL fraction is similar to low density environments (Castro-Rodriguez et al. 2003).

4 Spectroscopic Follow-Up

ICPNe are the only component of the ICL whose kinematics can be measured at this time. This is important since the high-resolution N-body and hydrodynamical simulations predict that the ICL is un-relaxed, showing significant substructure in its spatial and velocity distributions in clusters similar to Virgo.

The spectroscopic follow-up with FLAMES of the ICPN candidates selected from three survey fields in the Virgo cluster core was carried out by Arnaboldi et al. (2004). Radial velocities of 40 ICPNe in the Virgo cluster were obtained with the new multi-fiber FLAMES spectrograph on UT2 at VLT. The spectra were taken for a homogeneously selected sample of ICPNe, previously identified in three $\sim$ 0.25 deg^2 fields in the Virgo cluster core. For the first time, the 4959Å line of the [O III] doublet is seen in a large fraction (40%) of ICPNe spectra, and a large fraction of the photometric candidates with m(5007) < 27.2 is spectroscopically confirmed.

4.1 The Line-of-Sight Velocity Distributions of ICPNe in the Virgo Cluster Core

With these data, Arnaboldi et al. (2004) were able for the first time to determine radial velocity distributions of ICPNe and use these to investigate the dynamical state of the Virgo cluster. Fig. 5 shows an image of the Virgo cluster core with the positions of the imaged fields. The radial velocity distributions obtained from the FLAMES spectra in three of these fields are also displayed in Fig. 5. Clearly the velocity distribution histograms for the three pointings are very different.

In the FCJ field, the ICPNe distribution is dominated by the halo of M87. There are 3 additional outliers, 2 at low velocity, which are also in the brightest PNLF bin, and therefore may be in front of the cluster. The surface brightness of the ICL associated with the 3 outliers, e.g. the ICPNe in the FCJ field, amounts to $\mu_B \simeq 30.63$ mag arcsec^{-2}, in agreement with the surface brightness measurements of Ferguson et al. (1998) and Durrell et al. (2002) of the intracluster red giant stars.

The M87 peak of the FCJ velocity distribution contains 12 velocities with $\bar{v}_p = 1276 \pm 71$ km s^{-1} and $\sigma_p = 247 \pm 52$ km s^{-1}. The average velocity is consistent with that of M87, $v_{sys} = 1258$ km s^{-1}. The distance of the center of the FCJ field from the center of M87 is $15.'0 \simeq 65$ kpc for an assumed M87 distance of 15 Mpc. The value of σ_p is very consistent with the stellar velocity dispersion profile extrapolated outwards from $\simeq 150''$ in Figure 5 of Romanowsky & Kochanek (2001) and falls in the range spanned by their dynamical models for the M87 stars. The main result from our measurement of σ_p is that M87 has a stellar halo in approximate dynamical equilibrium out to at least 65 kpc.

In the CORE field, the distribution of ICPN line-of-sight (LOS) velocities is clearly broader than in the FCJ field. It has $\bar{v}_C = 1491 \pm 290$ km s^{-1} and $\sigma_C = 1000 \pm 210$ km s^{-1}. The CORE field is in a region of Virgo devoid of bright galaxies, but contains 7 dwarfs, and 3 low luminosity E/S near its S/W borders. None of the confirmed ICPNe lies within a circle of three times half the major axis diameter of any of these galaxies, and there are no correlations of their velocities with the velocities of the nearest galaxies where these are known. Thus in this field there is a true IC stellar component.

The mean velocity of the ICPN in this field is consistent with that of 25 Virgo dE and dS0 within 2° of M87, $< v_{\rm dE,M87} >= 1436 \pm 108$ km s^{-1} (Binggeli et al. 1987), and with that of 93 dE and dS0 Virgo members, $< v_{\rm dE,Virgo} >= 1139 \pm 67$ km s^{-1} (Binggeli et al. 1993). However, the velocity dispersion of these galaxies is smaller, $\sigma_{\rm dE,M87} = 538 \pm 77$ km s^{-1} and $\sigma_{\rm dE,Virgo} = 649 \pm 48$ km s^{-1}.

The inferred luminosity from the ICPNe in the CORE field is $1.8 \times 10^9 L_{B,\odot}$. This is about three times the luminosity of all dwarf galaxies in this field, $5.3 \times 10^8 L_{B,\odot}$, but an order of magnitude less than the luminosities of the three low-luminosity E/S galaxies near the field borders. Using the results of Nulsen & Böhringer (1995) and Matsushita et al. (2002), Arnaboldi et al. (2004) estimate the mass of the M87 subcluster inside 310 kpc (the projected distance D of the CORE field from M87) as $4.2 \times 10^{13} M_\odot$, and compute a dimensionless tidal parameter T for all these galaxies as the ratio of the mean density within the CORE field to the mean density of the galaxy. They find $T = 0.01 - 0.06$, independent of galaxy luminosity. Since $T \sim D^{-2}$, any of these galaxies whose orbit *now* comes closer to M87 than ~ 60 kpc would be subject to severe tidal mass loss. Based on the evidence so far, a tantalizing possibility is that the ICPN population in the CORE field could be debris from the tidal disruption of small galaxies on nearby orbits in the M87 halo.

In the SUB field the velocity distribution from FLAMES spectra is again different from CORE and FCJ fields. The histogram of the LOS velocities shows substructures related to M86, M84 and NGC 4388, respectively, and in Fig. 6 the projected phase space is shown. The association with the three galaxies is strengthened when we plot the LOS velocities of 4 HII regions (see Gerhard et al. 2002) detected with FLAMES in this pointing. The substructures in this

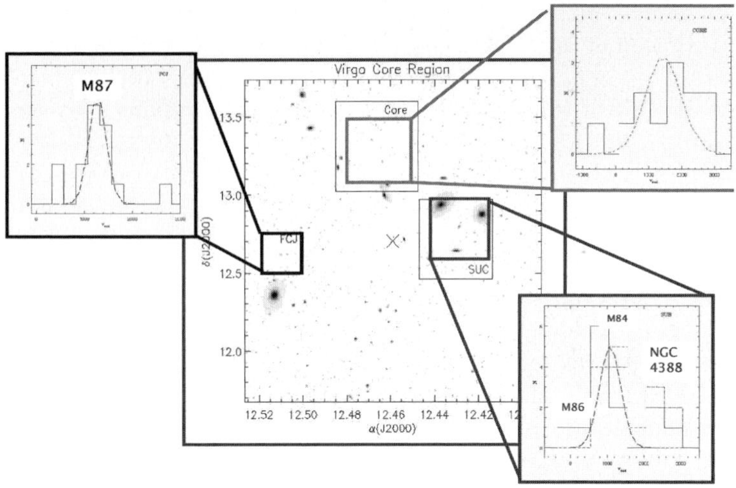

Fig. 5. ICPN radial velocity distributions in the three pointings (FCJ, CORE, and SUB) from Aguerri et al. (2004). In FCJ panel, blue dashed line shows a Gaussian with $\bar{v}_{rad} = 1276$ km s^{-1} and $\sigma_{rad} = 247$ km s^{-1}. In CORE, green dashed line shows a Gaussian with $\bar{v}_{rad} = 1436$ km s^{-1} and $\sigma_{rad} = 538$ km s^{-1}, for VC galaxies dE and dS0 within 2° of M87 (from Binggeli et al. 1987). In SUB, dashed histogram shows radial velocities from TNG spectroscopic follow-up (Arnaboldi et al. 2003). Dashed red line shows a Gaussian with $\bar{v}_{rad} = 1080$ km s^{-1} and $\sigma_{rad} = 286$ km s^{-1}. Dashed-dotted lines show the SUB-FLAMES spectra including those spectra for HII regions, which have radial velocities in M84 & NGC 4388 redshift ranges.

distribution are highly correlated with the galaxy systemic velocities. The highest peak in the distribution coincides with M84, and even more so when we add the LOS velocities obtained previously at the TNG (Arnaboldi et al. 2003). The 10 TNG velocities give $\bar{v}_{\rm M84} = 1079 \pm 103$ km s^{-1} and $\sigma_{\rm M84} = 325 \pm 75$ km s^{-1} within a square of $4R_e \times 4R_e$ of the M84 center. The 8 FLAMES velocities give $\bar{v}_{\rm M84} = 891 \pm 74$ km s^{-1} and $\sigma_{\rm M84} = 208 \pm 54$ km s^{-1}, going out to larger radii. Note that this includes the over-luminous PNe not attributed to M84 previously. The combined sample of 18 velocities gives $\bar{v}_{\rm M84} = 996 \pm 69$ km s^{-1} and $\sigma_{\rm M84} = 293 \pm 50$ km s^{-1}. Most likely, all these PNe belong to a very extended halo around M84 (see the deep image in Arnaboldi et al. 1996). It is possible that the somewhat low velocity with respect to M84 may be a sign of tidal stripping by M86.

5 Conclusions

While spatial structures have been observed in the ICPN number density distribution both in a single field (Okamura et al. 2002) and as field-to-field variations (Aguerri et al. 2004; Feldmeier et al. 2004a), substructures in velocity space still need to be investigated.

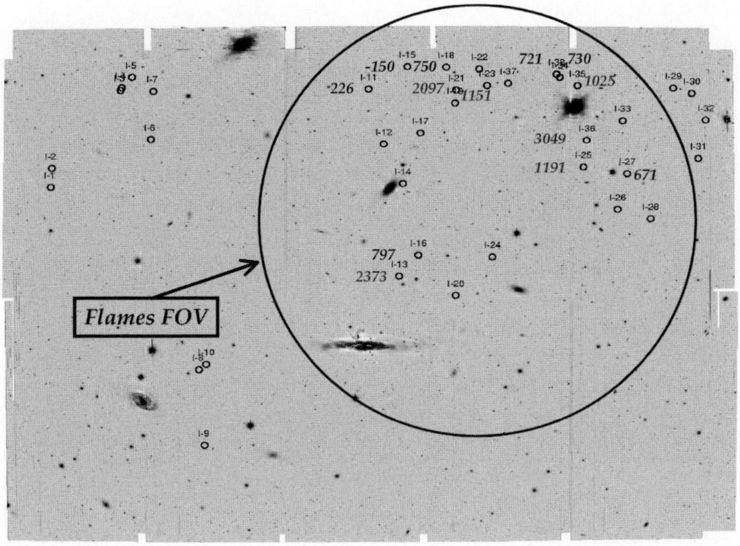

Fig. 6. ICPNe LOS radial velocity field in the SUB field.

Arnaboldi et al. (2004) were able for the first time to carry out the spectroscopic follow-up with the FLAMES spectrographs on UT2 at VLT. ICPNe with the velocity dispersion of the Virgo cluster are found in the CORE field 1.°2 from M87. These may have originated from tidal mass loss of smaller galaxies in the M87 subcluster halo. In a field 0.25 deg from M87, they detect an extended stellar halo of M87 in approximate dynamical equilibrium, but with few ICPNe. Finally, in a field near M84/M86, the ICPNe velocities are highly correlated with the galaxy velocities, showing that any well-mixed intracluster population is yet to form. Overall, the measured velocity distributions confirm the non-uniform dynamical structure and on-going assembly of the Virgo cluster.

Acknowledgements

M.A. would like to thank the organizing committee of the ESO workshop on Planetary Nebulae beyond the Milky Way for the invitation to give this presentation. This work has been done in collaboration with Ortwin Gerhard, Kenneth C. Freeman, and J. Alfonso Aguerri, Massimo Capaccioli, Nieves Castro-Rodriguez, John Feldmeier, Fabio Governato, Rolf-Peter Kudritzki, Roberto Mendez, Giuseppe Murante, Nicola R. Napolitano, Sadanori Okamura, Maurilio Pannella, Naoki Yasuda. M.A. wishes to thank ESO for the support of this project and the observing time allocated both at La Silla and Paranal Telescopes. M.A. wishes to thank the National Astronomical Observatory of Japan, for the observing time allocated at the Subaru Telescope. This work has been supported by INAF and the Swiss National Foundation.

References

1. Aguerri, J. A. L., et al., 2004, AJ, submitted
2. Arnaboldi, M. et al. 2004, ApJL, 614, 33
3. Arnaboldi, M. et al. 2003, AJ, 125, 514
4. Arnaboldi, M. et al. 2002, AJ, 123, 760
5. Arnaboldi, M. et al. 1996, ApJ, 472, 145
6. Bassino, L.P., et al. 2003, A&A, 399, 489
7. Bernstein, G.M., et al. 1995, AJ, 110, 1507
8. Binggeli, B., Popescu, C. C., & Tammann, G. A., 1993, A&A, 98, 275
9. Binggeli, B., Tammann, G.A., & Sandage, A. 1987, AJ, 94, 251
10. Calcáneo-Roldán, C., et al. 2000, MNRAS, 314, 324
11. Castro-Rodriguez, N., et al. 2003, A&A, 405, 803
12. Coté, P., et al. 2001, ApJ, 559, 828
13. Drinkwater, M.J., et al. 2003, Nature, 423, 519
14. Dubinski, J. 1998, ApJ, 502, 141
15. Durrell, P.R., et al. 2003, ApJ, 582, 170
16. Durrell, P.R., et al. 2002, ApJ, 570, 119
17. Feldmeier, J. J., Ciardullo, R., Jacoby, G. H., Durrell, P. R., 2004a, ApJ, 615, 196
18. Feldmeier, J.J., et al. 2004b, ApJ, 609, 617
19. Feldmeier, J.J., et al. 2003, ApJS, 145, 65
20. Feldmeier, J.J., et al. 2002, ApJ, 575, 779
21. Ferguson, H.C., Tanvir, N.R., & von Hippel, T. 1998, Nature, 391, 461
22. Gal-Yam, A.H. et al. 2000, ApJ, 536, 561
23. Gerhard, O., Arnaboldi, M., Freeman, K.C., Okamura, S. 2002, ApJ, 580, L121
24. Gonzalez, A. H., Zabludoff, A. I., Zaritsky, D., & Dalcanton, J. J. 2000, ApJ, 536, 561
25. Gregg, M.D., & West, M.J. 1998, Nature, 396, 549
26. Jordán, A., et al. 2003, AJ, 125, 1642
27. Matsushita, K., Belsole, E., Finoguenov, A., Böhringer, H, 2002, A&A, 386, 77
28. Murante, G. et al. 2004, ApJ, 607, L83
29. Napolitano, N.R. et al. 2003, ApJ, 594, 172
30. Nulsen, P. E. J., & Böhringer, H., 1995, MNRAS, 274, 1093
31. Okamura, S. et al. 2002, PASJ, 54, 883
32. Romanowsky, A. J., & Kochanek, C. S. 2001, ApJ, 553, 722
33. Schombert, J. 1986, ApJS, 60, 603
34. Smith, H.A., 1981, AJ, 86, 998
35. Sommer-Larsen, J., Romeo, A.D. & Portinari, L. 2004, MNRAS, submitted (astro-ph/0403282)
36. Trentham, N., & Mobasher, B., 1998, MNRAS, 293, 53
37. Uson, J.M., et al. 1991, ApJ, 369, 46
38. West, M.J., et al. 1995, ApJ, 453, L77
39. White, P.M., et al. 2003, ApJ, 585, 739
40. Willman, B., Governato, F., Wadsley, J., Quinn, T., 2004, MNRAS, 355, 159
41. Zwicky, F., 1951, PASP, 63, 61

Planetary Nebulae as Tracers of the Intergalactic Stellar Background: A Population Synthesis Theoretical Approach

Alberto Buzzoni[1] and Magda Arnaboldi[2]

[1] INAF – Osservatorio Astronomico di Bologna, Via Ranzani 1, 40127 Bologna (Italy)
[2] INAF – Osservatorio Astronomico di Torino, Via Osservatorio 20, 10025 Pino Torinese (Italy)

Abstract. We wish to assess the relationship between the population of planetary nebulae (PNe) and a given parent stellar population from a theoretical point of view. Our results rely on original population synthesis models used to estimate the expected luminosity-specific PN density and account for different evolutionary scenarios and star formation histories, as observed in galaxies in the near Universe. For a complete PN sample, we find that 1 PN/1.5 10^6 $L_\odot$ is a safe (IMF-independent) lower limit to the traced global bolometric luminosity of the parent stellar population. A tentative application to Virgo cluster data allows us to place a lower limit at $\sim 7\%$ for the global B luminosity of the cluster provided by "loose" intergalactic stars.

1 Introduction

Recent works on diffuse light in nearby clusters have renewed interest in planetary nebulae (PNe) as tracers of the parent stellar population. The efficient detection of these nebulae in the field, even at 70 kpc or more beyond the optical radius of bright galaxies (see e.g. Peng et. al 2004), makes them ideal candidates to trace the "loose" stellar population possibly filling the "empty" space among galaxies. PNe can be detected up to a distance of 20 Mpc in nearby clusters, as for instance for the Virgo or the Fornax cluster (see Arnaboldi et al. 2003; Feldmeier et al. 2003).

A major issue in this regard concerns the link between PN number density and the global (bolometric) luminosity of their parent stellar population. This ratio (the so-called "luminosity-specific PN density" or, shortly, the "α ratio", as defined first by Jacoby 1980) provides in fact direct information on the distinctive physical properties of the underlying stellar population as far as its observed PN component is concerned.

Population synthesis models can be especially useful in this sense, allowing us to tackle the problem from a fully theoretical point of view, thus overcoming most of the natural limits of the empirical calibrations (i.e. observational bias, incomplete sampling etc.). In this direction, we give here a brief progress report of our project aimed at calibrating "α" for different evolutionary scenarios and star formation histories as observed in the local galaxy sample and likely consistent

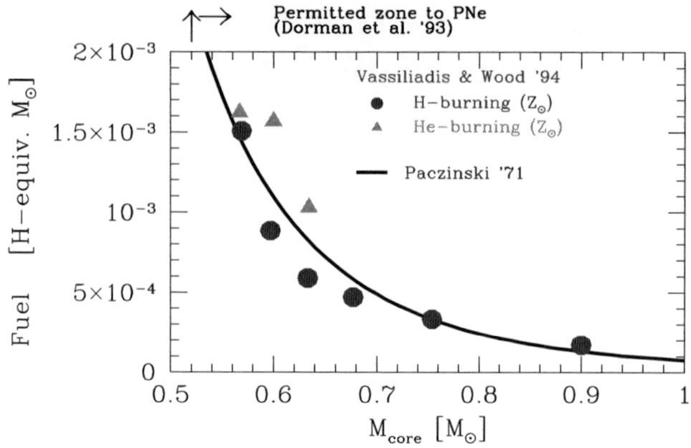

Fig. 1. Theoretical fuel consumption for stars along the Hot-PAGB evolution according to the models of Paczynski (1971; solid curve) and Vassiliadis and Wood (1994; solid dots and triangles). Fuel is expressed in Hydrogen-equivalent solar mass. The stellar core-mass range to produce PNe is marked, after Dorman et al. (1993).

with the expected properties of the diffuse stellar component of the intergalactic medium.

2 Theoretical Fundamentals

Theory of simple stellar populations (SSPs), as developed in Renzini & Buzzoni (1986) and Buzzoni (1989) (hereafter RB86 and B89, respectively), provides the basic reference framework for our task, allowing us to evaluate the value of "α" for coeval and chemically homogenous stellar aggregates.

Briefly, the expected number ($N_{\rm PN}$) of PNe for a SSP of total (bolometric) luminosity L_{tot} is:

$$N_{\rm PN} = \mathcal{B}\, L_{\rm tot}\, \tau_{\rm PN}. \tag{1}$$

In the equation, $\mathcal{B}$ is the so-called "specific evolutionary flux" (a nearly age- and IMF-independent parameter of the order of $2\,10^{-11}$ $[L_\odot^{-1}{\rm yr}^{-1}]$, see B89), and $\tau_{\rm PN}$ is the PN lifetime (i.e. the time for the nebula to be detectable in [O III] and/or Hα surveys).

The PN lifetime directly modulates with the stellar core-mass evolution, the latter depending on the nuclear fuel burnt at high temperature ($T_{\rm eff} \simeq 10^5$ K) along the Hot Post-AGB (H-PAGB) phase (this is a direct result of the so-called "Fuel Consumption Theorem" of RB86). Figure 1 summarizes the model predictions, after Paczynski (1971) and Vassiliadis and Wood (1994), displaying the H-PAGB fuel consumption for stars of different core masses.[1] In the plot,

[1] Note that *actual* PAGB core mass of stars can sensibly be lower than the *original* stellar mass when leaving the Main Sequence Turn Off point. This is a consequence of the mass loss processes that heavily affect stellar evolution along the red and asymptotic giant branch phases.

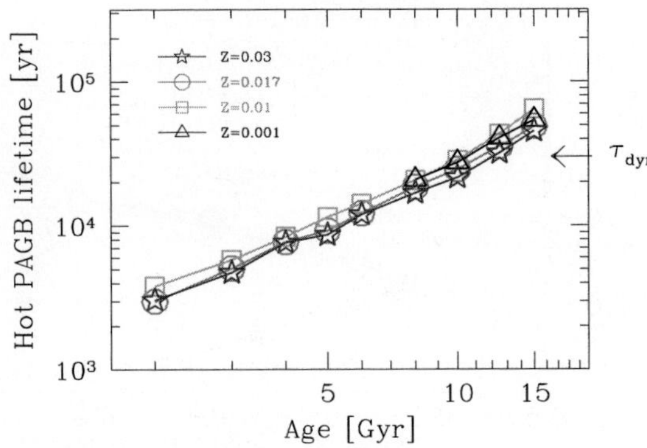

Fig. 2. Hot-PAGB core mass lifetime, for SSP evolution according to B89 synthesis models. A Salpeter IMF is assumed, and a Reimers mass loss rate coefficient $\eta = 0.3$. The explored metallicity spans the range from $Z \sim 1/20 \rightarrow 2\ Z_\odot$, as labeled in the plot. The PN dynamical timescale ($\tau_{\rm dyn}$) is also marked, for reference.

we express the fuel in terms of Hydrogen-equivalent solar mass (i.e. 1 g of H-equivalent mass $= 6\ 10^{18}$ ergs; cf. RB86).

The inferred H-PAGB lifetime for SSPs of different ages, according to B89, is reported in Fig. 2. It can be defined as $\tau_{\rm HPAGB} \simeq {\rm Fuel}/\ell_{\rm H-PAGB}$, where $\ell_{\rm H-PAGB}$ is the luminosity of the stellar core at the onset of the nebula ejection (see e.g. Paczynski 1971 for details). It is important to note that, in general, $\tau_{\rm HPAGB}$ is shorter than the dynamical timescale for nebula evaporation ($\tau_{\rm dyn} \simeq 30\,000$ yr; cf. Peimbert 1990); this implies that, at least for SSPs of young and intermediate age, $\tau_{\rm PN}$ should actually coincide with $\tau_{\rm HPAGB}$. Only for $t_{\rm SSP} \gtrsim 12$ Gyr we have that $\tau_{\rm PN} \simeq \tau_{\rm dyn}$, so that the luminosity-specific PN density could eventually be written as

$$\alpha = \frac{N_{\rm PN}}{L_{\rm SSP}} = \mathcal{B}\,\tau_{\rm PN} = \mathcal{B}\min\{\tau_{\rm HPAGB}, \tau_{\rm dyn}\} \qquad (2)$$

From eq. (2), and taking the B89 synthesis code as a theoretical reference, in Fig. 3 we report the expected luminosity-specific PN density for SSPs of different metallicity. Some physical age constraints exist for SSPs to produce PNe. On the one hand, for $t_{\rm SSP} \lesssim 10^8$ yr, SSP evolution is dominated by stars with $M_* \gtrsim 5\ M_\odot$ ending up as Supernovae. On the other hand, no PNe are expected to form from stars with actual mass $M_* \lesssim 0.52\ M_\odot$ (Dorman et al. 1993; see Fig. 1); this mass limit is likely to occur at some point along SSP evolution for ages between $10 \lesssim t_{\rm SSP} \lesssim 18$ Gyr, the exact details depending on the efficiency of the mass loss processes[2].

[2] The B89 SSP models parameterize mass loss via the coefficient η according to the Reimers (1975) standard notation. A value of $\eta = 0.3$ is adopted here, as an indicative value for the observed mass loss in the Galactic globular clusters (see B89 for details).

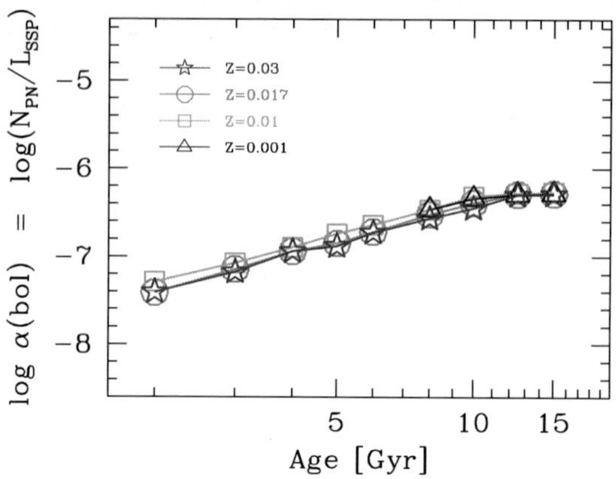

Fig. 3. Theoretical luminosity-specific PN density for SSPs of different metallicity. Models are according to B89, with the distinctive parameters as in Fig. 2.

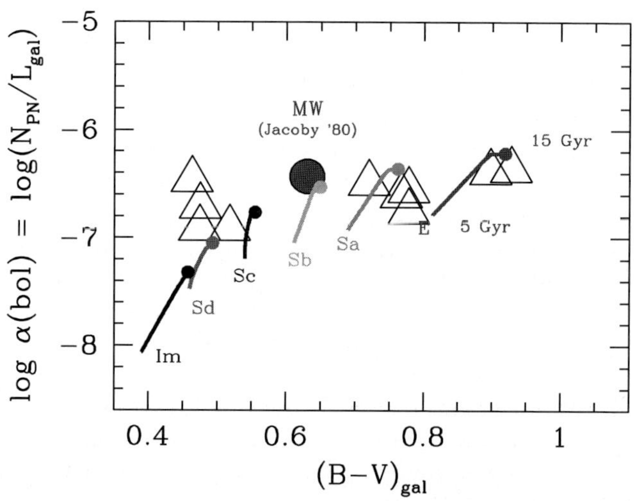

Fig. 4. Expected luminosity-specific PN density for template galaxy models (labeled with the corresponding Hubble type), according to Buzzoni (2002). Each curve tracks galaxy evolution from 5 Gyr to 15 Gyr (small solid dots), as labeled for the case of Ellipticals. Big triangles report the empirical estimate of $\alpha = N_{\mathrm{PN}}/L_{\mathrm{gal}}$ from the PN census of Jacoby (1980) for the galaxies in the Local Group. The case of the Milky Way is singled out in the plot by the big solid dot.

3 The Luminosity-Specific PN Density in Galaxies: Implications for the Intergalactic Medium

The theoretical estimate of the luminosity-specific PN density is important when we infer total bolometric luminosity of their parent stellar population in the intergalactic medium. So far, we have relied upon the empirical calibration of α on the bulge of M31, with all the uncertainties induced by the poorly understood effects of metallicity, age and star formation history.

Our new results, based on this improved theoretical approach, indicate that a luminosity-specific PN density

$$\alpha = 1 \text{ PN}/1.5\,10^6 \text{ L}_\odot \tag{3}$$

is a safe estimate for old SSPs. Quite interestingly, this value does *not* strongly depend on metallicity and, even more importantly, on the stellar IMF (as $\mathcal{B}$, in eq. (2), is nearly insensitive to the IMF details, as shown by RB86).

On the basis of the derived SSP evolution, and matching the Buzzoni (2002) template galaxy models, we have further elaborated our synthesis scheme computing in Fig. 4 the expected evolution of α for the different star formation histories of galaxies along the Hubble morphological sequence. A comparison with the observed PN population for the Local Group members (Jacoby 1980, including the relevant case of the Milky Way) is also reported in the figure.

As a tentative application of our results, we could match the recent observations of the Virgo cluster by Arnaboldi et al. (2003) in order to assess the luminosity contribution from the intergalactic stellar background. Some 37 PN candidates have been detected by Arnaboldi et al. ($\sim 25\%$ of which, it is claimed, possibly consist of spurious Lyα emitters at $z = 3.14$) across an area of 0.196 deg^2 on the cluster. As the *bona fide* PNe only span the brightest magnitude bin of the standard luminosity function we had to correct for sample incompleteness according to Ciardullo et al. (1989).

By adopting the reference range of α for 15 Gyr galaxy models of Fig. 4, an intergalactic surface brightness (in bolometric units) could be derived as $26.0 \lesssim \mu_{\rm bol} \lesssim 28.5$ mag arcsec^{-2}, depending on the assumed star formation history for the PN parent stellar population.[3] This implies a B surface brightness for intergalactic stars in the range $27.3 \lesssim \mu_{\rm B} \lesssim 30.3$, mag arcsec^{-2}. Compared to the smooth B surface brightness distribution of the bright-galaxy component of Virgo (i.e. $\mu_{\rm B} \simeq 27.4$ mag arcsec^{-2}, see Arnaboldi et al. 2003), this allows us to place a confident lower limit to 7% for the cluster global luminosity to be provided by "loose" intergalactic stars.

References

1. M. Arnaboldi et al.: Astron. Journal **125**, 514 (2003)
2. A. Buzzoni: Astrophys. Journal Suppl. Ser. **71**, 817 (1989) [B89]
3. A. Buzzoni: Astron. Journal **123**, 1188 (2002)
4. R. Ciardullo, G. H. Jacoby, H. C. Ford, J. D. Neill: Astrophys. Journal **339**, 53
5. B. Dorman, R. T. Rood, R. W. O'Connell: Astrophys. Journal **419**, 596 (1993)
6. J. J. Feldmeier, R. Ciardullo, G. H. Jacoby, P. R. Durrell: Astrophys. Journal Suppl. Ser. **145**, 65 (2003)

[3] For our calculations we adopt a distance modulus $(m - M) = 30.8$ mag for the Virgo cluster, and an absolute bolometric magnitude $M_{\rm bol,\odot} = 4.72$ for the Sun.

7. G. Jacoby: ApJ **42**, 1 (1980)
8. B. Paczynski: Acta Astr. **21**, 47 (1971)
9. M. Peimbert: Rev. Mexicana Astron. Astrof. **20**, 119 (1990)
10. E. W. Peng, H. C. Ford, K. C. Freeman: Astrophys. Journal **602**, 685 (2004)
11. D. Reimers: Mem. Soc. Roy. Sci. Liège, 6th Ser. **8**, 87 (1975)
12. A. Renzini A. Buzzoni: 'Global properties of stellar populations and spectral evolution of galaxies'. In *Spectral Evolution of Galaxies*, ed. C. Chiosi, A. Renzini (Reidel, Dordrecht 1986) pp. 195–235 [RB86]
13. E. Vassiliadis, P. R. Wood: Astrophys. Journal Suppl. Ser. **92**, 125 (1994)

The Survival of Planetary Nebulae in the Intracluster Medium

Eva Villaver[1,2] and Letizia Stanghellini[3]

[1] Space Telescope Science Institute, 3700 San Martin Drive, Baltimore, MD 21218, USA; villaver@stsci.edu
[2] Affiliated with the Space Telescope Division of the European Space Agency
[3] National Optical Astronomy Observatory, 950 N. Cherry Av., Tucson, AZ 85719, USA; letizia@noao.edu

1 Summary

In the last few years, quantitative estimates of the Intracluster (IC) stellar population have been inferred from the study of the Red Giant and Asymptotic Giant Branch (AGB) stars detected with the *HST* in the Virgo cluster [7,4]; Planetary Nebulae (PNe) from ground-based observations in the Virgo and Fornax clusters [9,5,6,3,1], and Supernovae (SNe) Ia discovered in archival *HST* images [8].

The stellar population stripped from galaxies in clusters evolves under the very extreme conditions imposed by the IC medium. Typically, the IC PNe have very high systemic velocities (up to $2000\,\mathrm{km\,s^{-1}}$) through a rarefied, extremely hot medium ($\sim 10^7$ K). This environment affects the formation and evolution of the AGB circumstellar envelopes and PN shells, and in general, any gas structure. We present numerical simulations aimed to test whether gaseous structures such as AGB envelopes and PNe shells survive under the extreme conditions of the IC medium, and, if so, how their evolution is affected.

2 The Numerical Simulations

The code setup is similar to the one used in [12] in the study of the interaction of Galactic PN with the interstellar medium (ISM). We have used a star with a main sequence mass of $1\,\mathrm{M_\odot}$. We follow the temporal evolution of its stellar wind during the AGB and PN phases [11,10]. The IC gas was assumed to be homogeneous with a temperature of 10^7 K and a density of $10^{-4}\,\mathrm{cm^{-3}}$. We considered a relative movement through the Virgo cluster of 1000 km $\mathrm{s^{-1}}$.

In Fig.1 we show the gas density (on a logarithmic scale) at 4.1×10^5 yr for a 1 $\mathrm{M_\odot}$ star evolving in the Virgo cluster. The main result of our simulations is that the delicate structure of the PN envelope survives, although, it is highly affected by its evolution under the extreme conditions of the IC medium. We find that ram pressure stripping plays a major role in decreasing the mass of the AGB circumstellar envelope and therefore the amount of gas that will be observed as a PN shell [12].

Instabilities develop in the upstream direction, breaking the main shell. As a result the gas ejected by the star is mixed through instabilities with the IC

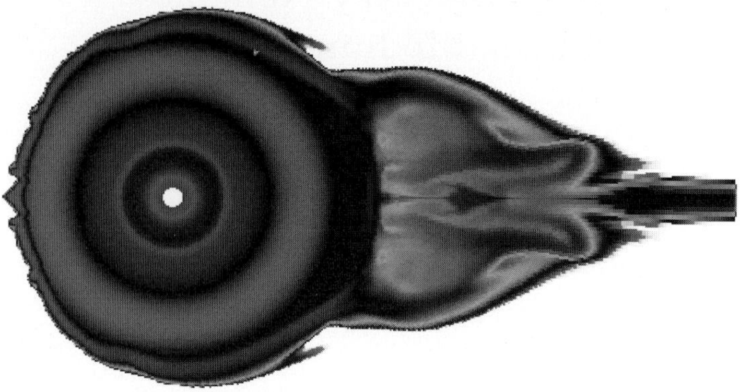

Fig. 1. Gas density at the end of the AGB evolution of a 1 $M_{\odot}$ star evolving under the conditions of the IC medium in Virgo. The intensity scale is logarithmic and the physical grid size 4×2 pc. Note that the relative velocity was assumed to be 1000 km s^{-1}, with the IC medium flowing into the grid from left to right.

material. In the downstream direction close to the symmetry axis, the flow turns back. Long tongues of gas, a consequence of the stripping process, are also visible (note the turbulent nature of the stripped gas). We expect the AGB envelopes and PN shells to be very sensitive to the IC medium conditions. An increase of the IC medium ram pressure will have a strong effect on the stripping of the shells. No IC PN were found in a survey in the Leo HI cloud [2], confirming the primordial nature of the intergalactic gas in Leo. We expect ram pressure stripping to be particularly efficient in the Leo HI cloud, due to its higher density; most of the gas left in the downstream direction of the movement will be removed, therefore possibly preventing the detection of bright PN.

References

1. Arnaboldi, M., et al. 2003, AJ, 125, 514
2. Castro-Rodriguez et al. 2003, A&A 405, 803
3. Ciardullo, R., Jacoby, G. H., Feldmeier, J. et al. 1998, ApJ, 492, 62
4. Durrell, P. R., Ciardullo, R., Feldmeier, et al. 2002, ApJ, 570, 119
5. Feldmeier, J., Ciardullo, R., & Jacoby, G. 1998, ApJ, 503, 109
6. Feldmeier, J., Ciardullo, R., Jacoby, G., et al. 2003, ApJS, 145, 65
7. Ferguson, H.C., Tanvir, N., & von Hippel, T. 1998, nature, 391, 461
8. Gal-Yam, A., Maoz, Guhathakurta & Filippenko 2003, AJ, 125, 1087
9. Mendez, R. H. et al. 1997, ApJ, 491, L23
10. Villaver, E., García-Segura, & Manchado, A. 2002, ApJ, 571, 880
11. Villaver, E., Manchado & García-Segura 2002, ApJ, 581, 1204
12. Villaver, E., García-Segura, & Manchado, A. 2003, ApJ, 585, L49

Part IX

Prospects

Extragalactic Planetary Nebulae: Observational Challenges and Future Prospects

Quentin A. Parker[1] and Richard Shaw[2]

[1] Macquarie University/Anglo-Australian Observatory, Sydney, Australia, qap@ics.mq.edu.au
[2] National Optical Astronomy Observatories, Tucson, USA, shaw@noao.edu

1 Introduction

The study of extragalactic planetary nebulae (EPN) is a rapidly expanding field. The advent of powerful new instrumentation such as the PN.Spectrograph (PN.S, Douglas, these proceedings) has led to an avalanche of new EPN discoveries both within and between galaxies. We now have thousands of EPN detections in a heterogeneous selection of nearby galaxies and their local environments, dwarfing the combined Galactic detection efforts of the last century. This alone brings its own problems of nomenclature and cataloging (Parker & Acker, these proceedings).

Key scientific motivations driving this rapid growth in EPN research and discovery have been the use of the PNLF as a standard candle (e.g. Ciardullo, Girardi, these proceedings), the use of EPN as dynamical tracers of their host galaxies and of dark matter (e.g. Napolitano, these proceedings), and as probes of Galactic evolution (e.g. Liu, Richer, these proceedings). This is coupled with the basic utility of PN as laboratories of nebula physics and the consequent comparison with theory, where population differences, abundance variations and star formation history (SFH) within and between stellar systems, informs both stellar and galactic evolution.

Here we pose some of the burning questions, discuss some of the observational challenges and outline some of the future prospects of this exciting, relatively new, research area as we strive to go fainter, image finer, see further and survey faster than ever before and over a wider wavelength regime.

2 Some Selected Questions

The ability to not only detect but to begin to study PN in external galaxies permits us to pose some new questions. What is the origin of the PNLF, how far can we push it and does it break down? How does the nature of a host galaxy affect SFH, chemical evolution and the PN population? Does host galaxy type play a role in nebular morphology? Can multi-galaxy studies provide clues to the progenitor/core mass relation for medium to low mass stars? What is the nature and extent of the intra-cluster stellar population and what can it tell us of the merger histories via detection of evidential streams? Finally what is the role of dust in PN formation and evolution?

3 The Role of Technology

The available technology and instrumentation both drives and limits what we can observationally achieve at any given epoch and in any wavelength domain. We need to develop the most viable strategies for tackling the above questions and push to influence the development of the facilities necessary to execute these strategies effectively. The PN spectrograph (PN.S) is a rare example of an instrument specifically designed to detect PN. Do we want more of these? Usually we have to make use of instruments that may not be ideally suited to our needs leading to compromise or adaptation of the science mission. We need to ask if the currently available technology is driving our research directions more than we are driving the technology. It behoves our community to take an active role in influencing the design and capability of the next generation of instrumentation to better fit requirements.

Most current EPN detection techniques perform on-band off-band imaging for a single (usually the [O III] 5007Å) line. Contamination is a serious issue. The ability to efficiently image several nebula lines in the rest frame of a galaxy over a wide field of view would be a significant step forward. Larger/multi aperture optical telescopes of 8m+ now outnumber the 4m class which are supplemented by new multi-wavelength capabilities from earth and space in the UV/IR/MIR and radio regimes. Higher resolution capabilities coupled with more efficient/powerful detectors, spectrographs, interferometers, Fibre/IFU/multi-slit technologies, narrow band capabilities and innovative techniques (e.g. CCD nod and shuffle and tunable filters) offers significant opportunities to advance EPN research.

Tunable filters offer much promise. They can rapidly image a series of diagnostic nebular lines over a wide field at a given redshift and can be used to scan through an entire depth of a cluster. The Taurus Tunable Filter on the AAT was a highly effective tool in this regard but although a powerful and unique capability it has now been de-commissioned. We need to ensure that the best facilities suited to our needs are maintained and enhanced.

Examples of ground-based instrumentation advances on 8m telescopes include use of optical fibres in MOS and IFU systems such as FLAMES/GIRAFFE (ESO VLT) and FMOS/Echindna (Subaru), enhanced IR imaging capabilities such as VISIR (VLT) and advances in active and adaptive optics systems (including MCAO, especially in the NIR). New space-based facilities such as Spitzer offer new multi-wavelength sensitivities, while planned future facilities such as SKA, OWL and the JWST will offer unique capabilities. We need to plan now to ensure that we achieve access to such facilities and influence their development.

4 The Observational and Theoretical Challenges

An ultimate aim is to obtain for EPN the observational detail we obtain for PN in our own Galaxy. Hence, aside from mere EPN detection, we would like to:
• detect/study their central stars (e.g. Villaver, these proceedings);

• detect/study their outer halos (now just possible for the LMC - Reid & Parker, these proceedings);
• detect highly evolved examples and determine PN morphologies and nebular abundances. Currently only possible to some degree for Magellanic Cloud PN via the HST (e.g. Stanghellini, these proceedings).
• obtain high S/N, decent resolution spectra permitting EPN nebular parameters to be determined across a wide wavelength range coupled with accurate radial and expansion velocities.

Faster computers and more powerful, precise and sophisticated N-body simulations and photoionisation code coupled to significant new observational EPN data would enable us to fine tune theory/models of diverse PN populations in widely different galaxy environments.

4.1 Going Fainter and Seeing Further

If we simply probe fainter via a combination of light gathering power (larger telescopes), system efficiency gains (e.g. by AO), sky/background suppression (e.g. narrow-band filters) and exposure times, we can push the PNLF to fainter limits in external galaxies at greater distances as the most luminous PN in more distant systems move into the detection threshold. Spectroscopically we can obtain higher S/N observations of more useful nebular diagnostic lines in our own Galaxy, LMC and Local Group and perhaps begin to sample strong emission lines in PN in more distant galaxies. Abundance determinations in systems > 10Mpc would become feasible, together with the detection of halos, additional morphological features and lower surface brightness PN.

4.2 Obtaining Finer Resolution

Improved resolution imaging and spectroscopy (across a wider wavelength domain) will provide the ability to discriminate finer morphological detail in EPN (currently only possible in the LMC via the HST) and permit the study of reaction interfaces between EPN and their local ISM. Measurements of accurate expansion velocities and systemic velocities in external systems would permit improved kinematical studies and the removal of contaminants in spectra that were overlapping in lower resolution systems.

4.3 Surveying Faster

We are truly entering an era of the data avalanche in astronomy as a consequence of large survey facilities, focal plane arrays, advanced MOS systems and other significant efficiency gains in coverage, resolution and sensitivity. The current AVO/Astrogrid initiatives to federate, incorporate and manage the massive petabyte catalogues now being created in astronomy have been set up to meet this challenge. These are coupled with vast storage and processing capabilities as technology and Moores law allows. We need to be able to handle/understand cross-correlate and disseminate our own EPN results in a timely fashion and make best use of these new AVO capabilities.

5 Problems and Issues Arising from this Meeting

The most significant problem is the extreme faintness of EPN even with line-imaging. This is coupled with poor resolution (EPN are unresolved in all but the closest galaxy neighbours with current technology). Another issue is the contamination of EPN samples when only single lines are imaged. However soon we will have tens of thousands of individual EPN crying out for further study and careful cataloguing. It is also clear that we badly need better and indeed any spectra of PN candidates in the intra-cluster medium. We also urgently need better simulations to leverage observations. We want to determine multi-dimensional PNLFs and directly detect EPN central stars and determine individual spectral and photometric properties. Such problems represent a considerable observational challenge. Only strategic planning, collaboration and technological developments are likely to be able to address these problems effectively.

6 Collaborations and Future Prospects

A community speaking with one voice can be heard more effectively and a community co-operating to achieve a common goal is more likely to succeed. The prohibitive cost of new instrumentation and the plans for 50m+ telescopes necessitates large multi-national collaborations. It is vitally important that our community has an influence when key instrumentation capabilities are chosen. Furthermore it may prove fruitful to inaugurate several key large-scale community wide collaborations to develop effective exploitation strategies to use these powerful new facilities and to tackle the really big issues such as indicated below.

While it will be some time before we can effectively study individual EPN in all but the closest galactic neighbours, currently available capabilities mean that there are realistic prospects of advancement in several key areas. Should we aim for a complete EPN census of Local Group galaxies, complete to ~ 5 mag below the L(5007Å) peak and a census of intra-cluster EPN to depths of several tens of Mpc? This will be invaluable in understanding the origins of the PNLF and what occurs in late type populations and how EPN inform host galaxy SFH. Can we perform morphological studies of complete samples of EPN in at least two local group galaxies, e.g. via HST/STIS or MCAO on 8m telescopes? Should we obtain a census of intra-cluster EPN in Virgo/Fornax together with velocity estimates? This will impact on our understanding of galaxy dynamics, the intra-cluster stellar population and the influence of dark matter.

Acknowledgements

The authors are grateful for the positive feedback from the participants during the course of the workshop and during the scheduled discussion.

Author Index

ESO ASTROPHYSICS SYMPOSIA
European Southern Observatory

Series Editor: Bruno Leibundgut

D.L. Clements, I. Pérez-Fournon (Eds.),
Quasar Hosts
Proceedings, 1996, XVII, 336 pages. 1997.

L.N. da Costa, A. Renzini (Eds.),
Galaxy Scaling Relations: Origins, Evolution and Applications
Proceedings, 1996. XX, 404 pages. 1997.

L. Kaper, A.W. Fullerton (Eds.),
Cyclical Variability in Stellar Winds
Proceedings, 1997. XXII, 415 pages. 1998.

R. Morganti, W.J. Couch (Eds.),
Looking Deep in the Southern Sky
Proceedings, 1997. XXIII 336 pages. 1999.

J.R. Walsh, M.R. Rosa (Eds.),
Chemical Evolution from Zero to High Redshift
Proceedings, 1998. XVIII, 312 pages. 1999.

J. Bergeron, A. Renzini (Eds.),
From Extrasolar Planets to Cosmology: The VLT Opening Symposium
Proceedings, 1999. XXVIII, 575 pages. 2000.

A. Weiss, T.G. Abel, V. Hill (Eds.),
The First Stars
Proceedings, 1999. XIII, 355 pages. 2000.

A. Fitzsimmons, D. Jewitt, R.M. West (Eds.),
Minor Bodies in the Outer Solar System
Proceedings, 1998. XV, 192 pages. 2000.

L. Kaper, E.P.J. van den Heuvel, P.A. Woudt (Eds.),
Black Holes in Binaries and Galactic Nuclei: Diagnostics, Demography and Formation
Proceedings, 1999. XXIII, 378 pages. 2001.

G. Setti, J.-P. Swings (Eds.)
Quasars, AGNs and Related Research Across 2000
Proceedings, 2000. XVII, 220 pages. 2001.

A.J. Banday, S. Zaroubi, M. Bartelmann (Eds.),
Mining the Sky
Proceedings, 2000. XV, 705 pages. 2001.

E. Costa, F. Frontera, J. Hjorth (Eds.),
Gamma-Ray Bursts in the Afterglow Era
Proceedings, 2000. XIX, 459 pages. 2001.

S. Cristiani, A. Renzini, R.E. Williams (Eds.),
Deep Fields
Proceedings, 2000. XXVI, 379 pages. 2001.

J.F. Alves, M.J. McCaughrean (Eds.),
The Origins of Stars and Planets: The VLT View
Proceedings, 2001. XXVII, 515 pages. 2002.

J. Bergeron, G. Monnet (Eds.),
Scientific Drivers for ESO Future VLT/VLTI Instrumentation
Proceedings, 2001. XVII, 356 pages. 2002.

M. Gilfanov, R. Sunyaev, E. Churazov (Eds.),
Lighthouses of the Universe: The Most Luminous Celestial Objects and Their Use for Cosmology
Proceedings, 2001. XIV, 618 pages. 2002.

R. Bender, A. Renzini (Eds.),
The Mass of Galaxies at Low and High Redshift
Proceedings, 2001. XXII, 363 pages. 2003.

W. Hillebrandt, B. Leibundgut (Eds.),
From Twilight to Highlight: The Physics of Supernovae
Proceedings, 2002. XVII, 414 pages. 2003.

P.A. Shaver, L. DiLella, A. Giménez (Eds.),
Astronomy, Cosmology and Fundamental Physics
Proceedings, 2002. XXI, 501 pages. 2003.

M. Kissler-Patig (Ed.),
Extragalactic Globular Cluster Systems
Proceedings, 2002. XVI, 356 pages. 2003.

P.J. Quinn, K.M. Górski (Eds.),
Toward an International Virtual Observatory
Proceedings, 2002. XXVII, 341 pages. 2004.

W. Brander, M. Kasper (Eds.),
Science with Adaptive Optics
Proceedings, 2003. XX, 387 pages. 2005.

A. Merloni, S. Nayakshin, R.A. Sunyaev (Eds.),
Growing Black Holes: Accretion in a Cosmological Context
Proceedings, 2004. XIV, 506 pages. 2005.

L. Stanghellini, J.R. Walsh, N.G. Douglas (Eds.)
Planetary Nebulae Beyond the Milky Way
Proceedings, 2004. XVI, 372 pages. 2006.